PROBABILISTIC INFERENCE AND STATISTICAL METHODS IN NETWORK ANALYSIS

PROBABILISTIC INFERENCE AND STATISTICAL METHODS IN NETWORK ANALYSIS

Edited by:

Olga Moreira

ARCLER
P R E S S

www.arclerpress.com

Probabilistic Inference and Statistical Methods in Network Analysis

Olga Moreira

Arcler Press
2010 Winston Park Drive,
2nd Floor
Oakville, ON L6H 5R7
Canada
www.arclerpress.com
Tel: 001-289-291-7705
001-905-616-2116
Fax: 001-289-291-7601
Email: orders@arclereducation.com

ISBN: 978-1-77361-555-4 (Hardcover)

DECLARATION

ABOUT THE EDITOR

Olga Moreira obtained her Ph.D. in Astrophysics from the University of Liege (Belgium) in 2010, her BSc. in Physics and Applied Mathematics from the University of Porto (Portugal). Her post-graduate travels and international collaborations with the European Space Agency (ESA) and European Southern Observatory (ESO) led to great personal and professional growth as a scientist. Currently, she is working as an independent researcher, technical writer, and editor in the fields of Mathematics, Physics, Astronomy and Astrophysics.

TABLE OF CONTENTS

SECTION II NEURAL NETWORKS

SECTION III GENE REGULATORY NETWORKS AND BIOLOGICAL SYSTEMS

SECTION IV BAYESIAN NETWORKS

SECTION V COMPLEX NETWORKS

LIST OF CONTRIBUTORS

Faiza Ghafoor
Department of Software Engineering, International Islamic University, Islamabad, Pakistan

Muaz A. Niazi
Department of Computer Science, COMSATS Institute of IT, Islamabad, Pakistan

Mostafa Shokoohi
Research Center for Modeling in Health, Institute for Futures Studies in Health, Kerman University of Medical Sciences, Kerman, Iran

Saharnaz Nedjat
School of Public Health, and Knowledge Utilization Research Center (KURC), Tehran University of Medical Sciences, #12, Nosrat East North Kargar, Tehran, Iran

Reza Majdzadeh
School of Public Health, and Knowledge Utilization Research Center (KURC), Tehran University of Medical Sciences, #12, Nosrat East North Kargar, Tehran, Iran
Endocrinology and Metabolism Research Institute, Tehran University of Medical Sciences, Tehran, Iran

Victor Ströele
Postgraduate Program in Computer Science/PPGCC, Federal University of Juiz de Fora (UFJF), Juiz de Fora, MG 36036-900, Brazil

Renato Crivano
Systems Engineering and Computation Program/COPPE, Federal University of Rio de Janeiro (UFRJ), Rio de Janeiro, RJ 21941-972, Brazil

Geraldo Zimbrão
Systems Engineering and Computation Program/COPPE, Federal University of Rio de Janeiro (UFRJ), Rio de Janeiro, RJ 21941-972, Brazil

Jano M. Souza
Systems Engineering and Computation Program/COPPE, Federal University of Rio de Janeiro (UFRJ), Rio de Janeiro, RJ 21941-972, Brazil

Fernanda Campos
Postgraduate Program in Computer Science/PPGCC, Federal University of Juiz de Fora (UFJF), Juiz de Fora, MG 36036-900, Brazil

José Maria N. David
Postgraduate Program in Computer Science/PPGCC, Federal University of Juiz de Fora (UFJF), Juiz de Fora, MG 36036-900, Brazil

Regina Braga
Postgraduate Program in Computer Science/PPGCC, Federal University of Juiz de Fora (UFJF), Juiz de Fora, MG 36036-900, Brazil

Weizhe Zhang
School of Computer Science and Technology, Harbin Institute of Technology, Harbin 150001, China

Xiaoqiang Li
School of Computer Science and Technology, Harbin Institute of Technology, Harbin 150001, China

Hui He
School of Computer Science and Technology, Harbin Institute of Technology, Harbin 150001, China

Xing Wang
School of Computer Science and Technology, Harbin Institute of Technology, Harbin 150001, China

Damien R. Farine
Department of Zoology, Edward Grey Institute of Field Ornithology, University of Oxford, South Parks Road, Oxford OX1 3PS, UK
Department of Anthropology (Evolutionary), University of California Davis, 1

Shields Avenue, Davis, CA 95616, USA
Smithsonian Tropical Research Institute, Ancon, Panama

Hal Whitehead
Department of Biology, Dalhousie University, 1355 Oxford St, Halifax, NS, Canada B3H 4J1

IldefonsMagrans de Abril
Okinawa Institute of Science and Technology, Graduate University, Japan
ARAYA, Inc., Tokyo, Japan

JunichiroYoshimoto
Nara Advanced Institute of Science and Technology, Japan

KenjiDoya
Okinawa Institute of Science and Technology, Graduate University, Japan

Richard F. Betzel
Department of Bioengineering

Danielle S. Bassett
Department of Bioengineering
Department of Electrical and Systems Engineering, University of Pennsylvania, Philadelphia, PA 19104, USA

Va^n Anh Huynh-Thu
Department of Electrical Engineering and Computer Science, Systems and Modeling, University of Lie`ge, Lie`ge, Belgium
GIGA-Research, Bioinformatics and Modeling, University of Lie`ge, Lie`ge, Belgium

Alexandre Irrthum
Department of Electrical Engineering and Computer Science, Systems and Modeling, University of Lie`ge, Lie`ge, Belgium
GIGA-Research, Bioinformatics and Modeling, University of Lie`ge, Lie`ge, Belgium

Louis Wehenkel
Department of Electrical Engineering and Computer Science, Systems and Modeling, University of Lie`ge, Lie`ge, Belgium

GIGA-Research, Bioinformatics and Modeling, University of Lie`ge, Lie`ge, Belgium

Pierre Geurts
Department of Electrical Engineering and Computer Science, Systems and Modeling, University of Lie`ge, Lie`ge, Belgium
GIGA-Research, Bioinformatics and Modeling, University of Lie`ge, Lie`ge, Belgium

Roozbeh Manshaei
Electrical and Computer Engineering Department, Ryerson University, Toronto, ON, Canada M5B 2K3

Pooya Sobhe Bidari
Electrical and Computer Engineering Department, Ryerson University, Toronto, ON, Canada M5B 2K3

Mahdi Aliyari Shoorehdeli
Electrical and Computer Engineering Department, K.N. Toosi University of Technology, Tehran 16315-1355, Iran

Amir Feizi
Department of Chemical and Biological Engineering, Systems and Synthetic Biology Group, Chalmers University, 41296 Gutenberg, Sweden

Tahmineh Lohrasebi
National Institute of Genetic Engineering and Biotechnology (NIGEB), Tehran 14965/161, Iran

Mohammad Ali Malboobi
National Institute of Genetic Engineering and Biotechnology (NIGEB), Tehran 14965/161, Iran

Matthew Kyan
Electrical and Computer Engineering Department, Ryerson University, Toronto, ON, Canada M5B 2K3

Javad Alirezaie
Electrical and Computer Engineering Department, Ryerson University, Toronto, ON, Canada M5B 2K3

Marco Scutari
Department of Statistics, University of Oxford, 24-29 St Giles', Oxford, OX1 3LB, UK

Pietro Auconi
Private Practice of Orthodontics, Roma, Italy

Guido Caldarelli
IMT School for Advanced Studies, Piazza San Francesco 19, 55100, Lucca, Italy
Istituto dei Sistemi Complessi CNR, Unita Sapienza, Dip. Fisica, P.le A. Moro 2, 00185, Rome, Italy
London Institute for Mathematical Sciences, 35a South St, Mayfair, London, W1K 2XF, UK

Lorenzo Franchi
Dipartimento di Chirurgia e Medicina Traslazionale, Universita degli Studi di Firenze, Firenze, Italy
Thomas M. Graber Visiting Scholar, Department of Orthodontics and Pediatric Dentistry, School of Dentistry, The University of Michigan, Ann Arbor, MI, USA

Xin Zou
School of Natural and Built Environments, University of South Australia, Adelaide, SA 5095, Australia

Wen Long Yue
School of Natural and Built Environments, University of South Australia, Adelaide, SA 5095, Australia

Enrique Hernández-Lemus
Computational Genomics Department, National Institute of Genomic Medicine, Mexico City, México
Complexity in Systems Biology, Center for Complexity Sciences, National Autonomous University of México, Mexico City, Mexico

Jesús M Siqueiros-García
Ethics, Legal and Social Studies Department, National Institute Genomic Medicine, Mexico City, México

Marc Timme
Network Dynamics, Max Planck Institute for Dynamics and Self-Organization (MPIDS), D-37077 Gottingen, Germany
Bernstein Center for Computational Neuroscience (BCCN) Gottingen, ¨ D-37077 Gottingen, Germany
Faculty of Physics, Institute for Nonlinear Dynamics, University of Gottingen, ¨ D-37077 Gottingen, Germany

Jose Casadiego
Network Dynamics, Max Planck Institute for Dynamics and Self-Organization (MPIDS), D-37077 Gottingen, Germany

LIST OF ABBREVIATIONS

AC	Activator criterion
ABS	Anti-lock braking system
CBD	Central Business District
CV	Coefficient of variation
CI	Coincidence Index
CPT	Conditional Probability Table
CNs	Connection networks
CME	Continuous medical education
CNN	Convolutional Neural Network
CC	Cross correlation
DPI	Data Processing Inequality
DFU	Diabetic foot ulcers
DAG	Directed acyclic graph
DALY	Disability Adjusted Lost Years
DBN	Dynamic Bayesian Network
EI	Educational Influential
EC	Effective connectivity
ET	Effectivity threshold
EM	Expectation-Maximization
EKF	Extended Kalman filter
FC	Functional connectivity
GNW	Gene Net Weaver
GRN	Gene regulatory network
GEE	Generalized estimation equation
GLMMs	Generalized linear mixed models
GTE	Generalized transfer entropy
GRNs	Genetic regulatory networks
GECIs	Genetically encoded calcium indicators

GEVIs	Genetically encoded voltage indicators
GPS	Global positioning system
GGM	Graphical Gaussian model
GUI	Graphical User Interface
HMM	Hidden Markov Models
IID	Independent and identically distributed
ILP	Inductive logic programming
IG	Information gain
IT	Information theory
IQR	Interquartile range
JE	Joint entropy
LASSO	Least absolute shrinkage and selection operator
MAP	Maximum a posteriori
MF	Maximum flow
ML	Maximum likelihood
MAE	Mean absolute error
MI	Mutual information
NHTSA	National Highway Traffic Safety Administration
NBDA	Network-based diffusion analysis
OSS	Open-source software
PC	Partial-Correlation
PF	Particle Filter
PR	Precision-recall
PA	Preferential attachment
RENM	Rational Erdös number model
ROC	Receiver operating characteristic
RC	Repressor criterion
RBNFNs	Rule-based neurofuzzy networks
SNA	Social network analysis
SNS	Social Network Sites
SM	Statistical mechanics
TE	Transfer entropy
UKPDS	United Kingdom prospective diabetes study
UHI	Urban heat island

PREFACE

This edited book, *Probabilistic Inference and Statistical Methods in Network Analysis*, is a collection of contemporary open access articles which highlight the development of computational methods for constructing social and biological networks; detecting the topological structure of a network and identifying important nodes within.

This book features two classes of computational methods currently used in network analysis: (a) model-free methods based on statistical and information theory measures such as centrality, correlation, cross-correlation, and partial-correlation, mutual information, joint entropy, and transfer entropy; and (b) generative model-based methods.

The open access articles are organized around the following research domains:

- **Social Networks:** Chapters 1 to 5 focus on the behavioral analysis of social networks, featuring centrality-based methods used in the detection communities, and in the identification of leaders. Ghafoor & Niazi (2016), for instance, analyze the behavioral influence and interests of people participating in large social networks by measuring the betweenness, closeness, and eigenvector centralities in their data. Shokoohi, Nedjat & Majdzadeh (2013) apply density, centrality and reciprocity measures for analyzing the knowledge transfer in the health and medical providers community. Ströele et al. (2018) introduce measurement models based on maximum flow and resistance distance for evaluating the interactions and relationships in social networks. Zhang et al. (2014) propose a network opinion leader recognition method based on Markov Logic Networks. Farine & Whitehead (2015) discuss network analysis as a toolbox for describing and investigating a social system of species or population.
- **Neural Networks:** Chapters 6 to 7 focus on the inference of neural connectivity by applying information theory and statistical measures. Main methods are discussed in Abril et al. (2018). Betzel & Bassett (2017) discuss the prospects of generative models for the analysis of the neural networks. The use of cost-integrated and generalized entropy

measures as topological metrics of interest are investigated by Ginestet et al. (2011) and Gajowniczek et al. (2018).

- **Gene Regulatory Networks (GRNs) and Biological Systems:** Chapters 8 to 9 focus on the development of computational methods for recovering GRNs from gene expression data. This includes the tree-based ensemble methods developed by Huynh-Thu et al. (2010) ; a hybrid rule-based neuro-fuzzy algorithm developed by Manshaei et al. (2012). Gill, Datta & Datta (2010) describes a methodology for differential network analysis. Allen et al. (2012) provide a comprehensive evaluation of statistical methods based on correlation and information-theory measures for constructing large-scale GRN and for detecting highly connected genes within a network.
- **Bayesian Networks:** Chapters 10 and 11 demonstrate potential applications of Bayesian networks for data analysis and statistical inference. Bayesian networks are used for studying the interaction of various malocclusion maxillofacial features during growth and treatment (Scutari et al., 2017); determining the cause of road accidents in road accidents (Zou & Yue,2017); inferring mesoscale structures in networks (Tunç & Verma, 2015).
- **Complex Networks:** Chapters 12 and 13 are dedicated to the analysis of complex networks according to their topological structures (Hernández-Lemus and Siqueiros-García, 2013) and dynamics (Timme & Casadiego, 2014).

The intended audience of this edited book will mainly consist of researchers and graduate students in the Natural and Computer Sciences. The book is also of particular interest to scientists and engineers in areas such as machine learning, data mining, information theory, computational neuroscience, and biological systems. It is suitable for readers with basic knowledge of statistical inference, differential equations, calculus, algebra, graph theory, scientific modelling and computer simulation.

SECTION I
SOCIAL NETWORKS

CHAPTER

1

USING SOCIAL NETWORK ANALYSIS OF HUMAN ASPECTS FOR ONLINE SOCIAL NETWORK SOFTWARE: A DESIGN METHODOLOGY

Faiza Ghafoor[1] and Muaz A. Niazi[2]

[1] Department of Software Engineering, International Islamic University, Islamabad, Pakistan

[2] Department of Computer Science, COMSATS Institute of IT, Islamabad, Pakistan

CITATION (APA): Ghafoor, F., & Niazi, M. A. (2016). Using social network analysis of human aspects for online social network software: a design methodology. Complex Adaptive Systems Modeling, 4(1), 14. (19 pages)

ABSTRACT

Background

Online social networks share similar topological characteristics as real-world social networks. Many studies have been conducted to analyze the online social networks, but it is difficult to link human interests with social network software design.

Purpose

The goal of this work is to propose a methodology involving the analysis of human interactions for use in designing online social network software.

Methods

We propose a novel use of social network analysis techniques to elicit requirements in order to design better online Social network-based software. The validation case study involved the collection of real-world data by means of a questionnaire to perform a network design construction and analysis. The key idea is to examine social network to help in the identification of behaviors and interests of people for better software requirements elicitation.

Results

The validation case study demonstrates how unexpected centrality measures can emerge in real world networks. Our case study can thus conducted as a baseline for better requirement elicitation studies for online social network software design. This work also indicates how sociometric methods may be used to analyze any social domain as a possible standard practice in online social network software design. Overall, the study proved the effectiveness of the proposed novel methodology for the design of online social network software. The methodology specifically improves upon traditional methods for software design by involving social network modeling and analysis to first study the behavior and elicit requirements to develop more resilient online social network sites.

Keywords: Online social network, Social network analysis, Centrality measures, Software requirements, elicitation, Software Engineering

INTRODUCTION

Online social networks, also known as Social Network Sites (SNS) are web-based services allowing users to communicate with each other - whether for personal or professional usage. Boyd and Ellison have defined SNS in their paper as "Web Based Services that allow individuals to (1) construct a public or semi-public profile within a bounded system (2) to articulate a list of other users with whom they share a connection and (3) view and traverse their list of connections and those made by others within the systems" (Ellison 2007). These social network sites not only allow individuals to meet strangers, but also provide them the opportunity to regularly interact with their connections living across the world.

Over time, however, it is clear that many social network sites have been launched and closed due to various reasons ranging from security issues to unsatisfactory and missing features and from the launch of new and more interesting social network sites. Still, the current software design methodologies do not have any means to ensure the effective requirements elicitation for SNS. As a result, many SNS projects have failed due to hitherto unidentified reasons (Ellison 2007). We believe that the complexity of larger groups of end users is something which is currently not catered for in existing software design methodologies. The broad application of social network sites and the virtual presence of users make it challenging to elicit user requirements.

Requirement elicitation has always been considered a long and complicated process—especially when it comes to web-based services—such as web-based information system (Yang and Tang 2003), web-based social networks, or in general, any software custom-designed for online social-network software for interaction in communities. Yang and Tang have discussed these difficulties of requirement elicitation in web-based services and have presented a three stage model of requirement elicitation:

- When we come to web-based services, users are generally diverse.
- It is very hard to identify key users in such a large population.
- User requirements are always changing and even differ from culture to culture.
- Users may grow in future.

- There are no traditional methods to gather requirements (Interviews or observations) for web-based services as they become impractical if we consider the facts we have discussed above.

Sorenson and Skouby have proposed the approach of requirement elicitation by using interviews (Sorensen and Skouby 2008). They however have also subsequently discussed the disadvantages of using interviews. The biggest problem is the limited number of users to be interviewed. In another paper, Tang and Yang (2006) have discussed the drawbacks of taking survey-based studies of online users. They note that a key problem is getting no response back from the intended members of the network.

Therefore, by considering all these previous studies, it is quite clear that online social networks are unlike traditional software. They involve building software for a much larger and much diverse group of users from the ones who can be practically interviewed. As a result, traditional software engineering approaches such as interview-taking and conducting surveys are not enough to ascertain the specific requirements related to behavior of the end-users-requirements which, if not elicited, can result in possible failure of the software. One key reason, Online SNS can fail is because traditional software requirement elicitation techniques do not have any means of discovery of human behavioral influences. The key novel idea in the proposed methodology is to employ the use of social network modeling and analysis (Niazi and Hussain 2012) to understand the behaviors and interests of people for better elicitation of software requirements.

Social networks can be analyzed by analyzing structural characteristics of the network. The key identifying features of social networks are the users and their connections (Howard 2008; Oinas-Kukkonen et al. 2010). These connections make the structure of the network. While not without their inherent problems, as noted by Batool and Niazi (Batool and Niazi 2014), structural and topological characteristics have been used in various studies to understand the nuances of human behavior in social networks (Shapiro and Varian 2013). Online social network sites can also be analyzed in detail by applying social network analysis (Bonchi et al. 2011). Previous researchers have found many similarities between offline and online social networks. E.g. both type of networks can be scale-free and, at times, some of them can also follow the power law distribution (Chun et al. 2008). Several measures have been suggested yet to analyze the structural characteristics of the network. One of these is commonly known as the centrality measures (Hanneman et al. 2001). The centrality measures are important because they tell us about the most influential and powerful node in the network. Many centrality measures have been proposed yet; the most famous ones are degree centrality, closeness centrality, eccentricity and betweenness centrality (Hanneman and Riddle 2005). The fifth, most popular one is eigenvector centrality measure which has been proposed by Bonacich (Bonacich 1972). This centrality measure are usually used to find the most influential person in the network. Still researchers have noted that using these measures mechanically may not be the most effective means of understanding the interaction structure of the social network (Batool and Niazi 2014).

There are numerous studies that have been conducted to understand the social influences on behaviors of people by using social network analysis (Goyal et al. 2010; Blansky et al. 2013; Hill et al. 2010; Christakis and Fowler 2007; Christakis and Fowler 2008; Mednick et al. 2010). However, the goal of this work is to propose a novel software requirements elicitation methodology involving the analysis of human interactions for use in designing online social network software. Using a network of 498 girls living in a university hostel, we have analyzed the role of the strength of friendship relationship (for example best friend, friend and acquaintance) besides their everyday life habits (1) whether strength of relationship with different friends affects their sleeping habit (2) whether a girl's eating habits are associated with her social contacts (3) whether girls take influence of each other while doing exercise or not. We believe that the conducting of social network modeling and analysis has allowed for a better understanding of the hidden behaviors of the large-scale group.

This design methodology gives us the idea to understand human aspects for online social networks by using social network analysis.

The outline of the rest of the paper is as follows: we first present background related to the failure of online social networks or SNS. In the methodology section, we give the details of the proposed design methodology. We also present the social network analysis techniques used. In the results section, we discuss the analysis results and a detailed discussion on the implications of this proposed design methodology to understand human aspects for online social networks by using social network analysis (SNA).

FAILURE AND TIMELINE OF ONLINE SOCIAL NETWORK SITES

Here, we present the timeline and failures of online SNS.

Six Degree

It was founded by Andrew Weinreich in 1997 (New York) by using the concept of six degrees of separation. In few years, this Six Degree attracted almost 1 million users (Heidemann et al. 2010; Hanneman and Riddle 2005).

Failure of Six Degree

But within 3 years, this site had to face failure due to poorly developed web technology (Mednick et al. 2010) whereas the founder of this site had accepted this failure with the statement "It was simply ahead of its time" (Ellison et al. 2007). Despite the failure, six degree has proved as a trendsetter for future online SNS.

Live Journal

Live Journal was a SNS based in San Francisco California. It was founded by Brad Fitzpatrick in April 1999. By using this site, users were able to keep a diary, a blog, or a journal with another blogging company. In 2007, six apart sold Live Journal to another Russian company. In 2009, they laid off some employees and moved product development and design to Russia.

Failure of Live Journal

In 2007, 2010, and 2012 this site was blocked by different countries due to several issues such as extremist propaganda perpetuated in its blogs etc (Ellison 2007).

Ethnic Sites

In the following years, many other online sites had been launched such as Asian Avenue, Mi Gente, and Black Planet. These can be classified as ethnic community sites.

Failure of Ethnic Sites

These sites had tried to fill the gap of technical functionality but they also failed in a few years possibly due to limited friendship options. Limited options did not allow users to connect with the maximum possible nodes. This way, the question of better designing of social network sites still remained unsolved (Ellison 2007).

Ryze

In 2001 (San Francisco) Adrian Scott founded Ryze as a business network. Ryze can therefore be considered to serve as a trendsetter and an example model for future business networks (e.g., Linked In).

Failure of Ryze

Ryze too had to shut down in a few years. To the best of our knowledge, the exact reasons of the failure of this site are still unidentified (Ellison 2007).

Friendster

In 2002, Jonathan Abrams founded "Friendster"—primarily as a competitor to online dating sites. It was created on the assumption that friends of friends would make better romantic partners (Boyd 2004). Friendster allowed the access to the profiles within 4 degrees of separation (Boyd and Heer 2006). Till 2004, Friendster enjoyed considerable popularity and was possibly the biggest site of its time. In 2011, Friendster was relaunched and a number of registered users reached over 115 million. In 2012, it remained notably popular in Indonesia. In 2015, Friendster finally suspended their services.

Failure of Friendster

Due to its extremely popular nature, the site began to face technical and social issues. These problems included unscalability in part due to servers and databases not being able to handle the interactions of such a large group of people. Secondly, people often ending up a social dilemma by having to deal with undesired contacts. Furthermore, fake users started to become a part of the system, thereby becoming a viable threat to the privacy of the users. In the meanwhile, expulsion of a group from Friendster for not complying with the regulations resulted in a chain reaction that ended in a large departure of end-users (Ellison et al. 2007).

MySpace

In 2003, MySpace was launched in California. It was able to attract frustrated Friendster users. At the start, music bands from the Los Angeles started using MySpace for advertising VIP passes for popular clubs. MySpace put in a better effort by continuously adding features based on users demand (Ellison 2007). It also allowed users to personalize their pages.

Failure of MySpace

In 2005, News Corporation bought MySpace for $580 million but later sold it for $35 million. The reported reason for this loss was issues pertaining to safety such as sexual interactions between adults and minors. Although some researchers reported it as an exaggerated reason (Boyd and Heer 2006), the end result was a complete failure.

After gathering all the identified reasons of the failures of online social network sites, we propose the idea of modifying traditional methods of software requirements elicitation by the addition of social network modeling and analysis of the expected group of end users.

Launch Time of the Major Online SNS

See Table 1 for details of the timeline for various online SNS.

Table 1: Launch time of famous online social networks

Launch time	Online social networks
1997	Six Degrees
1999	Live Journal, Asian Avenue, Black Planet
2000	Lunar Storm, My Gente
2001	CyWorld, Ryze
2002	Fotolog, Friendster, skyblog
2003	Couch surfing, Linked In, MySpace, Tribe.net, Last.Fm, Hi5
2004	Orkut, Dogster, Flicker, Piczo, Mixi, Facebook (Harvard only), Dodgeball, Care2 (relaunch), Catster, Hyves
2005	Yahoo!360, Youtube, Xanga (relaunch), Cyworld, Bebo (relaunch), Facebook (high schools), Ning, Asian Avenue, Black Planet (relaunch)
2006	QQ (relaunch), Facebook (corporate network), Windows live space, CyWorld (U.S), Twitter, MyChurch, Facebook (everyone)

Timeline

See Fig. 1.

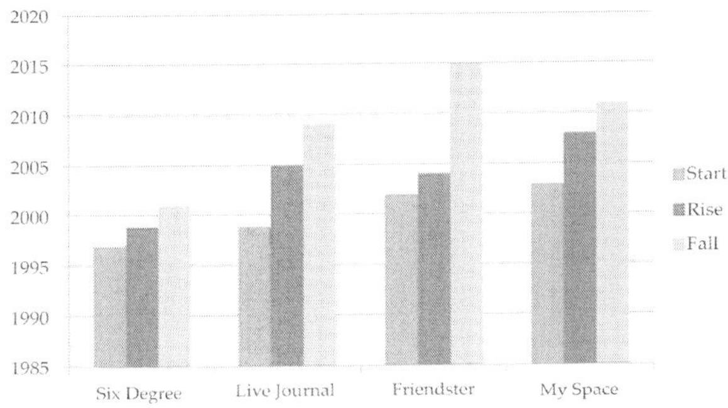

Figure 1: Start, rise, and fall of famous online social networks.

Previous Offline Behavioral Studies

In the past, social network analysis has been enormously used to analyze the behavioral influence and the spread of habits or information across the network. Christakis et al. (2011) have widely presented their research over the spread of behavior across the networks by using social network analysis and statistical analysis. Specifically they have observed the spread of obesity (Christakis and Fowler 2007), spread of happiness in human society (Fowler et al. 2009), diffusion of emotional states either positive or negative (Hill et al. 2010), social contagion of sleep behavior and drug use behavior (Mednick et al. 2010), spread of smoking behavior (Christakis and Fowler 2008), spread of alcohol consumption (Rosenquist et al. 2010) etc. In all these studies they have taken a large social network of people and applied centrality measures and statistical analysis such as logistic regression and generalized estimation equation (GEE) procedures on longitudinal data. In all these studies they have focused on the spread of behavior.

To analyze the spread of academic success, Hiroki Sayama (Blansky et al. 2013) has also used social network analysis, linear regression, and correlation and observed the relatively higher influence of intermediate level friendship for the spread of academic success in school students.

Use of Behaviorial influence in the Proposed Approach

We have used the approach of analyzing behavioral influences in offline social networks to identify the hidden requirements of a person representing a specific social domain or society. As we have discussed earlier that targeting each individual for requirement elicitation in online social networks (OSN) is impossible. The approach of analyzing behavioral influences suggest the idea of taking requirements from an influential person (who can be identified by using centrality measure) who can possibly influence his/her contacts or befriend similar people regarding their choices and interests. In this way, we can analyze the software design requirements of the society by targeting minimum people.

PROPOSED DESIGN METHODOLOGY

This section introduces the set of activities related to the design of online social network software.

The proposed design methodology for online social network software is based on the concept of a traditional method of software development design process. Our idea encapsulates the traditional method of software development in addition to social network analysis of human aspects.

The methodology diagram Fig. 2 represents several phases of the development of online social network software design.

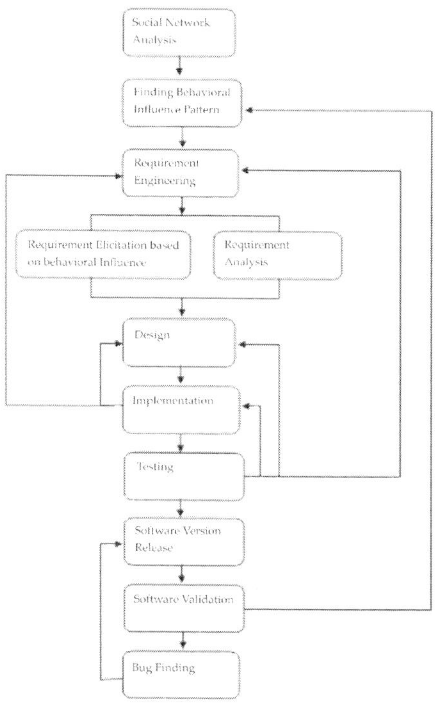

Figure 2: Proposed design methodology diagram, the proposed idea encapsulates the traditional software development processes in addition with social network analysis of human aspects.

- The first activity of the proposed design methodology takes the social network analysis of human aspects into account.
- Once the network analysis is performed, identification of influential people is the next task.
- After finding influential people, the process of requirement engineering would be performed. Requirements gathering from the influential people of the society or that specific social domain would be performed. After requirements gathering, requirements would be analyzed.
- After requirement engineering process, design phase will start.
- After the design phase of online social network software, implementation, testing, version release, software validation processes would be performed.
- In software validation process, the software would be validated against our identified influential people's requirements.

Methods

Source Data

All the data was obtained from the girl's hostel of International Islamic University, Islamabad by developing the questionnaire related to their choices. Girls were asked to list their three best friends and rate them, according to the strength of the relationship. Girls freely recalled their friends and listed their names. We targeted their friends and collected their data. For this purpose, we delivered 600 Questionnaires in the hostels of International Islamic University Hostels and got almost 498 responses.

Social Network Analysis

The diagram in Fig. 3 presents the sequence of techniques of SNA performed to understand the human behaviors and their influence.

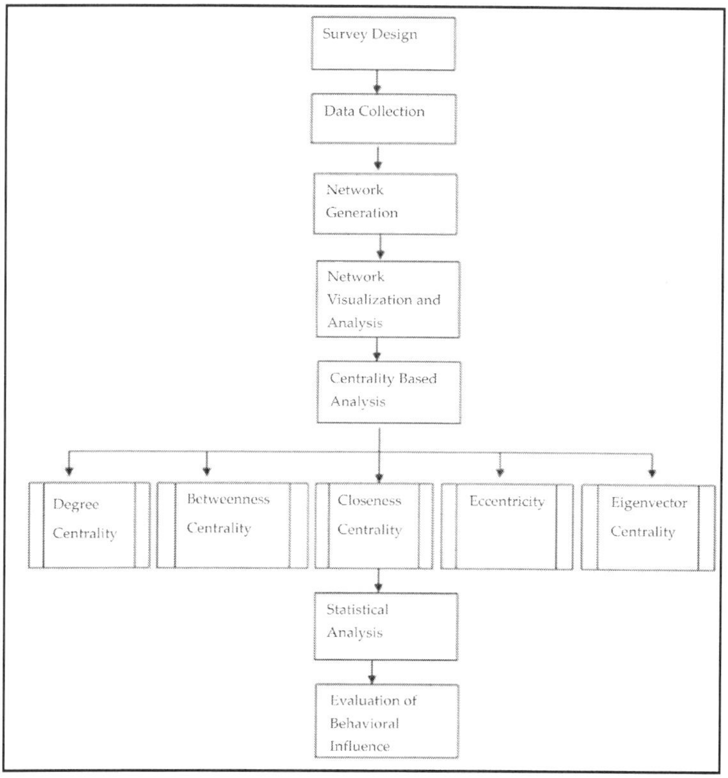

Figure 3: Social network analysis techniques, the *diagram* presents the sequence of the network analysis techniques we have performed to conduct the study. First of all the data was gathered and then a network was generated for the analysis. Centrality measures have been calculated to find the prominent actors in the network. After analyzing prominent actors, statistical analysis was conducted in order to validate the results of social network analysis.

- At the first step of our case study, we design a survey in order to collect data. We target 600 people and get 498 responses.
- After data collection, we first generate a network of friends and represented friends as "nodes" in the network and their relationship as "edges" (directed links) and then manipulate it by deletion of unnecessary nodes (nodes representing girls having no friends), insertion of necessary nodes (targeting missing friends whose presence are necessary to establish the links between friends in the network) and then develop a social network ready for the network analysis.

- Next, we apply visualization and social network analysis on the network.
- We also measure centralities in order to identify the prominent actors and their positions. We have also observed the power law behavior of these centrality measures. After performing centrality based analysis on the whole network, we find a unique node in the network which we have termed as the focal node in our study. This node can be considered as the most influential node having highest eigenvector centrality. Then we extract a cluster of the most influential node and her neighbors (friends) located at the one degree of separation.
- In the end, we have performed statistical analysis to validate the behavioral influence of a focal node in a real life.

The main idea of this study design is to understand the behavioral influence and interests of people for better software requirements.

Methodology used for Analysis of Social Network

First of all, the social network of friends is developed. The main components of the network are friends and their links as nodes and edges respectively. Attributes of the edges represent the strength of the relationship between friends. The basic idea in this social network analysis is that friends according to their position in the social network take the influence of other friends by whom they are surrounded in a social network. Social network analysis is often referred as a structural analysis because this approach is used for exploring structures of the networks. First of all, we made an analysis of the whole network.

Structural Characteristics of the Network

In the following section, we have described different structural properties of the network.

Network size

Network size of our network is 498 nodes connected by 1226 edges. To calculate the network size is the first step to analyze any network in order to get interesting structural properties of the network.

Cluster size

A cluster is a group of people who are strongly connected to each other but sparsely connected to the rest of the people, 64 Clusters are observed in the network.

Degree

The degree is the number of connections of a node in a network. The average degree of our network is 4.9196, where one node represented the highest degree of 24.

See Fig. 4a for degree partition of a network.

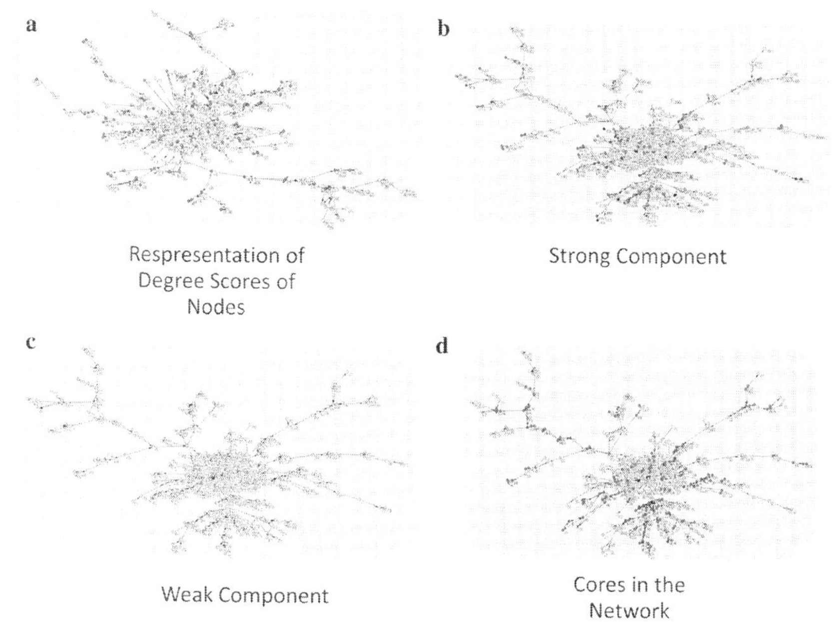

Figure 4: Network analysis and visualization. a This *figure* represents specific degree measure of every node such as the nodes with *yellow* color are labeled with *1* which indicates nodes with 1 degree, the nodes with *green* color have 2 degree, *red* have 3, *blue* have 4, *pink*have 5, *white* have 6 and so on. b This *figure* represents 210 strong components. *Digits*labeled on the nodes (*circles*) show the component number. c This *figure* represents 1 weak component in *yellow color*. d This *figure* represents 6 cores in the network, *yellow*colored nodes belong from core 1, *green* from core 2, *red* from core 3, *blue* from core 4, *pink* from 5 and *white* colored nodes represent core 6.

Components

Components of a network show the level of connectivity of a network, our data set has 1 Weak and 210 Strong components.

See Fig. 4b for strong components and Fig. 4c for weak components.

Cores

Cores are the denser regions of the network. These are helpful in identifying the denser parts of the network. 6 Cores are observed on the network (see Fig. 4d).

These structural properties of the network help us in identifying the denser and strongly connected cluster in the network. In the strongly connected cluster of the nodes (representing people) we can identify friends having strong bonds among friends.

Centrality-based Analysis

To measure the location of each actor, these centrality measures are required. Centrality is a metric for one node, whereas centralization metric tells us about the whole network.

Degree Centrality

A number of links a node has presents the degree centrality of a node. Directed networks are analyzed by using the metrics of in degree and out degree. In degree is referred as the number of links coming towards nodes. Regarding our friendship network, in degree correlates with the popularity of girl among her friends. Out degree is referred as the number of links going out from the node. In the friendship network out degree correlates with the social nature of the girl in a network.

Degree centrality (Freeman 1979) can be calculated by using Freeman's formula of Degree centrality:

$$C_D(v) = \frac{K_v}{n-1} = \sum_{j \in G} \frac{a_{vj}}{n-1} \qquad (1)$$

where Kv is the number of connections of a node and n is the total number of nodes in the network.

We found one unique (focal) node in a network having the highest degree of 24.

Betweenness Centrality

According to the Freeman's approach betweenness centrality is defined as the proportion of times that the node acts as a bridge between two different nodes for sending information. So this is the most favored position of the actors in the network. So the actor offering shortest (geodesic) pathway between other pairs of actors tend to have high betweenness centrality (Hanneman and Riddle 2005).

Betweenness centrality can be calculated using the Freeman's formula:

$$C_B(v) = \sum_{s \neq v \neq t} \frac{\delta st(v)}{\delta st} \tag{2}$$

where δst represents the sum of the shortest path, where "s" and "t" can be considered as two different nodes. δst(v) represents the sum of paths that intersect node v.

Our focal nodes betweenness centrality is 0.05.

Closeness Centralization

Closeness centrality referred as how far a node from all other nodes in the network (Hanneman and Riddle 2005). If we consider the distance of node as a farness of the node. The closeness would be "the inverse of the farness". The less total distance of a node, lesser will be the closeness centrality. A node having low closeness centrality considered to be more central in the network. Closeness centrality can be calculated using the Freeman's formula:

$$C_C(v) = \sum \frac{1}{dist(v,t)} \tag{3}$$

Here v and t are the nodes.

Our focal node's closeness centrality is 0.31.

Eigenvector Centrality

Bonacich's eigenvector centrality is used to calculate the centrality of a node as a function of the centrality of its neighbors. Eigenvector centrality focuses on the fact that a node having the connection to high centrality nodes are more important than the node having links with low centrality nodes (Boudin 2013). This is a measure for finding the most influential node in a network.

$$\lambda v = Av \quad (4)$$

λ is constant, v is the eigenvector and A is the adjacency matrix. Highest eigenvector value is 1 in our network and only one unique (focal) node is observed with 1 eigenvector centrality.

After doing this network analysis, we took that unique node as a focal person and analyzed her friends (best friend, friend or just an acquaintance) and their social influence over each other. We used the Kamada–Kwai algorithm to get the clear image of the focal person and her neighboring friends at the one degree of separation.

(Focal person/node is the term, we are using here for the most influential person according to the network analysis report; person with the highest degree and the highest eigenvector centrality).

Statistical Analysis

To validate the results of structural analysis of the social network we have performed statistical analysis over attribute data of the nodes. Our goal was to evaluate whether a focal person's habits are associated with her friends in a real life. To test this hypothesis, we took a cluster of friends and analyzed the habits of the focal person and her friends by using the statistical tool named "R".

RESULTS AND DISCUSSION

In the following subsection, results of the study are presented in detail.

Network Analysis and Visualization

Our complete network consists of 498 nodes with 1226 edges. Figure 4a shows all degree partition of a network. 24 is the highest degree of one unique node and the average degree of a network is calculated as 4.9196.

Figure 4b shows the strong component in the network. The size of the largest strong component of a network is 18 vertices having 3.614 of the total network. We used Kamada–Kwai algorithm for the spatialisation of the network.

Figure 4c shows the weak component consists of a whole network. The size of the weak component is 498 nodes which are 100 % of a whole network.

Figure 4d shows cores of the network. Our network consists of 6 cores.

Centrality based Visual Analysis

For Centrality based visual analysis, we used Gephi software with "Frutcherman Rheingold layout". Increasing size of the node correlates with the highest centrality value. Figure 5a shows partition of the network, on the basis of degree. The one having purple color with the biggest size is the focal node having the highest degree in a network. Figure 5b shows the visual display of the ranking (size) of the nodes according to the high betweenness centrality and partitioned (colored) on the basis of degree. Figure 5c shows the ranking (size) of nodes with high closeness centrality. Though the network is partitioned (colored) on the basis of degree. Our focal node in purple color is not showing the high closeness centrality. Figure 5d shows the ranking (size) of the nodes with high eigenvector centrality. Again the network is partitioned (colored) on the basis of degree. Our focal node (purple colored) is showing highest eigenvector centrality value in the network, which is the indication of the most influential node.

a

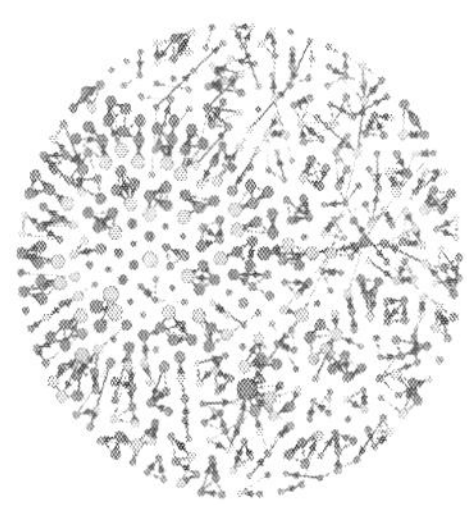

Degree Centrality

b

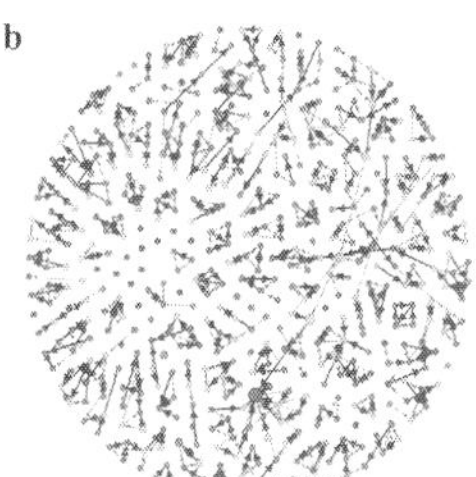

Betweenness Centrality

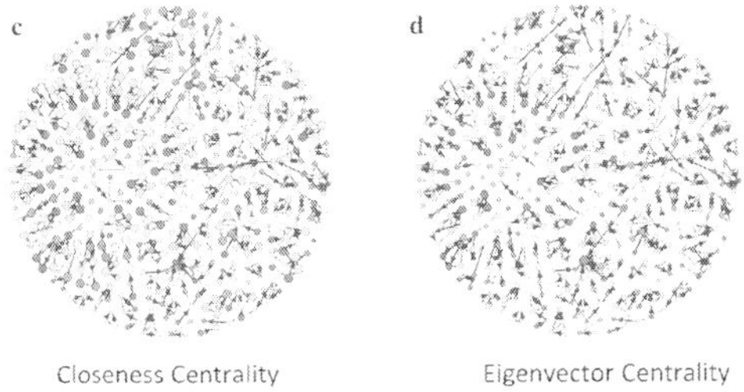

Figure 5: Centrality-based visual analysis: focal node is in *purple* color in all networks, size correlates with centrality measures, whereas colors represent the degrees of the nodes in each network. a The one having *purple* color with the biggest size are the focal node having the highest degree in a network. b A visual display of the ranking (size) of the nodes with the highest betweenness centrality and partitioned (*colored*) on the basis of degree. c The ranking (size) of nodes with high closeness centrality. d The ranking (size) of the nodes with high eigenvector centrality.

After having the complete report of the network, we started focusing the node having the highest degree and eigenvector centrality. These measures have proved this node as the most influential and popular node (person) in the network. We looked for the neighbors of the focal person located at the one degree of separation. We have extracted the cluster including focal person and her neighboring friends located at the one degree of separation from the entire network using "Pajek software". We have used Kamada–Kwai algorithm to get the clear image of the neighboring friends and the focal person. Figure 6a, b shows the focal person with her neighbors in the network.

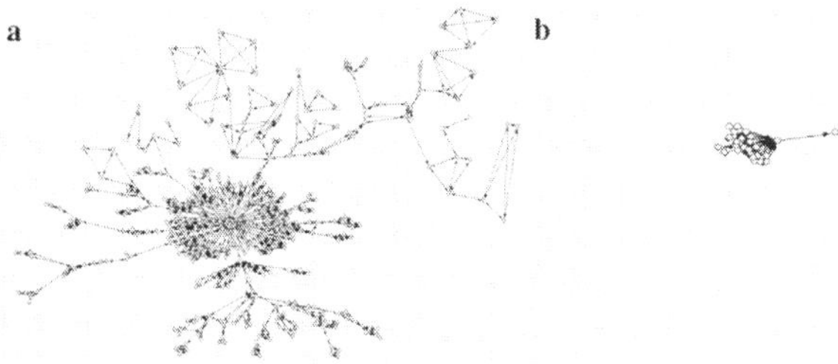

Figure 6: Neighbors of the focal node. a Node with the *blue* color is a focal node having the highest degree and eigenvector centrality, *yellow* colored nodes are the neighbors of the node at one degree of separation, and all the other nodes are in *gray color*. b *Blue* colored node is a focal node; *yellow* colored nodes are the neighbors of the focal node with one degree of separation.

(Focal node is the term, we are using here for the most influential node/ person according to the network analysis report; the node with the highest degree and the highest eigenvector centrality).

Focal Node and the Neighboring Nodes in a Network

Power Law Distribution of Centrality Measures of the Focal Node and the Neighboring Nodes

Power Law distributions are typically fitted to empirical data when the frequency of an event varies with the power of some attribute of that event however, Clauset et al. (Clauset, 2009) have noted the problems in fitting these distributions on data. In this case we have seen if the frequency of nodes varies as a power of centrality measures.

These plots in Fig. 7 represent the correlation of centrality metrics and the frequency of nodes in the cluster of the focal node (focal person) and the neighboring nodes (friends) located at one degree of separation.

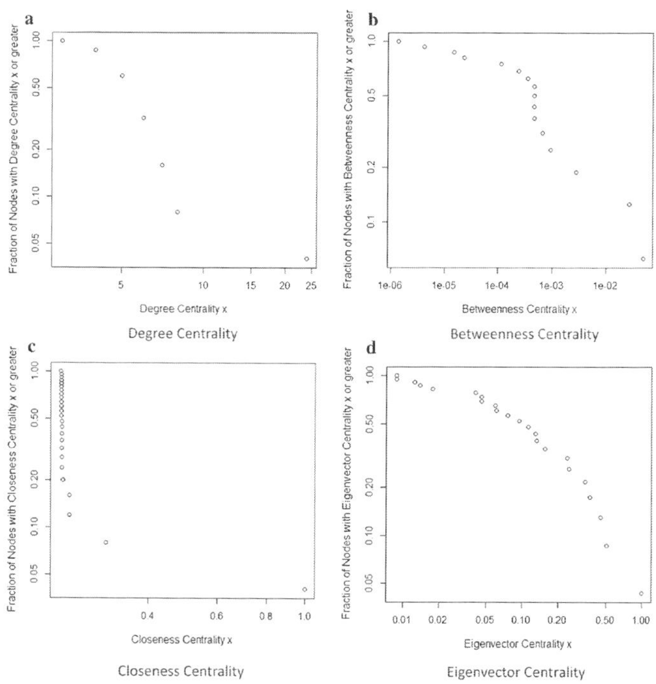

Figure 7: Centrality plots. a Our focal node has a maximum degree centrality 24. b Our focal node betweenness centrality is 0.05. c Our focal node's closeness centrality is 0.31. d Our focal node's eigenvector centrality is 1.

Behavioral Analysis Results

Statistical analysis over the attribute data of the focal node and her neighboring nodes (friends) is performed to analyze the behavioral influences. We took contact type as a function of different variables such as their sleeping habit, the frequency of physical activity in a week, fast food intake and fresh fruits and juices intake to analyze whether the focal person and her contact type (Best friend, friend, acquaintance) share same interests or not.

Figure 8 show the strip charts of the habits of focal person and her contacts. The goal of statistical analysis of their behavior is to understand either people living in the same community take the influence of their friends or not. Additionally, we have tried to validate our network analysis result which has indicated 1 unique node as the most influential node in the network (which we have termed as the focal node in our study).

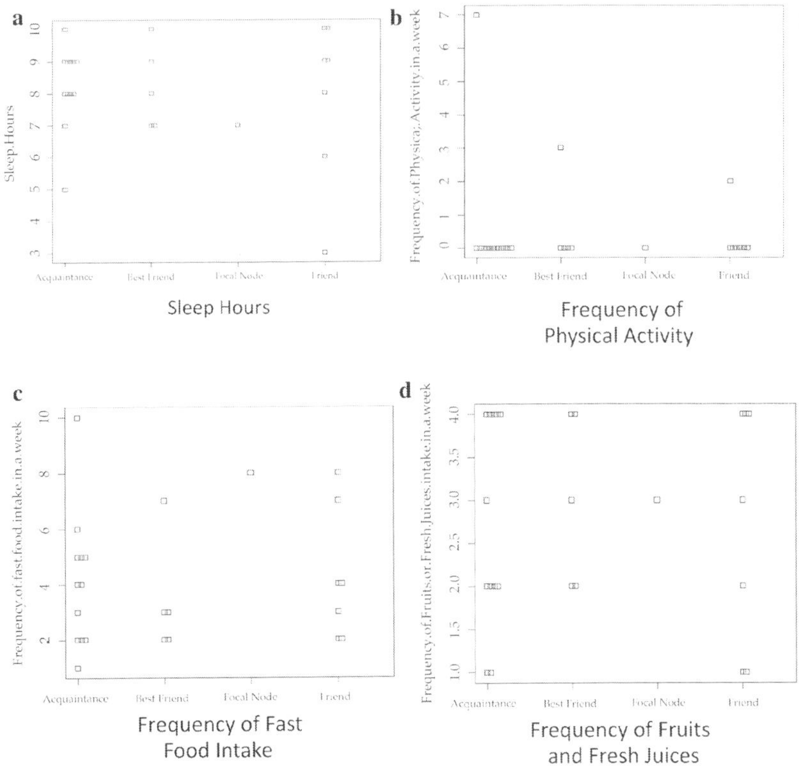

Figure 8: Behavioral analysis using strip chart. a Focal node sleeps 7 h in a day. b Physical activity of a focal node in a week is 0. c Focal node's fast food intake is 8 times in a week. d Focal node's intake is 3 times in a week.

These results showed a slight similarity in their interests.

"R" statistical tool has been used to perform statistical analysis.

Discussion

Next Generation of Social Media and Understanding of Behavioral and Cultural Motives

Adam Smith in "The theory of Moral Sentiments" noted that to understand the social phenomenon, one must incorporate the multitude of psychological and cultural motives. Pent et al. (Cebrian et al. 2016) in their paper have referred the Smith and suggested that "modeling the observable processes and the underlying motivational dynamics can pay tribute to the Smith's nuanced understanding of human nature and this way next generation of social media can be designed".

Manifestation of Actual Real Life Personalities in Online Social Networks

Gosling et al. (2011) have stated that how offline personalities reflect in the online social domain. To map out the connection between original personality and online social network behavior, they have presented two studies and examined how real personality traits are expressed in Facebook. Their correlation results showed the number of links between Facebook information and respondents actual personality. So they concluded that social and real personality processes are parallel in virtual and non-virtual environment. They suggested that future research should examine how other processes such as social influence and other aspects of personality (attitude, value etc.) are manifested in virtual and non-virtual environment.

Lately in 2015 (Dunbar et al. 2015), latest research has been conducted on Facebook, twitter, and offline social network data sets. They have found sociological similarities between online and offline environments, as implied by the data. In addition, they have also found similar structural characteristics between offline (non-virtual) and online networks (virtual environments).

In another research Lewis et al. (2012) discussed the problem that, do people pick friends who are similar to them or do they take the influence of their friends over time? By using Facebook activity data of college students they analyzed the choice of students regarding their music and movie interests. They picked the people who were already friends in their real life too. Their finding suggests that friends share their interest not because they influence each other but also because they become friends due to similar nature. They have stated that selection and influence play a very important role in relationships and tastes expressed online not only show their psychological preferences but also their presentation of actual self.

By considering these reported ideas, we have proposed this design methodology involving the social network analysis techniques on the offline social network in order to elicit requirements for designing the online social network software. Similar structural properties of network and psychological preferences of people participating in virtual and non-virtual environments suggests the idea of taking an efficient approach of analyzing real life interests of influential people in order to establish future online social network sites.

If we consider the fact of the success of Facebook, and get back to the history this fact is inevitable that Facebook too was first launched for a specific social domain (Harvard only) and then it became famous worldwide in coming years.

Summary of Discussion

From 1997 to 2004, many famous online social networks have been launched and had to face drastic failures. The reasons behind many of these failures have remained unidentified. The solution to these problems needs to be worked on. To the best of our knowledge, no one has tried to analyze the behaviors of people participating in the network before launching the website. All the websites were launched on the trial basis and the fate decided their success and failures.

By considering the failed ideas of taking interviews and online surveys of social networks. The idea of offline social network analysis is proposed in our study. By using this approach, we can elicit requirements for use in the design of social network based software. The main idea is to use SNA to understand the behaviors (in terms of influence) and interests of people for better software requirements. Our goal is to conduct an example case study acts as a baseline for better software design.

In this study, we have performed the social network analysis of the offline network to gather the information about human behaviors and their influence over each person participating in a network. The structural characteristics of the whole network have been studied and then the centrality measures are calculated of the cluster of friends.

By applying these measures and then after statistically analyzing the attribute data of the people under study, we have deduced that social network analysis in combination with statistical analysis techniques is an effective method to analyze the interest of people participating in the large social network. The best implication of this design methodology is to link the behaviors of people to online networks in order to perceive the patterns of basic human interests and behaviors where targeting single human for requirement gathering seems impossible.

This case study is considered as just a proof of analyzing hidden requirements, as a standard practice in software design.

CONCLUSION

In this paper, we have proposed a design methodology by using social network analysis of human aspects for online social network software.

This methodology presents an idea of analysis of human behaviors, participating in large social networks. We have used the approach of analyzing human choices in offline social networks to identify the hidden requirements of a person representing a specific social domain or society. As we have discussed earlier that targeting each individual for requirement elicitation in OSN is impossible, the proposed approach of analyzing behavioral influences suggests the idea of taking surveys to develop models for analysis using social network methods from the intended group of end users. The presented case study has demonstrated the effectiveness of the proposed approach.

REFERENCES

1. Batool K, and Niazi MA (2014) Towards a methodology for validation of centrality measures in complex networks. PloS one 9(4):e90283
2. Blansky D, Kavanaugh C, Boothroyd C, Benson B, Gallagher J, Endress J, Sayama H (2013) Spread of academic success in a high school social network. PloS ONE 8(2):e55944
3. Bonacich P (1972) Factoring and weighting approaches to status scores and clique identification. J Math Soc 2(1):113–120
4. Bonchi F, Castillo C, Gionis A, Jaimes A (2011) Social network analysis and mining for business applications. ACM Trans Intell Syst Technol (TIST) 2(3):22
5. Boudin F (2013) A comparison of centrality measures for graph-based keyphrase extraction. In: International Joint Conference on Natural Language Processing (IJCNLP), p 834–838
6. Boyd DM (2004) Friendster and publicly articulated social networking. ACM Press, New York
7. Boyd D, Heer J (2006) Profiles as conversation: networked identity performance on Friendster. In: Proceedings of the 39th annual Hawaii international conference on system sciences, 2006. HICSS'06, vol 3, p 59c
8. Cebrian M, Rahwan I, Pentland AS (2016) Beyond viral. Commun ACM 59(4):36–39
9. Christakis NA, Fowler JH (2007) The spread of obesity in a large social network over 32 years. N Engl J Med 357(4):370–379
10. Christakis NA, Fowler JH (2008) The collective dynamics of smoking in a large social network. N Engl J Med 358(21):2249–2258
11. Chun H, Kwak H, Eom Y-H, Ahn Y-Y, Moon S, Jeong H (2008) Comparison of online social relations in volume vs interaction: a case study of Cyworld. In: Proceedings of the 8th ACM SIGCOMM conference on internet measurement, p 57–70
12. Dunbar RIM, Arnaboldi V, Conti M, Passarella A (2015) The structure of online social networks mirrors those in the offline world. Soc Netw 43:39–47
13. Ellison NB (2007) Social network sites: definition, history, and scholarship. J Comput Med Commun 13(1):210–230
14. Ellison NB, Steinfield C, Lampe C (2007) The benefits of Facebook

"friends:" social capital and college students' use of online social network sites. J Comput Med Commun 12(4):1143–1168

15. Fowler JH, Christakis NA, Steptoe, Roux D (2009) Dynamic spread of happiness in a large social network: longitudinal analysis of the Framingham heart study social network. Br Med J 3:23–27
16. Freeman LC (1979) Centrality in social networks conceptual clarification. Soc Netw 1(3):215–239
17. Gosling SD, Augustine AA, Vazire S, Holtzman N, Gaddis S (2011) Manifestations of personality in online social networks: self-reported Facebook-related behaviors and observable profile information. Cyberpsychol Behav Soc Netw 14(9):483–488
18. Goyal A, Bonchi F, Lakshmanan LVS (2010) Learning influence probabilities in social networks. In: Proceedings of the third ACM international conference on web search and data mining, p 241–250
19. Hanneman RA, Riddle M (2005) Introduction to social network methods. University of California Riverside, Riverside
20. Hanneman Robert A, Riddle Mark, Robert A (2001) Social network analysis. University of California, Riverside, pp 1–154
21. Heidemann J, Klier M, Probst F (2010) Identifying key users in online social networks: a PageRank based approach
22. Hill AL, Rand DG, Nowak MA, Christakis NA (2010) Emotions as infectious diseases in a large social network: the SISa model. Proc R Soc B Biol Sci 277(1701):3827–3835
23. Howard B (2008) Analyzing online social networks. Commun ACM 51(11):14–16
24. Lai W (2016) Fitting power law distributions to data
25. Lewis K, Gonzalez M, Kaufman J (2012) Social selection and peer influence in an online social network. Proc Natl Acad Sci 109(1):68–72
26. Mednick SC, Christakis NA, Fowler JH (2010) The spread of sleep loss influences drug use in adolescent social networks. PloS ONE 5(3):e9775
27. Niazi, Muaz A., and Amir Hussain. Cognitive Agent-based Computing-I: A Unified Framework for Modeling Complex Adaptive Systems Using Agent-based & Complex Network-based Methods. Springer Science & Business Media, 2012

28. Oinas-Kukkonen Harri, Lyytinen Kalle, Yoo Youngjin (2010) Social networks and information systems: ongoing and future research streams. J Assoc Inform Syst 11(2):3
29. Rosenquist JN, Murabito J, Fowler JH, Christakis NA (2010) The spread of alcohol consumption behavior in a large social network. Ann Intern Med 152(7):426–433
30. Shapiro C, Varian HR (2013) A strategic guide to the network economy. Harvard Business Press, Information Rules
31. Sorensen L, Skouby KE (2008) Next generation social networks-elicitation of user requirements. In: IEEE 19th international symposium on personal, indoor and mobile radio communications, 2008. PIMRC 2008, p 1–5
32. Tang J-H, Yang H-L (2006) Emergent user roles and perceived requirements in a social-oriented community. Libr Rev 55(8):508–519
33. Yang H-L, Tang J-H (2003) A three-stage model of requirements elicitation for web-based information systems. Ind Manag Data Syst 103(6):398–409

CHAPTER 2

A SOCIAL NETWORK ANALYSIS ON CLINICAL EDUCATION OF DIABETIC FOOT

Mostafa Shokoohi[1] , Saharnaz Nedjat[2] and Reza Majdzadeh[2,3]

[1] Research Center for Modeling in Health, Institute for Futures Studies in Health, Kerman University of Medical Sciences, Kerman, Iran

[2] School of Public Health, and Knowledge Utilization Research Center (KURC), Tehran University of Medical Sciences, #12, Nosrat East North Kargar, Tehran, Iran

[3] Endocrinology and Metabolism Research Institute, Tehran University of Medical Sciences, Tehran, Iran

ABSTRACT

Introduction

Identification of Educational Influentials (EIs) in clinical settings helps considerably to knowledge transfer among health and medical practice provid-

CITATION (APA): Shokoohi, M., Nedjat, S., & Majdzadeh, R. (2013). A social network analysis on clinical education of diabetic foot. Journal of Diabetes & Metabolic Disorders, 12(1), 44. (7 pages)

ers. The aim of this study was identifying EIs in diabetic foot ulcers (DFU) by medical students (clerks, interns and residents) and providing their relational pattern in this subject.

Methods

Subjects were medical students at clerk, intern and resident levels in a local educational hospital. A standard questionnaire with four domains (knowledge, communication, participation and professional ethics) was used for identifying EIs. Students introduced those people with these characteristics who referred them for DFU. Respective communication networks were drawn as intra-group (such as resident-resident) and inter-group (such as intern-resident) networks and quantitative criteria of density, in-degree and out-degree centrality and reciprocity were measured.

Results

The network density of clerks-residents (0.024) and interns-residents (0.038) were higher than clerks-attends (0.015) and interns-attends (0.05); indicating that there were more consulting interactions in former networks than the latter. Degree centrality in residents-related networks (clerks-residents=2.3; interns-residents=2.6) were higher than attends-related ones (clerks-attends=1.1; interns-attends=1.7), while they were not statistically significant. However, In-degree centralization, which indicating a degree of variance of the whole network of ingoing relationships, in attends-related networks was greater than resident-related networks.

Conclusion

Resident were consulted with almost as same as attends on DFU. It showed that residents were playing a remarkable role in knowledge transfer and they can be considered as EIs in this clinical setting. It seemed that the availability was the main reason for this key role.

Keywords: Knowledge Transfer, Degree Centrality, Social Network Analysis, United Kingdom Prospective Diabetes Study, Diabetic Foot Ulcer

INTRODUCTION

There are more than 3 million people over 20 years old with diabetes in Iran. One of the major complications is diabetic foot ulcers (DFUs) and the sever consequence is foot amputation. Disability Adjusted Lost Years (DALY) of type 2 diabetes in Iran was estimated at 4.7 years per 1000 [1] and burden of diabetic foot ulcers and organ amputation were assessed to 5848 and 1573 lost years, respectively [2]. In 2002, approximately 100,000 deaths occurred in Iran due to diabetes [3]. The United Kingdom prospective diabetes study (UKPDS) indicated that 20% of diabetic people suffer from diabetic neuropathy within 4 years after diagnosis of diabetes, which reaches to 50% after 15 years [4]. Lavery et al. in a study on Mexican–Americans and non-Hispanic Whites found that foot amputation is 5.9 per 1,000 ones annually [5].

Health care providers and medical staff including physicians and nurses are the key groups, which could prevent deteriorating problems of DFUs by providing appropriate health care for the patients [6, 7, 8]. Studies in Iran have shown that the knowledge, attitude and practice of physicians in the field of diabetes care, especially in the fields of patients" education, treatment and care of complications, was insufficient [9]. Even studies in developed countries including USA and Netherlands have shown that 30-40% of the patients did not receive health care based on best available evidence, and about 20-25% of the provided cares were insufficient or even potentially harmful [10, 11, 12].

Educational Influential (EI) concept and its importance have been introduced to medical education few decades ago. Hiss et al. were among the first people who found the importance of this issue and developed the method for identification of EI in the medical field [13]. Various studies have shown that these people were distinct from other colleagues due to their specific characteristics and had gained an influential and strong role in knowledge transfer and diffusion of innovations. They can lead others to the quality of care and subsequently affect the patients' health [14, 15, 16].

Studies indicated that the most trusted and influential source of information for physicians were their colleagues who communicating with them. Relationships between individuals within a network is as an important source for clinical knowledge and decision making [17, 18]. This knowledge significantly influence decision making and management of the patients [19, 20, 21]. Pathman et al. proposed a process for implementation of guidelines including awareness, intellectual agreement, decision, adoption and adherence to the guideline or its integration in the routine practice [22]. Traditional continuous medical education (CME) programs generally stopped in the level of awareness or agreement so that they did not reach to the adopt stage. Thus, nowadays considering the adoption stage is an important target in the CMEs. EIs are those people who influence on others' opinions, attitudes, beliefs, motivations, judgments and behaviors. They can affect adoption of innovation by other practitioners. Knowledge transfer would be easier by identification of these people since in this way, knowledge reached more effectively to the recipients [23].

One way to determine influential individuals within a network, e.g. a hospital, is the social network analysis (SNA). The interpersonal relationships and knowledge transfer patterns within networks can be investigated by using this analysis. Social network consists of a set of actors and their relationships [24]. Tindall and Wellman defined the SNA as analyzing the social structure and its impacts [25]. This analysis can be used for studying the influence of networks' structure on physicians' attributes, dissemination of medical knowledge and implementation of the guidelines [26, 27, 28].

The aim of this study was identifying the health care providers' social network in a sample hospital in Iran and knowledge transfer relationships regarding DFU. The specific objective was comparing the role of attends and residents in the aforementioned network.

METHODS

Subjects

This study was conducted in one of the educational hospitals with about 400 beds in one of the medical sciences universities in Iran. The hospital had five wards including general, surgery, pediatrics, obstetrics and dermatology with approximately 80 faculty members. 140 students participated in this study including 70 clerks, 45 interns and 25 residents.

Questionnaire

Students were asked to introduce whom they were consulted with for solving their problem (questions) on DFU care. Hiss et al. developed a sociometric instrument for identifying EIs for the first time. It included three domains:

knowledge ability, communication skills and humanism. However, authors of the present study developed an identical sociometric questionnaire which published elsewhere and compared characteristics of EI in developed and developing countries [29]. In the later questionnaire, which was used in the present study, the EIs can be identified by four criteria including high level of knowledge, communication skills, taking into account of the stakeholders and following professional ethics.

Analysis

In the present study, the network was assumed as egocentric. It means that when density is to be calculated, ego nodes and their direct relations are ignored and only relationships between other nodes are considered. Four groups of indicators were estimated in the present study including density, centralization (including in-degree and out-degree), degree centrality and Reciprocity (including dyad and arc).

Density

It is a description of general situation of links or interconnection of points in the network [30]. This indicator shows the ratio of the existing relations between individuals to a maximum number of possible relationships between them which is measured as a value between zero and one. In a network, the more actors exist in the relationship; the network density will be higher [31, 32, 33]. Network density may show the information flow among the nodes.

Centralization

Centralization implies influential people within a network. Network centralization score ranged between zero to one, that is, zero means every network individual have relationship with every other member, and one means that all members are connected to only one individual. In the other words, the centralization implies the degree of asymmetry in the network. When some individuals have more connections than others, it is expected to have a high centralization score. Centralization has two sub measurements in a directed network: 1) *In-degree centralization*: degree of variance of the total network of ingoing relationships in comparison with an ideal star network (which, theoretically, implies the most centralization). Centralization provides an analysis of centrality of consulted individuals. If it is close to one, it means that few individuals of the network are consulted by the rest of the individuals. 2) *Out-degree centralization*: degree of variance of the total network of outgoing relationships in comparison with an ideal star network. It shows consulted individuals' situation, and its high level indicates that small individuals of the network do most of consulting of others [31, 33].

Degree Centrality

The mean number of (in/out) going relationships of a person. People who receive many links may be prominent, while subjects who have many connections to others may be influential. The measure refers to direct connections to an individual only [34].

Reciprocity

Which indicating relationship or interaction between the two individuals symmetrically, that is, if actor A asks consultation from actor B, if actor B asks consultation from actor A too [33]. This indicator was calculated in two ways: 1) *dyad* (or *hybrid*); which means there are several pairs or reciprocated relationship between the actors. This method is obtained from the number of pairs of communication between two individuals within the network divided by existing communications in the network. 2) The *arc* is the other one; i.e. instead of focusing on the number of existing communications, it is focused on the maximum possible number of communications within the network, by focusing on all possible communications within the network.

All of these indicators were calculated for three kinds of communication networks in the following subgroups: A) *Intergroup Communication Networks*: including six subgroups on 1. Clerks with interns, 2. Clerks with residents, 3. Clerks with attends, 4. Interns with residents, 5. Night interns with attends, and 6. Residents with attends. B) *Intra-group Communication Networks*: including three subgroups on 1. Clerks with clerks, 2. Interns with interns, and 3. Residents with residents.

UCInet 6 software for windows was used for determining the centrality criteria, providing maps for indicating interpersonal relationships and visual identification of EIs as well as data analysis. NetDraw 2.41 (Network

Visualization Software) (a UCInet subprogram) was used for drawing network diagrams [35, 36].

The present study was approved by the ethics committee of Tehran University of Medical Sciences (TUMS). An informed consent was obtained from all students who participated to the study.

RESULTS

Figures 1, 2 and 3 present both intergroup and intra-groups connections of the networks. As shown in Figure 1, residents and attends were similarly introduced by clerk as the source of consultation on the DFU. Figure 2 displays connections between interns with residents (2A) and attends (2B). Interns introduced both attends and residents as EIs people. Figure 3A illustrates the relationships within residents when they were seeking for getting the consultation on DFU, it means that residents were not only consulting to the other students, including clerks and interns, but also they were seeking information about DFU from other residents (peers), in addition to the attends (Figure 3B).

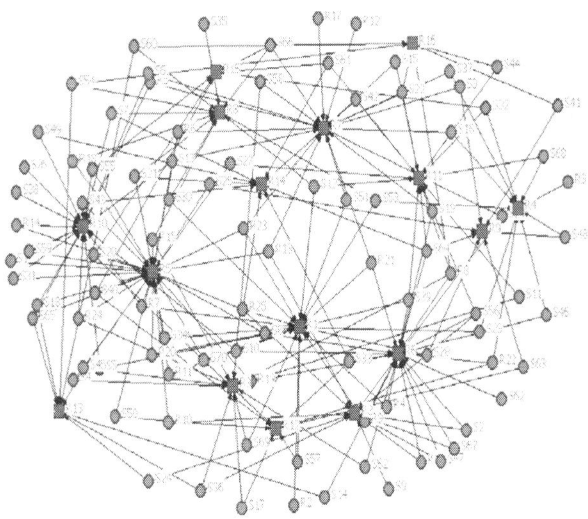

A

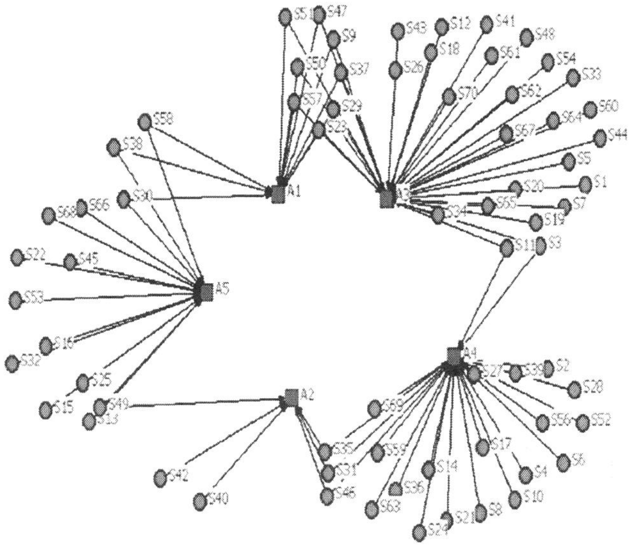

B

Figure 1: Clerks-residents network (A) vs. clerks-attends network (B).

A

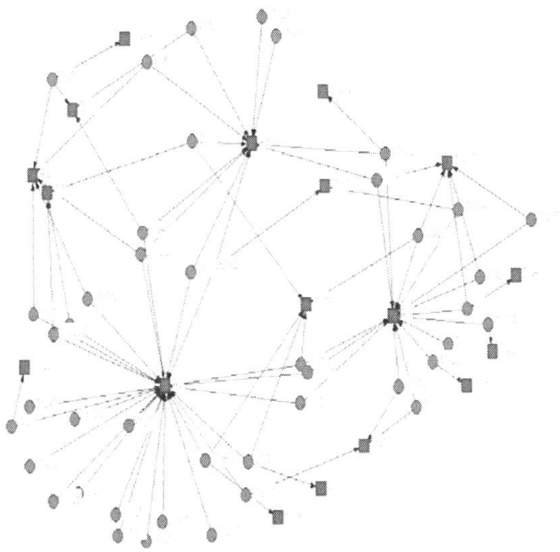

B

Figure 2: Interns-residents network (A) vs. interns-attends network (B).

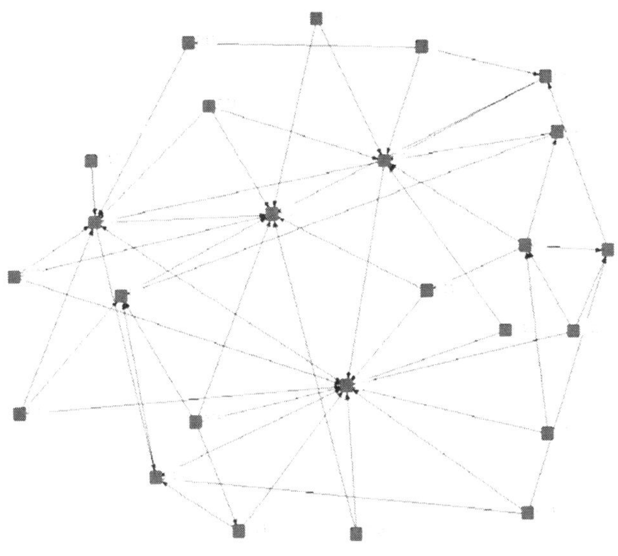

A

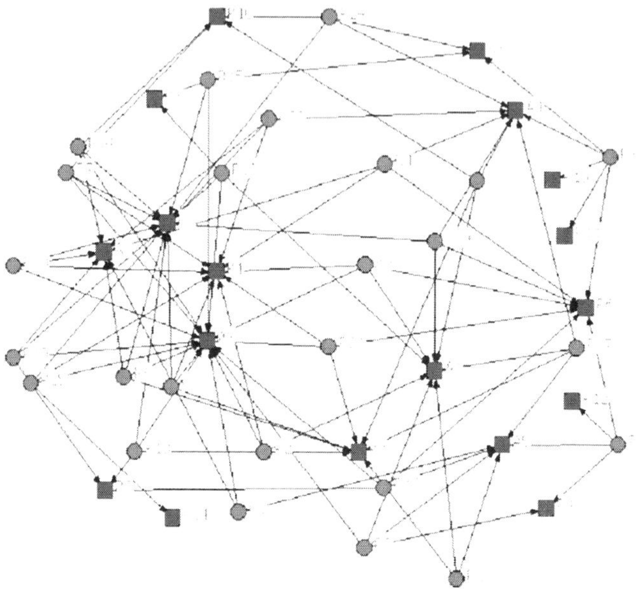

B

Figure 3: Residents-residents network (A) and residents-attends network (B).

Table 1 shows quantitative attributes of the "Intergroup Communication Networks" according to the different subgroups of students. In-degree centralization of the network in attends-related subgroups was not much higher than resident-related subgroups. This indicator for clerks-attends and interns-attends were 42.2% and 52.6%, while for clerks-residents and interns-residents were 31.9% and 38.7%. There were not also considerable differences between attends- and residents-related networks in terms of out-degree centralization of network, which in clerks-attends and clerks-residents, it was 1.1% and 1.8%, and for intern-attends and interns-residents networks equal to 2.6% and 2.0%. However, in- and out-degree centralization belonged to the network of residents-attends subgroups was 45.3 and 6.1%, respectively. Clerks-interns network had the least score in Table 1.

Table 1: Description of inter-subgroup networks

	Inter subgroups networks					
	Clerks with interns	**Clerks with residents**	**Clerks with attends**	**Interns with residents**	**Interns with attends**	**Residents with attends**
In-degree centralization of network*	7.44%	31.9%	42.25%	38.7%	52.6%	45.3%

In-degree centrality, mean±SD (min, max)	0.58±1.74 (0, 9) *	2.3±5.9 (0, 32)	1.1±4.9 (0, 32)	2.6±5.9 (0, 29)	1.7±4.5 (0, 27)	3.3±4.1 (0, 16)
Out-degree centralization of network	1.25%	1.79%	1.16%	2.01%	2.62%	6.1%
Out-degree centrality, mean±SD (min, max)	0.58±0.65 (0, 2)	2.3±0.73 (0, 4)	1.1±0.53 (0, 2)	2.6±0.89 (0, 4)	1.7±0.93 (0, 3)	3.3±1.5 (0, 5)
Density	0.005±0.07	0.024±0.15	0.015±0.12	0.038±0.19	0.05±0.18	0.11±0.31

* The definitions are presented in the body text.

The mean degree centrality in resident-related networks was almost higher than attends-related ones. The mean degree centrality for clerks-residents and interns-residents was 2.3 and 2.6, while for clerks-attends and interns-attends was 1.1 and 1.7, respectively. The maximum degree centrality belonged to resident-attends network, while the minimum was to clerks-interns network.

The mean network density of clerks-residents (0.024) and interns-residents (0.038) were higher than that in clerks-attends (0.015) and interns-attends (0.05) networks, indicating that there were more consulting interactions in former networks than the latter. The maximum mean density was related to residents-attends network (0.11±0.31). The reciprocity indicator was zero for all of the subgroups based on both of dyad and arc methods (not showed in the table).

Table 2 shows the "Intra-group Communication Networks" indicators. The estimated indicators for residents-residents communication network were higher compared to other communication networks (interns - interns and clerks - clerks). Values for in-degree, out-degree and density for resident-resident communication network were 46.7%, 7.8% and 0.091, respectively. These three indicators were zero for clerk- clerk communication network. The mean degree centrality in residents-residents and interns-interns networks was 2.2±0.98 and 0.57±0.69, respectively. Reciprocity indicator for resident-resident subgroup was 3.7% and 7.2% according to the dyad and arc.

Table 2: Description of intra-subgroup networks

	Intra subgroups networks		
	Residents with residents	**Interns with interns**	**Clerks with clerks**

In-degree centralization of network (%)	46.8%	10.2%	0%
In-degree centrality (Mean±SD) (min, max)	2.2±3.3 (0, 13)	0.57±1.2 (0, 5)	0.0±0.0 (0, 0)
Out-degree centralization of network (%)	7.8%	3.3%	0%
Out-degree centrality (Mean±SD) (min, max)	2.2±0.98 (0, 4)	0.57±0.69 (0, 2)	0.0±0.0 (0, 0)
Density	0.091±0.28	0.013±0.11	0.0±0.0
Reciprocity			
Dyad or hybrid	3.7%	0%	0%
Arc	7.2%	0%	0%

DISCUSSION

This study indicated that the communication degree and the distribution of connections in residents-related networks and attends- related ones were nearly similar to each other. In the other words, not only the residents were in relationship with each other, but also the other subgroups including interns and clerks communicated with the residents. This relationship was nearly similar to the communications that the clerks and interns had with their attending.

There are four approaches for identifying EIs: sociometric methods, key informant methods, self-designing methods and observation methods. Most of the studies for identification of EIs used the sociometric instrument, which we used in the present study. A recent systematic review showed that nine trails out of twelve clinical trials used this method [37].

Relatively low density of these networks may be attributed to the type of network, because when density was measured in egocentric method (unlike density measurement based on socio-centric method), ego nodes (those who were consulted with and communicated with other nodes) were not considered. This characteristic of ego-centric network was regarded as a privilege in SNA methods since it sought for finding relationship of other network individuals with each other. In a qualitative study aiming at identification of obstacles in using evidence based medicine and fail to answer their clinical questions, it was shown that several factors including shortage of enough time, lack of access to electronic resources, not having skills required to search electronic resources, lack of tracking unanswered questions, lack of priority for some questions and lack of personal initiative in tracking questions were the barriers. More importantly, team dynamics and institutional culture were considered as the main obstacles. It was indicated that dynamism or team work among medical students especially residents and hospital dominant culture could considerably help transfer of knowledge [38]. In the present study, one reason for justifying low density in communication subgroups, especially in resident - attend and resident - resident subgroups, might be due to the fact that this group of students searched sources other than their colleagues (books, papers, …) for finding the answers for their questions related to the complicated cases such as DFU.

Rappolt performed a qualitative study on 45 family physicians in order to identify sources of clinical information gathering and found that the first source was informal consultants with peers (44%) due to easy accessibility and approachability of them. The second and the third sources for gathering clinical information were asking peer experts (24%) and searching related literature (22%), respectively [17]. In addition, it could be mentioned that the responsibility burden on residents was often higher than the other groups which could justify high communication of this group with others in comparison to other networks.

Reciprocity could show communications among the residents. According to arc indicator, it was 7.2%, implying interaction between the individuals within resident - resident network. However, this indicator was zero for intern - intern and clerk - clerk subgroups, showing lack of interaction among these subgroups. It was clear that the later groups were at the initial states of service provision and possible clinical stresses did not allow them to take risk and rely on each other. However, the residents, who were at higher levels of clinical decision making, consulted with each other. Thus, experience can be considered as an important factor in seeking for consultation from colleagues.

In a survey study conducted at a teaching hospital in Boston, Keating et al. showed that the clinical information had a directed from highly experienced colleagues and accessible colleagues based on the location and schedule. Other predictors to take the clinical consultations among those physicians was having the same gender [39].

Considering in-degree centrality in clerk - attend, clerk - resident, intern - resident, and intern - attend, resident, it is clear that not only the attends could be the consultation source, but also the residents played the role of EI in this setting. High values for these indicators, whether in average or in the range level, suggested the existence of influential individuals within the related networks.

Flodgren et al. in a systematic review found that EIs independently or in conjunction with other interventions could improve evidence-based practice successfully. Effectiveness varied within and between the studies due to high heterogeneities of the collected studies. Therefore, there was no clear way to show and optimize the effectiveness of EIs [40].

Network approaches and science can provide some concepts and techniques to scrutinize the connections between people in any system. Although, network approaches' application in medical care research is fairly new, in many scientific disciplines such as neurosciences, molecular life sciences, and public health, it has been used. Recent research including studies of patient care teams in treatment of diabetes and chronic heart failure in primary care [32], opinion networks of long-term care specialists [41], and connectedness of health care professionals in the treatment of Parkinson [33]. A social network approach may be particularly relevant if actors (people) have flawed knowledge on their performance options and the disease-related outcomes.

ACKNOWLEDGEMENT

The authors would like to thank medical students (clerks, interns and residents) to participate in the present study. A part of the present study was the Master's thesis of the first author (MS) and was financially supported by the Tehran University of Medical Sciences.

AUTHORS' CONTRIBUTIONS

Research concept: RM, MS; developing the design: RM, MS and SN; data gathering: MS; analyzing the data: RM, MS and SN, reporting and preparing the manuscript: RM, MS and SN. All authors read and approved the final manuscript.

REFERENCES

1. Abol Hassani F, Mohajeri Tehrani MR, Tabatabaei O, Larijani B: Burden of diabetes and its complications in Iran in year 2000. *Iran J Diabetes and Lipid Disord* 2005, 5: 35–48.
2. Naghavi M: *A Study on Burden of Disease, Accidents, and Risk Factors in the Islamic Republic of Iran*. Tehran: Ministry of Health and Medical Education; 2008.
3. Delavari AR: *National Program for Control and Prevention of Diabetes*. Tehran: Center for Disease Control; 2004.
4. Tight blood pressure control and risk of macrovascular and microvascular complications in type 2 diabetes: UKPDS 38: UK Prospective Diabetes Study Group. *Bmj* 1998, 317: 703–713.
5. Lavery LA, Armstrong DG, Wunderlich RP, Tredwell J, Boulton AJM: Diabetic foot syndrome: Evaluating the prevalence and incidence of foot pathology in Mexican Americans and Non-Hispanic Whites from a diabetes disease management cohort. *Diabetes Care* 2003, 26: 1435–1438.
6. Aalaa M, Tabatabaei Malazy O, Sanjari M, Peimani M, Mohajeri-Tehrani MR: Nurses' role in diabetic foot prevention and care; a review. *J Diabetes Metab Disord* 2012, 11: 2–6.
7. Aliasgharpour M, Dehghan-Nayeri N: The care process of diabetic foot ulcer patients: a qualitative study in Iran. *J Diabetes Metab Disord* 2012, 11: 2–9.
8. Madanchi NT-MO, Pajouhi M, Heshmat R, Larijani B, Mohajeri-Tehrani MR: Who are diabetic foot patients? A descriptive study on 873 patients. *J Diabetes Metab Disord* 2013, 12: 2–6.
9. Peimani M, Tabatabaei-Malazy O, Heshmat R, Amiri Moghaddam S, Sanjari M, Pajouhi M: Knowledge, Attitude and Practice of physicians in the field of diabetes and its complications; A pilot study. *Iranian Journal of Diabetes and Lipid Disorders*2010, 9: 1–7.
10. Berardis G, Pellegrini F, Franciosi M, Belfiglio M, Di Nardo B, Greenfield S, *et al.*: Are Type 2 diabetic patients offered adequate foot care? The role of physician and patient characteristics. *J Diabetes Complications* 2005, 19: 319–327.
11. Litaker D, Flocke SA, Frolkis JP, Stange KC: Physicians' attitudes and preventive care delivery: insights from the DOPC study. *Prev Med* 2005, 40: 556–563.

12. Mason J, O'Keeffet C, Mclntosht A, Hutchinson A, Booth A, Young RJ: A systematic review of foot ulcer in patients with type 2 diabetes mellitus. *I: Prevention. Diabetic Medicine* 1999, 16: 801–812.
13. HissRG,MacdonaldR,DavisWK:Identificationofphysicianeducational influentials (EI's) in small community hospitals. In *Proceedings of the Seventeenth Annual Conference on Research in Medical Eduaction.* Washington, DC; 1978:283–288.
14. Lippet R, Watson J, Westley B: *The Dynamic of Planned Change*. New York: Harcourt, Brace and World, Inc; 1958.
15. Wenrick JW, Mann FC, Morris W, Reilly A: Informal Education for Practicing Physicians. *J Med Educ* 1971, 46: 299–305.
16. Wright FC, Ryan DP, Dodge JE, Last LD, Law CH: Identifying educationally influential specialists: issues arising from the use of "classic" criteria. *J Contin Educ Health Prof* 2004, 24: 213–226.
17. Rappolt S: Family physicians' selection of informal peer consultants: implications for continuing education. *J Contin Educ Health Prof* 2002, 22: 113–120.
18. Rogers EM: *Diffusion of innovations*. New York: Free Press; 1995.
19. Covell DG, Uman GC, Manning PR: Information needs in office practice: are they being met? *Ann Intern Med* 1985, 103: 596–599.
20. Cullen R: The medical specialist: information gateway or gatekeeper for the family practitioner. *Bull Med Libr Assoc* 1997, 85: 348–355.
21. Williamson JW, German PS, Weiss R, Skinner EA, Bowes F 3rd: Health science information management and continuing education of physicians. A survey of U.S. primary care practitioners and their opinion leaders. *Ann Intern Med* 1989, 110: 151–160.
22. Pathman DE, Konrad TR, Freed GL, Freeman VA, Kock GG: The awareness – to- adherence model of the steps to clinical guideline compliance: the case of pediatric Vacine Recommendations. *Med Care* 1996, 34: 873–889.
23. Nissen ME: An extended model of knowledge-flow dynamics. *Commun Assoc Inf Syst* 2002, 8: 251–266.
24. Hanneman R, Riddle M: *Introduction to social network methods*. Riverside, CA: University of California, Riverside; 2005. (published in digital form at) http://faculty.ucr.edu/~hanneman/nettext/Introduction_to_Social_Network_Methods.pdf

25. Tindall D, Wellman B: Canada as social structure: Social network analysis and Canadian sociology. *Can J Sociol* 2001, 26: 265–308.
26. Fattore G, Frosini F, Salvatore D, Tozzi V: Social network analysis in primary care: The impact of interactions on prescribing behaviour. *Health Policy* 2009, 92:141–148.
27. Gifford D, Holloway R, Frankel M, Albright CL, Meyerson R, Griggs RC, Vickrey BG: Improving adherence to dementia guidelines through education and opinion leaders. A randomized, controlled trial. *Ann Intern Med* 1999, 131: 237–246.
28. Soumerai SB, McLaughlin TJ, Gurwitz JH, Guadagnoli E, Hauptman PJ, Borbas C, *et al.*: Effect of local medical opinion leaders on quality of care for acute myocardial infarction: a randomized controlled trial. *JAMA* 1998, 279: 1358–1363.
29. Shokoohi M, Nedjat S, Golestan B, Soltani A, Majdzadeh R: Can Criteria for Identifying Educational Influentials in Developed Countries be Applied to Other Countries? A Study in Iran. *J Contin Educ Health Prof* 2011, 31: 95–102.
30. Freeman LC: ity in social networks I: conceptual clarifi cation. *Soc Networks* 1979, 1: 215–239.
31. Scott JTA, Crosson JC, Orzano AJ, Stroebel C, DiCicco-Bloom B: Social Network Analysis as an Analytic Tool for Interaction Patterns in Primary Care Practices. *Ann Fam Med* 2005, 3: 443–448.
32. Weenink JW, van Lieshout J, Jung HP, Wensing M: Patient Care Teams in treatment of diabetes and chronic heart failure in primary care: an observational networks study. *Implement Sci* 2011, 6: 2–9.
33. Wensing M, van der Eijk M, Koetsenruijter J, Bloem BR, Munneke M, Faber M: Connectedness of healthcare professionals involved in the treatment of patients with Parkinson's disease: a social networks study. *Implement Sci* 2011, 6: 67.
34. Barbara Z, Hlebec V: Reliability of Measures of ity and Prominence. *Soc Networks* 2005, 27: 73–88.
35. Borgatti SP, Everett MG, Freeman LC: *Ucinet for Windows: Software for Social Network Analysis*. Harvard, MA: Analytic Technologies; 2002.
36. Borgatti SP, Foster PC: The network paradigm in organizational research: a review and typology. *J anagement* 2003, 29: 991–1013.

37. Doumit G, Gattellari M, Grimshaw J, O'Brien MA: Local opinion leaders: effects on professional practice and health care outcomes. *Cochrane Database Syst Rev*2007,24(1):CD000125. 39
38. Green ML, Ruff TR: Why Do Residents Fail to Answer Their Clinical Questions? A Qualitative Study of Barriers to Practicing Evidence-Based Medicine. *Acad Med* 2005, 80: 176–182.
39. Keating NL, Ayanian JZ, Cleary PD, Marsden PV: Factors Affecting Influential Discussions among Physicians: A Social Network Analysis of a Primary Care Practice. *J Gen Intern Med* 2007, 22: 794–798.
40. Flodgren G, Parmelli E, Doumit G, Gattellari M, O'Brien MA, Grimshaw J, Eccles MP: Local opinion leaders: effects on professional practice and health care outcomes. *Cochrane Database Syst Rev* 2007, 1: CD000125. DOI:10.1002/14651858.CD000125.pub3
41. Clark MA, Linkletter CD, Wen X, *et al.*: Opinion networks among long-term care specialists. *Med Care Res Rev* 2010, 67: 102S-125S.

CHAPTER 3

RATIONAL ERDÖS NUMBER AND MAXIMUM FLOW AS MEASUREMENT MODELS FOR SCIENTIFIC SOCIAL NETWORK ANALYSIS

Victor Ströele[1], Renato Crivano[2], Geraldo Zimbrão[2], Jano M. Souza[2], Fernanda Campos[1], José Maria N. David[1] and Regina Braga[1]

[1] Postgraduate Program in Computer Science/PPGCC, Federal University of Juiz de Fora (UFJF), Juiz de Fora, MG 36036-900, Brazil

[2] Systems Engineering and Computation Program/COPPE, Federal University of Rio de Janeiro (UFRJ), Rio de Janeiro, RJ 21941-972, Brazil

CITATION (APA): Ströele, V., Crivano, R., Zimbrão, G., Souza, J. M., Campos, F., David, J. M. N., & Braga, R. (2018). Rational Erdös number and maximum flow as measurement models for scientific social network analysis. Journal of the Brazilian Computer Society, 24(1), 6. (17 pages)

ABSTRACT

In social network analysis, the detection of communities—composed of people with common interests—is a classical problem. Moreover, people can somehow influence any other in the community, i.e., they can spread information among them. In this paper, two models are proposed considering information diffusion strategies and the identification of communities in a scientific social network built through these two model concepts. The maximum flow-based and the Erdös number-based models are proposed as a measurement to weigh all the relationships between elements. A clustering algorithm (*k*-medoids) was used for the identification of communities of closely connected people in order to evaluate the proposed models in a scientific social network. Detailed analysis of the obtained scientific communities was conducted to compare the structure of formed groups and to demonstrate the feasibility of the solution. The results demonstrate the viability and effectiveness of the proposed solution, showing that information reaches elements that are not directly related to the element that produces it.

Keywords: Scientific social network analysis, Information diffusion, Measurement models, Clustering algorithm

INTRODUCTION

Web evolution and the increasing availability of data allow researchers to study the ways in which connections between people are established, and how they evolve over time. Social networks emerged to represent people and their relationships, and thereafter, many efforts have been made to analyze these networks, contributing to a better understanding of the social structures [45]. A relationship is defined as a specific contact or connection type between pairs of actors. Relationships may be direct, when an actor provides information and the other receives it directly, or indirect, when information reaches its destination through intermediate actors.

Since actors are directly or indirectly related, social networks play a key role in information diffusion, increasing the spread of new information and different points of view [2]. Facebook, for example, was very powerful in the Arab Spring in 2010 [20] and Twitter in the US presidential elections in 2008 [22]. More recently, these social networks were also powerful in the 2013 Brazilian protests [5, 46]. All these events are highly linked to social networking platforms, which contributed to disseminate information at specific moments.

Considering (i) information diffusion in social networks [17], (ii) some social network analysis studies ([16, 50] show that even two indirectly related people may be influenced by each other; and (iii) in previous studies from our research group in scientific social networks [41, 40], the authors proposed models for scientific social network analysis that include information diffusion strategies, throughout indirect relationships among actors [17, 27].

This paper aims to describe models capable of exploiting information diffusion in scientific social networks, i.e., how information propagates through people who have different potentials for spreading information (different tie weight). Our first objective is to build a scientific social network which allows developing studies to consider information diffusion in the network itself. It aims to analyze social network considering that all elements can somehow influence one another. *Influence* is related to the individual's ability to affect other people in a social network community. In general, in a scientific context, influential researchers tend to create or strengthen ties, as they can propagate their knowledge, reaching a larger number of people. Our second objective, with the construction of a scientific social network, is to consider the proposed models and the identification of scientific communities using the clustering algorithm *k*-medoids.

As contributions, the authors can highlight the development of two evaluation measures to analyze how information diffusion occurs between pairs of researchers, namely *maximum flow*and *weighted resistance distance*. These measures consider that information travels throughout all possible paths between two researchers, like cascade analysis in information diffusion studies [27]. As a result, the influence that researchers have over each other is calculated.

Another contribution is the modeling of a scientific social network using those two evaluation measures and the use of a clustering algorithm to find scientific communities. The resulting groups are compared in order to validate the *weighted resistance distance* and to evaluate which measure produces the most homogeneous groups. In homogeneous groups, people of similar scientific interest are engaged. In this paper, maximum flow and weighted resistance distance are used as similarity measures.

The paper has five sections, besides its introduction. The "Scientific social networks and information diffusion" section presents the background. The "Measurement models to define tie weight" section discusses the proposed measurement models. The "Case study" section introduces a case study and analyzes the feasibility of the proposed models. The "Final remarks" section presents the conclusions and future work.

SCIENTIFIC SOCIAL NETWORKS AND INFORMATION DIFFUSION

The growth of social networks is due to the Web evolution, and researchers are dealing with its challenges. Some works include mining newsgroups [50], predicting the popularity of links [35, 48, 52] and videos [8], and discovering useful information and patterns from data streams in sensor networks [25].

In addition, structure analysis of social networks helps in identifying critical regions and people [42]. Identifying them is a complex but essential task, as these elements are responsible for collaboration among all other peers in the network, and they are potential elements in information diffusion. Some studies have been conducted to gather information from social networks since appropriate aggregation of multiple social networks could offer a better opportunity for deep user understanding [38, 39].

Scientific social networks are specific types of social networks where two scientists are considered connected if they have co-authored a paper [23, 32]. In the real world, the nodes of a social network tend to be tightly connected, forming groups of people who work together, named *communities*. This effect also occurs in scientific social networks where researchers who have high concentration of common edges define *scientific communities* [33, 34]. Recovering these communities has been a challenging task, and many studies on social network analysis have been developed in order to identify them [48, 50].

In general, social network analysis involves the definition of a model to represent the relationship weights among researchers. According to [2], the higher the edge width/weight which defines the relationship between two individuals, the higher the influence that they exert on each other. Studies have shown that information spreads more easily between elements that have tighter relationships. However, looser relationships also play a key role in information diffusion [16, 24].

By means of a network topology, the behavior of its elements is used with several objectives. There are studies that attempted to model developers' networks to explore the coordination performance of open-source software (OSS) [19] or the behavior of developers in OSS communities [21]. Both apply information diffusion measures to help understand those complex networks. This kind of analysis has been carried out in the health scenario, for instance, for the identification of influential members [37] or communities [11].

This study differs from others in that it addresses (i) the identification of relationships and their quantification and its use by a clustering algorithm to identify *scientific communities*, (ii) considering *information diffusion* concepts, (iii) in relationship weight definition of *scientific social networks* in terms of direct and indirect influence, as well as tight and loose existing connections.

MEASUREMENT MODELS TO DEFINE TIE WEIGHT

In some cases, social network analysis is carried out by a quantitative evaluation that indicates the link weight between related elements [16, 24, 44]. Based on such quantitative analysis, the weight of links between authors is defined, allowing not only the use of other metrics (degree, closeness, betweenness, PageRank, etc.) [43] but also the discovery of communities [50, 32, 14, 34, 49].

However, quantitative analysis alone is not enough to evaluate the effect that indirectly related elements produced on each other. Many studies are being developed in order to evaluate information diffusion [2, 17]. As aforementioned, in these studies, all the elements of the social network have some influence on information diffusion.

Previous studies have shown that there is influence between peer researchers even though they are not directly related [41]. This indirect influence can be calculated in different ways. In this study, a comparison between the maximum flow-based model and the resistance distance-based model was made to evaluate the communication potential between nodes in a scientific social network.

Maximum Flow-based Model

The modeling process of the scientific social network was divided into three steps as follows: number of common relationships between researchers, age of the relationships that link these researchers, and loss of knowledge when the relationship between researchers is indirect [41].

Common Relationships between Researchers

To generalize this model, Eq. (1) considers that there may be different types of relationships between researchers, such as project participation, co-authored publications, advisory work, and technical productions. This study took into account only the co-authorship relationship since the database used has no other relationship type.

$$TR(\alpha)^{a,b} = \sum_{i=1}^{t} \left[\alpha_i \left(\frac{CR_i^{a,b}}{P_i^a + P_i^b} \right) \right], \tag{1}$$

where $TR(\alpha)^{a,b}$ means the weight of relationships between researchers *a* and *b* weighted by the number of relationships of type *i* of these researchers, preventing relationships with the same frequency from having the same weight. In this equation, α_i is the weight of relationship type *i*, *t* is the total number of relationship types, $CR_i^{a,b}$ is the number of common relationships of type *i* between researchers *a* and *b*, and P_i^a and P_i^b mean the total relationships of type *i* of researchers *a* and *b*, respectively. The result of Eq. (1) is normalized by the natural logarithm and min-max normalization (Eq. (2)), and a connected graph is obtained.

$$TR(\alpha)_N^{a,b} = \frac{\ln\left(TR(\alpha)^{a,b}\right) - \min(\ln(TR(\alpha)))}{\max(\ln(TR(\alpha))) - \min(\ln(TR(\alpha)))} \tag{2}$$

where $TR(\alpha)$ is a matrix with all relationship values of each researcher pair.

Relationship's Age

Another factor considered in a relationship is its age. Relationship's age is useful to indicate whether the relationship reflects a link of present elements or if it is just a connection that existed in the past. In this sense, a penalty function was added to Eq. (1) considering the relationship in terms of years, resulting in Eq. (3).

$$TR(\alpha.\rho)^{a.b} = \sum_{i=1}^{t} \sum_{j=0}^{d-1} \left[\rho(RY + j)\alpha_i \left(\frac{CR_i^{a,b}}{P_i^a + P_i^b} \right) \right], \tag{3}$$

where $TR(\alpha,\rho)^{a,b}$ is the weight of the links between researchers *a* and *b* considering the base year of the relationship; *d* is the relationship duration in years. Exact relationships (e.g., co-authorship) have lasted 1 year, $d=1$ and it contributes only once to the relationship weight between the researchers. On the other hand, ongoing relationships have equal duration considering the number of years in effect (e.g., projects), so if the authors worked, for example, in the same project for 3 years, then $d=3$, and it will be considered more than once in the relationship weight. RY is the relationship year, i.e., the year in which the two researchers worked together or when they started to work (co-authored, participated in the same project, and others); t is the total number of link types, α_i is the relationship weight i, and ρ is as defined in Eq. (4).

$$\rho(y) = e^{1/(\mathrm{BY} - y + 1)} \tag{4}$$

where BY is the base year, and y is the year to be considered in the penalty, as previously defined. Figure 1 represents a graph of the penalty function.

Figure 1: Graph of Eq. 4.

After Eq. (3), an $M \times M$ symmetric weighted matrix is obtained representing the relationship weight between each pair of researchers, where M is the number of researchers. As relationship weight represents the similarity between researchers, this matrix was named similarity matrix, as represented in Eq. (5).

$$\mathrm{SM} = \begin{cases} \mathrm{TR}(\alpha, \rho)^{a,b} & \text{if researcher 'a' links to researcher 'b'} \\ 0 & \text{otherwise.} \end{cases} \tag{5}$$

Content Loss in Long Relationships

Communication potential (relation weight) between two nodes is calculated using the maximum flow algorithm, which evaluates all possible paths between two researchers and consolidates the largest flow of knowledge that can be transmitted. This strategy is interesting since it considers all communication possibilities; however, it disregards the number of intermediate elements to define the relationship weight. Figure 2 illustrates this situation, in which A and B represent two elements of distinct groups, where it is necessary to bind element C to one of these two groups. This simple graph shows that there are two ways from C to A and B. Element C is directly linked to A, and C is linked to B through D and E. Using the maximum flow (MF) calculation, obtained as $MF_{CA} = MF_{CB} = 4$, i.e., 2 from the edge between A and C, and from the other path allows passing only 2 through the edge that links E and B. So, maximum flow between C and A is equal to the maximum flow between C and B. However, C is directly related to A, indicating a better communication between them.

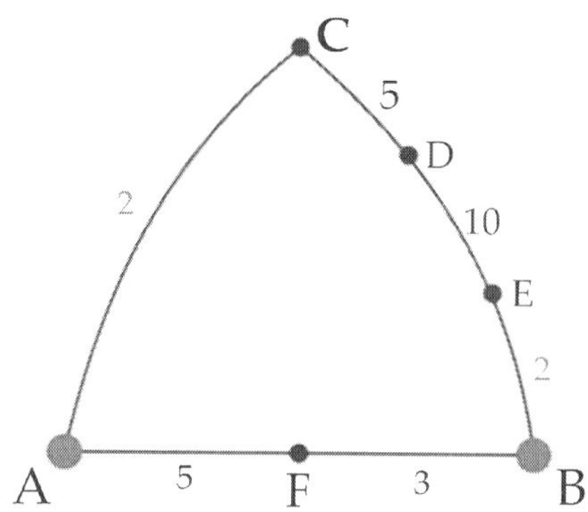

Figure 2: Maximum flow analysis.

Formally, the network maximum flow problem can be interpreted as a graph flow problem. The social graph with flow is represented by $G = (X, U, \mathbf{f})$, in which **f** is a vector of dimension $m+1$ and can be written in the form $\mathbf{f} = (f_0, f_1, \ldots, f_m)$.

Vector f is the flow in graph G, and each of the components indicates the value of the flow between the elements of G. The social graph is represented in this study by a non-oriented graph, so the maximum flow coming out of x_i to x_j is equal to the maximum flow of x_j to x_i, for all $x_i, x_j \in X$.

Consider G as the social graph represented by SM defined in Eq. (5), X is the set of researchers, U is the set of relationships between these researchers, and f is the set of maximum flows between each pair of researchers. Thus, for any x_i, $x_j \in X$, the maximum flow between these two researchers will be equal to $f_w = \sum_{v:(i,v)\in E} \overline{f}_{iv}$, where i is the source node, $v \in X \backslash \{i, j\}$, $\overline{f}_{iv}$ ivrepresents the amount of flow passing from source i to node v and $0 \leq w \leq m$.

The maximum flow was calculated using the Edmonds-Karp algorithm [13]. This algorithm is a variation of the Ford-Fulkerson maximum flow algorithm [15]. The main difference between these two approaches is that the Edmonds-Karp algorithm targets the maximum flow between two elements considering the shortest paths. Thus, it is guaranteed that the algorithm will converge on a finite number of iterations, even for non-oriented graphs.

As the Edmonds-Karp algorithm considers the shortest paths, an adaptation was made in the algorithm so that the maximum flow was penalized with a percentage according to the path length.

In this context, it was considered that information is received by the researcher with a knowledge loss of N%, where N is the number of intermediate nodes between source and receiver. Thus, assuming that the maximum knowledge flow between A and B is MaxFlow^{a,b}, the new relationship weight between them is given by Eq. (6).

$$\mathrm{TR}(\alpha, \rho, N) = \mathrm{MaxFlow}^{a.b} \times (1 - N/100) \tag{6}$$

At the end of the multi-relational scientific social network modeling process, the *maximum flow matrix* with resistance is achieved, which will be used by the clustering algorithm. Algorithm 1 shows the step-by-step of the maximum flow-based model.

Algorithm 1 Maximum Flow-based Model

Input: X: Set of authors; α: weight of relationships set; n: number of intermediate nodes
Output: MF: Max-Flow Matrix returned by Algorithm

```
Begin procedure
MF ← 0;
SM ← 0;                                          // Similarity Matrix
pubDate ← null;                                  // Publication date
ρ ← 0;                                           // Penalty value
    for each integer i in X do
        P ← getPublicationsOf(X_i);              // Return a set of publications of researcher X_i
        for each integer j in P do
            Y ← getAuthorsOf(P_j);               // Return a set of authors of publication P_j
            pubDate ← getDateOf(P_j);
            ρ ← e^(1/(CURRENT_DATE − pubDate + 1));
```

```
            for each integer k in Y do
                a ← getIndexOf(Xi);                      // Get researcher Xi corresponding index in the SM
                b ← getIndexOf(Yi);                      // Get researcher Yi corresponding index in the SM
                SM(a, b) ← SM(a, b) + TR(α,ρ)^(Xi,Yi) ;
            end for
        end for
    end for
    MF ← EdmonsKarpAdapted(SM, n);
end procedure
```

Erdös Number-based Model

Michael Barr proposed the rational Erdös number model (RENM) as a distance measure (Michael [31]). The idea was to consider a social network of researchers who have published together, as being akin to an electric circuit of resistances. To each person, a rational number is assigned representing the total resistance from that node to the center of the network—in this case, the famous mathematician Paul Erdös. If two researchers coauthored one paper, there would be a 1-Ω resistor between them. If they had two coauthored papers, then there would be two 1-Ω resistors connected in parallel, which, by the laws of electricity, are equivalent to a 0.5-Ω resistance. In Michael Barr's proposal, a paper that is written by more than two authors should be represented by a new node in the graph, and $N/4$ Ω resistors should be placed connecting that node to each one of the N authors involved.

The RENM was intended as a distance measure to Paul Erdös, who coauthored many articles and is somehow related to most of the mathematical community. But the idea of representing a social network as an electric circuit of resistances can be applied in other realities, and it is possible to calculate not only distances to a center but also distances between each pair of people. If the distance between each node of a graph is to be calculated, then an algorithm can be used to identify groups of tightly connected nodes and to identify elements that have the shortest distance to all other elements of their groups, i.e., identify the medoids.

Weighted Resistance Distance

Most of the complexity in this model is related to the calculation of the total resistance between any two arbitrary nodes of an electric circuit of resistors. If our problem is simpler, and all resistors could have a nominal value of 1 Ω (as in Fig. 3), then a better-studied situation will be dealing with the calculation of resistance distance in the graphs.

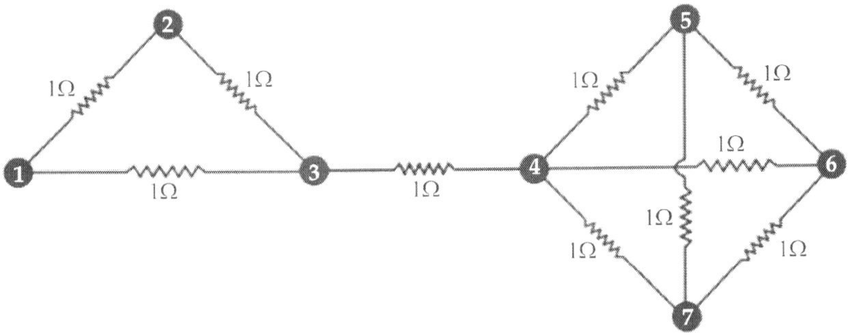

Figure 3: A simple graph where all edges represent 1 Ω resistors.

To solve the resistance distance, there is a method named the determinantal formula, which was proposed by Bapat [4].

To understand the determinantal formula, first, it is necessary to understand how the Laplacian matrix, denoted by $L=L(G)$, of a graph is formed. For each of the nodes i and j, $L_{ij}=-1$ if i and j are connected or 0 otherwise. If $i=j$, L_{ij} is the number of first neighbors of i. For the circuit above, L would be constructed as follows:

$$L(G) = \begin{pmatrix} 2 & -1 & -1 & 0 & 0 & 0 & 0 \\ -1 & 2 & -1 & 0 & 0 & 0 & 0 \\ -1 & -1 & 3 & -1 & 0 & 0 & 0 \\ 0 & 0 & -1 & 4 & -1 & -1 & -1 \\ 0 & 0 & 0 & -1 & 3 & -1 & -1 \\ 0 & 0 & 0 & -1 & -1 & 3 & -1 \\ 0 & 0 & 0 & -1 & -1 & -1 & 3 \end{pmatrix}$$

The determinantal formula states that the resistance distance between two nodes can be obtained by (i) building the Laplacian matrix of the graph, (ii) removing both the lines and the columns related to these two nodes, (iii) calculating the determinant, and (iv) dividing it by the number of spanning trees (Eq. (7)).

$$r_{ij} = \frac{\det(L(i,j))}{t(G)} \tag{7}$$

Since the number of spanning trees $t(G)$ can be obtained by omitting the ith line and the i^{th}column of L and calculating the determinant, another view of the determinantal formula can be seen in Eq. (8).

$$r_{ij} = \frac{\det(L(i,j))}{\det(L(i))} \tag{8}$$

Unfortunately, our preliminary studies showed that if all resistances are equal, no good groups are then identified. Thus, it is necessary to find an appropriate mathematical method to solve the circuit and obtain the effective resistance between each two authors.

The standard method to calculate the resistance between two points in an electrical network is to solve all equations provided by the first and second laws of Kirchhoff, and also by the Ohm's law [30]. Although being feasible, it would lead to very complex and tedious calculations. The formula for the determinant of the resistance matrix was derived by Bapat [3] and is shown to reduce to the formula obtained by Xiao and Gutman in the unweighted case [47]. Consequently, it was possible to calculate the effective resistance even if the circuit had different valued resistors.

Multiple Terminals Delta-star Transformation

According to Michael Barr's original proposal, new nodes should be created to represent publications involving more than two people. Even though it is possible, the creation of more nodes would imply a bigger matrix, increasing the processing time of the algorithm that calculates the resistance distances. Therefore, this study worked with an alternative approach, which produces the same result, but does not raise the number of nodes.

The approach consists of a generalization of the well-known delta-star transformation on resistive circuits [28], and it was only possible because in this case, all resistances between the center point (publication) and the terminals (authors) are equal. So, when more than two people coauthored a publication, the star-like array of $N/4\ \Omega$ resistors was replaced by a network of same valued resistors connecting each author to every other author, just like in a complete graph.

To calculate the precise value of the resistors that should be used, the determinantal formula for resistance was applied together with Cayley' formula [7, 9] for the number of spanning trees in complete graphs: $t(K_n) = n^{n-2}$, resulting in Eq. (9), where n is the number of researchers.

$$r_n = \frac{n}{2} \times \frac{n^{n-2}}{\det(L(i,j))} \tag{9}$$

The Model in Action

A scenario consisting of three publications was considered to validate the proposal. The first one was coauthored by three researchers named 1, 2, and 3; the second one by researchers 3 and 4; and the last one by researchers 5, 6, 7, and 8. The application of multiple terminals delta-star transformation formula, described above, leads to the electric circuit of Fig. 4.

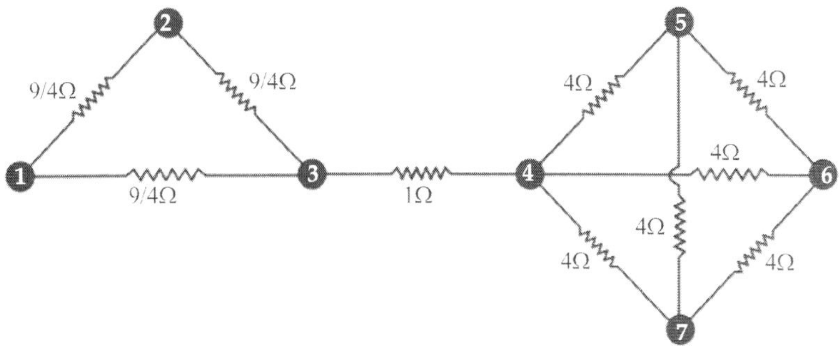

Figure 4: A social network compliant with the RENM paradigm.

The weighted Laplacian matrix of conductance for this RENM network can be written as follows:

$$L(G) = \begin{pmatrix} 0.89 & -0.44 & -0.44 & 0 & 0 & 0 & 0 \\ -0.44 & 0.89 & -0.44 & 0 & 0 & 0 & 0 \\ -0.44 & -0.44 & 1.89 & -1 & 0 & 0 & 0 \\ 0 & 0 & -1 & -1.75 & -0.25 & -0.25 & -0.25 \\ 0 & 0 & 0 & -0.25 & 0.75 & -0.25 & -0.25 \\ 0 & 0 & 0 & -0.25 & -0.25 & 0.75 & -0.25 \\ 0 & 0 & 0 & -0.25 & -0.25 & -0.25 & 0.75 \end{pmatrix}$$

and by applying the determinantal formula to compute the resistances between each two authors, the following values can be found:

$$r(G) = \begin{pmatrix} 0 & 1.5 & 1.5 & 2.5 & 4.5 & 4.5 & 4.5 \\ & 0 & 1.5 & 2.5 & 4.5 & 4.5 & 4.5 \\ & & 0 & 1 & 3 & 3 & 3 \\ & & & 0 & 2 & 2 & 2 \\ & & & & 0 & 2 & 2 \\ & & & & & 0 & 2 \\ & & & & & & 0 \end{pmatrix}$$

After obtaining the resistances between all pairs of researchers, we can run the *k*-medoids algorithm [18, 26] using the resistances produced by the previous calculations as distance measures between each author in the social network. For the above example, and considering $K=2$, the first group was formed by elements 1, 2, 3, and 4 and the second one by 5, 6, and 7, showing that the model is, in some way, feasible. Algorithm 2 shows the step-by-step of the Erdös number-based model.

Algorithm 2 Erdös Number-based Model

Input: P: Set of Publications
Output: RM: Resistances Matrix

Begin procedure
$RM \leftarrow 0$; $resistors \leftarrow 0$;
$circuit \leftarrow 0$; // *Social network compliant with the RENM paradigm*
$localL \leftarrow 0$; // *Laplacian Matrix of complete graph (star-like)*
$L \leftarrow 0$; // *Weighted Laplacian matrix of conductance for the RENM network*
for each *integer i* **in** *P* **do**
  $authors \leftarrow$ getCoauthorsOf(P_i) // *Returns a set of authors of publication P_i*
  if $|authors| > 2$ **then**
    $n \leftarrow |authors|$;
    $localL \leftarrow$ getLaplacianMatrix(n); // *Builds Laplacian Matrix for a complete graph with n nodes*
    $resistors \leftarrow \frac{n}{2} \times \frac{n^{n-2}}{\det(L(i,j))}$;
  else
    $resistors \leftarrow 1$;
  end if
  addResistors($authors$, $resistors$, $circuit$); // *Adds the authors and the resistors among then in the circuit (RENM network)*
end for
$L \leftarrow$ buildWeightedLaplacian($circuit$); // *Builds the weighted Laplacian matrix of conductance for the RENM network*
$nAuthors \leftarrow |L(i)|$;
for each *integer* $0 \leq i < nAuthors$ **do**
  for each *integer* $i < j < nAuthors$ **do**
    $RM(i,j) \leftarrow \frac{\det(L(i,j))}{\det(L(i))}$;
  end for
end for
end procedure

CASE STUDY

The case study was conducted in order to evaluate the measurement models proposed to define tie weight. The scope of the evaluation, based on the GQM method [29], was described as follows: "*To analyze* the scientific communities generated by clustering techniques and/or information diffusion in social networks *for the purpose of* evaluating scientific communities homogeneity and compare the two proposed approaches *in relation to* information diffusion *from the point of view* of the researchers *in the context of* scientific communities obtained based on researchers' information diffusion potential in a scientific social network."

Based on the scope of the case study, the main research question and three secondary ones were defined:

- How are scientific communities organized considering individual influences and measurement models that quantify information diffusion among researchers from a scientific social network?
- RQ1: Which information diffusion approach produces more homogeneous scientific communities?
- RQ2: Does the use of cluster analysis retrieve real scientific communities considering the activities developed by the researchers?
- RQ3: Are researchers from the same scientific community connected through direct or indirect relationships?

In view of the above research questions, the case study was a suitable choice as a research method, considering that a contemporary phenomenon was evaluated, in its "real-world context," according to Yin [51].

Dataset and Case Study Process

Data for the construction of scientific social network were selected from DBLP,[1] one of the databases commonly used in scientific studies on social networks [10, 50]. For this case study, the data of five out of eight high-quality Brazilian institutions were extracted (COPPE/UFRJ, PUC-RJ, UFPE, UFRGS, and UFMG). Altogether, 169 researchers were analyzed from the area of Computer Science.

Figure 5 shows the DBLP data used in this study in the form of a social network. The nodes were highlighted according to its degree. As can be seen, there are many relationships between the pairs of researchers, indicating that these researchers have co-authored more than once.

Figure 5: DBLP scientific social network.

In summary, the network contains the following: number of relationships, 1401; clustering coeficient, 0.325; and network density, 0.035. In order to explore some complex network metrics, we plotted graphs. Figure 6 shows the betweenness centrality distribution of the scientific network. The betweenness centrality of a node reflects the amount of control that this node exerts over the interactions of other nodes in the network [45]. This measure favors nodes that join communities, rather than nodes that lie inside a community. As can be seen in the network of Fig. 5, only three nodes have more control over other nodes.

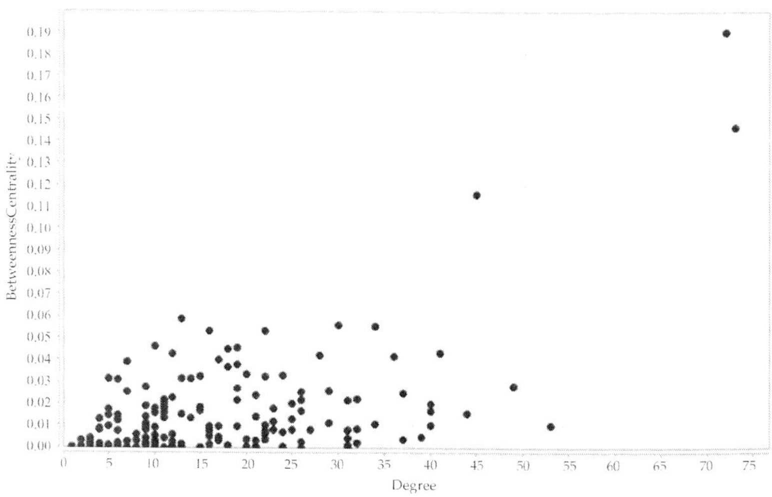

Figure 6: Betweenness centrality of the network.

Figure 7 shows the closeness centrality distribution of the scientific network. Closeness centrality is the measure of how fast information spreads from a given node to other reachable nodes in the network [45]. We can observe that, excepted by the two nodes with high degree, there is no other node that stands out from the others considering this metric.

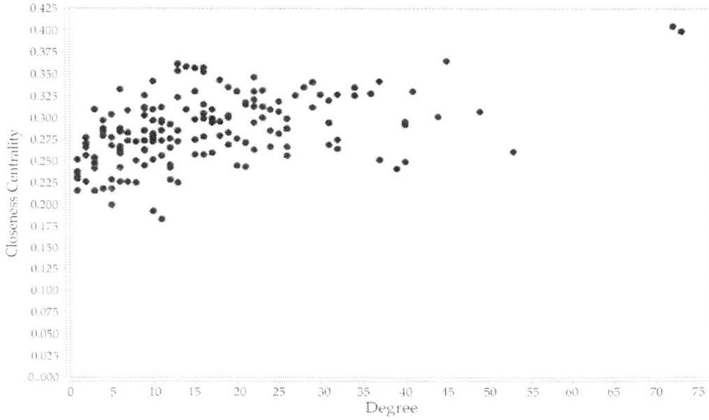

Figure 7: Closeness centrality.

The neighborhood connectivity distribution gives the average of the neighborhood connectivities of all nodes n with k neighbors for k=0,1,.... Figure 8 shows the neighborhood connectivity distribution for the network presented. As the neighborhood connectivity distribution is an increasing function in k, edges between highly connected nodes prevail in the network, i.e., nodes with more neighbors tend to have these neighbors more connected. On the other hand, nodes with few neighbors tend to have their neighbors less connected.

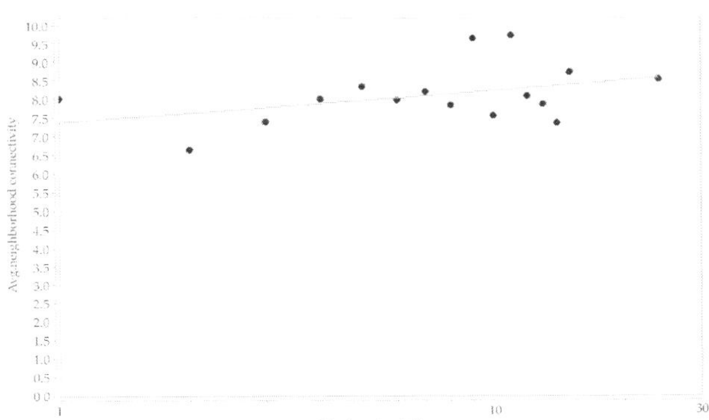

Figure 8: Neighborhood connectivity distribution of the network, with a fitted line.

This case study follows the process illustrated in Fig. 9 with the following steps:

- Data extraction from DBLP and construction of the social graph where nodes represent researchers and edges represent their co-authoring relationships. This social graph was used by both proposed models and to compare them.
- After construction of this social graph, the relationship weight (tie weight) between each pair of researchers was defined. The weight was defined by the two measurement models, previously described.
- As a result of the application of these models, two matrices representing the communication potential between pairs of researchers were obtained.
- In order to identify the scientific communities, the *k*-medoids clustering algorithm was applied to each of these matrices.
- The obtained measurements were compared based on the quality of the clusters generated by each of them.

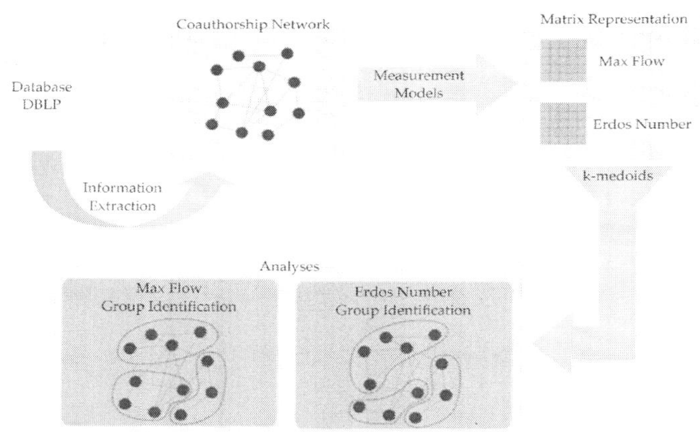

Figure 9: Case study process steps.

k-medoids

The clustering algorithm developed aims to group researchers who have the greatest potential of communication with each other. To assist the development basis of the algorithm, the *k*-medoids algorithm was used [18].

As in the k-medoids algorithm, the proposed algorithm defined that k elements randomly represented the medoids. In the second step, each element of the data set is associated with the group (medoid) in which this element has the greatest potential for communication. In the third step, the medoids of each group are set once again, and the elements are regrouped.

The definition of medoids is based on the internal communication of each researcher group. The relationship weight of each researcher internal to the group is added, and the researcher who has the largest sum is considered the medoid of the group. After defining the new medoids, the algorithm comes back to the second step until there are no changes in the structure of the groups.

One of the biggest difficulties of some clustering techniques, including k-medoids, is in defining the ideal number of groups. Cluster analysis aims to identify homogeneous groups so that the sum of differences within the group (intragroup) is minimized, and the sum of differences among groups (intergroup) is maximized [1]. The groups are validated by evaluating which set of groups has the best grouping structure.

There are several techniques that can be used to assist in defining this number, such as PBM index, intragroup distance, and intergroup distance [6, 36].

The sum of the intragroup differences is a good measure to evaluate the homogeneity of the obtained groups. This measure assesses the position of objects in the variables space within their respective groups, so it will indicate whether the objects of the same group are close to the medoid. This sum is defined by Eq. (10), where k is the number of groups, m is the number of elements in the group i, x_j is an element of group i, and $\overline{x}_i$ is the medoids of group i. The value of k, which produces the smallest sum of intragroup distances, indicates the ideal number of groups to the problem that is being solved.

$$\text{Intragroup} = \sum_{i=0}^{k} \sum_{j=0}^{m} \left\| x_j - \overline{x}_i \right\|. \tag{10}$$

Another inherent difficulty in the k-medoids algorithm is the initial setting of medoids because the result of the clustering process depends on the initial selection of these elements. Therefore, to select the best group, it is necessary to define the number of groups and the best set of elements that compose the initial medoids. In the next section, the details for setting these parameters and the results of clustering algorithm for the two proposed measurement models will be presented.

Experiments and Results Analysis

In this section, many experiments were conducted to analyze, compare, and evaluate the effectiveness of the two proposed models and answer the research questions.

Number of Groups

To answer RQ1, it was necessary to set the number of groups and evaluate the best clustering, so the k-medoids algorithm was applied to the data generated by the two proposed measurement models. The value of k was evaluated in the range between 20 and 80 groups, and the values of seeds for random definition of medoids ranged from 1 to 199.

At this stage, the internal costs of the groups were analyzed for each seed. However, the sum of the internal maximum flow and the sum of the internal resistance (models proposed in the "Measurement models to define tie weight" section) produce values quantitatively and semantically different. Thus, it was necessary to define a criterion to calculate the internal distance of the groups so that the two methods could be compared.

As illustrated in Fig. 9, both models are based on a social graph where the relationship weight is defined by the scientific work done by them together. Since this graph is the same for both models, the intragroup distance (Eq. (10)) will be calculated based on it. It is worth noting that into the obtained groups may have elements that are not directly related to the medoid, i.e., there are elements that relate indirectly to the medoid. Thus, the distance between an element and the related medoid was calculated based on Dijkstra's algorithm [12], which calculates the shortest path between nodes in a graph.

Dijkstra's algorithm was used by the groups obtained for each seed in order to calculate the intragroup distance. As a result, for each seed, the internal average distances of the groups and the variation of this average were calculated, as can be seen in Fig. 10. The values in the graphs are ordered from smallest to largest average, i.e., from best to worst seed. As can be seen in Fig. 10a, the maximum flow-based model has the best result using seed 100, because this seed had the lowest intragroup average distances. On the other hand, seed 3 showed best results for the Erdös model (Fig. 10b).

The seeds that produced more homogeneous groups were selected for a more detailed study of the intragroup distances variation of the groups.

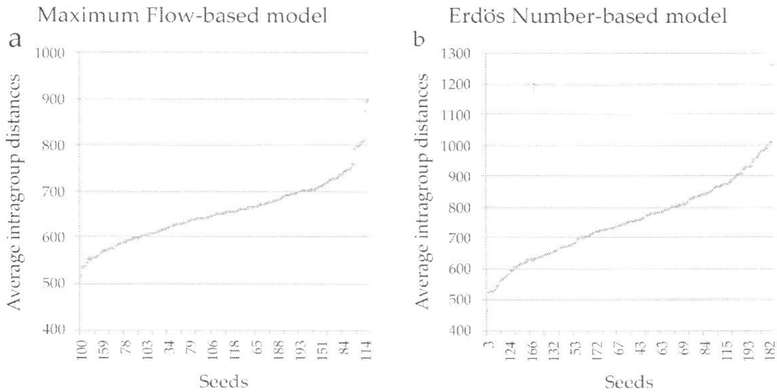

Figure 10: Average variation of intragroup distances for each seed in (a) Maximum Flow and (b) Erdös Number models.

With the aim of determining the best k value for each of the proposed models, a more detailed analysis was carried out considering seeds 100 and 3, as shown in Fig. 11. Considering it, $k=43$ was chosen because there was a decrease in the intragroup distance value in this partitioning and the values decreased slowly after that. So, for the maximum flow-based model, it was determined that the optimal number of groups is 43 and the best seed is 100. The same analysis was conducted, as illustrated in Fig. 11, and $k=47$ was chosen for the same reasons described above. Therefore, for the Erdös number-based model, it was determined that the optimal number of groups is 47 and the best seed is 3.

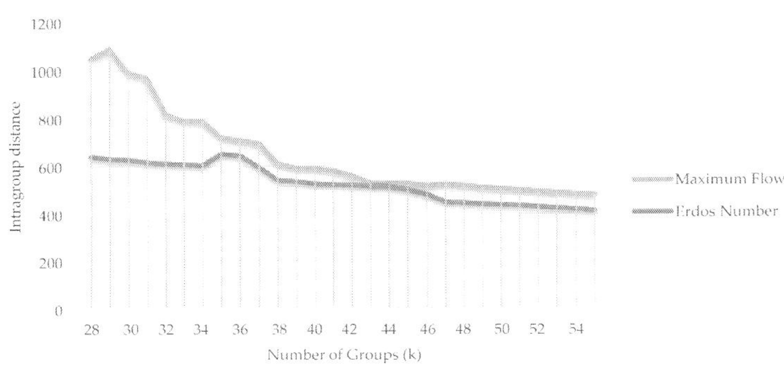

Figure 11: Variation of intragroup distances for the maximum flow-based model (seed 100) and the Erdös number-based model (seed 3).

Average Intragroup Distances

To evaluate the behavior of the two proposed models, a detailed analysis was performed focusing on changes in the intragroup distances. Considering the boxplot graph, it is worth stressing the following three points: the first middle point (the median) and the middle points of the two halves. These three points divide the entire data set into quarters, named "quartiles" (Q1, Q2, and Q3). The distance between the first (Q1) and third (Q3) quartiles is a simple dispersion measure that represents the range containing the data on the average. This distance is named interquartile range (IQR), represented by the box on the chart. IQR can be used as a measure of how the values are spread out. If IQR is small, data dispersion is lower, i.e., the data are more homogeneous.

The maximum flow model has a lower interquartile range than the Erdös model, which suggests that the clusters produced from its data are more homogeneous. On the other hand, the Erdös model showed a larger interquartile range and with higher values. This issue indicates that the clusters of this model are less homogeneous since there is a greater dispersion of the average intragroup distances and the average value is likewise higher. The graphs in Figs. 12 and 13represent the variation of the intragroup distances for each group considering all seeds, i.e., for each value of *k*, the clusters were generated using all seeds (1–199), and the intragroup distance variation is in the *k*-group boxplot. As expected, a smaller number of groups produces a greater variation average of the distances. Thus, with an increase in the number of groups, there is a decrease in these mean values.

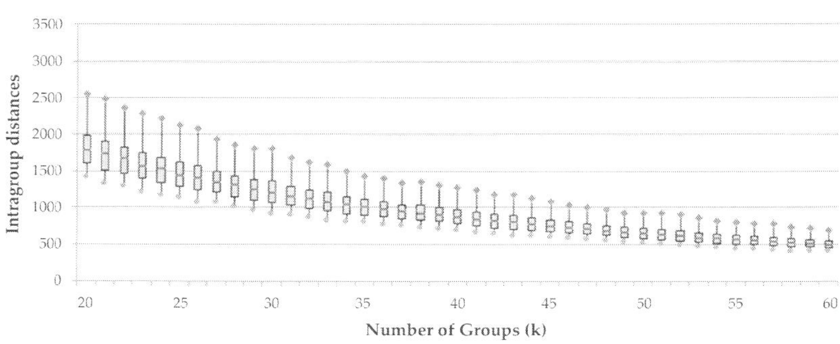

Figure 12: Variation of intragroup distances per group (maximum flow model).

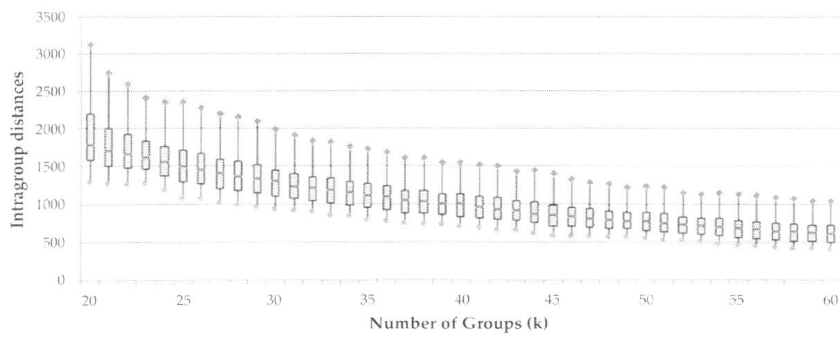

Figure 13: Variation of intragroup distances per group (Erdös number model).

The maximum flow model has its interquartile range (variation around the median) always smaller than that of the Erdös model. This is another indication that the maximum flow model produces more homogeneous groups than the Erdös model does, regardless of the number of clusters.

Group Elements

To answer RQ2, another important analysis related to the element distribution in the groups was conducted. The purpose of the analysis was to check which model could distribute more evenly the elements between groups, reducing the number of groups with a single element. The graph in Fig. 14 shows this distribution.

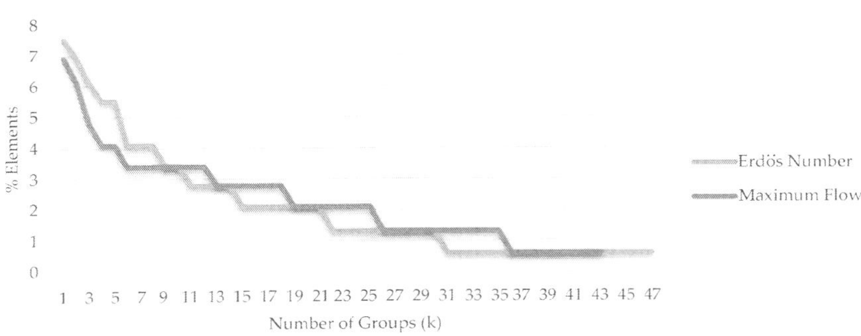

Figure 14: Elements distribution in the groups.

According to Fig. 14, there is a tendency that larger groups produced by the maximum flow model have fewer elements than the larger groups produced by the Erdös model. Moreover, the maximum flow model tends to increase the smaller groups, while the Erdös model produces a greater number of groups with fewer elements. While the Erdös model has the lowest average intragroup distances, as shown in Fig. 10b, this model produces many groups with few elements. The maximum flow model tends to produce groups with a better distribution of elements.

Figure 15 shows the obtained result from the Erdös model, each rectangle indicates the researcher number, and the ones in parentheses indicate the group number to which they belong. The large areas behind researchers represent the institutions to which they are affiliated. It was possible to validate the cluster generated by the Erdös model by comparing it to that obtained by the maximum flow model, previously validated in Ströele et al. [41].

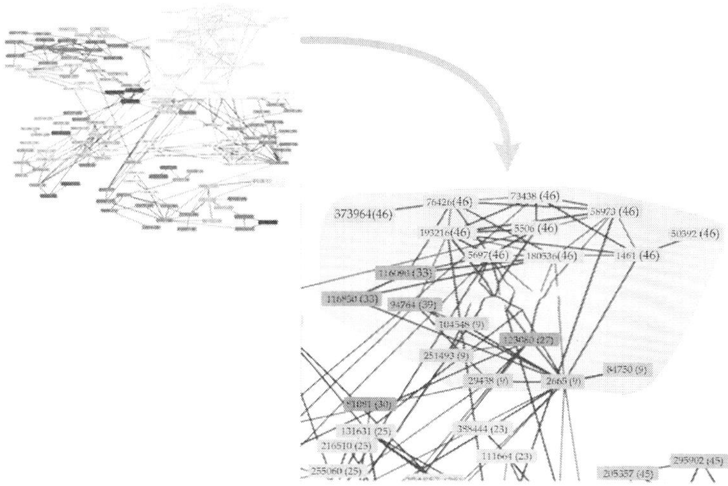

Figure 15: General and detailed view of the generated groups for the Erdös number-based model.

An important feature of the measurement models is the possibility of considering all the alternative paths between two elements to define their relationship weight. This feature answers RQ3, as can be seen in Fig. 15, where elements 50392 and 373964 are in group 46 due to their strong relationship with medoid 73438, and the relationship among them is defined by intermediary elements.

After the case study, it can be stated that both approaches produce good clusters, but the maximum flow-based approach produced more homogeneous groups with better distribution of the researchers among them. Moreover, the results showed that researchers participating in the same community could be indirectly related.

FINAL REMARKS

All elements in a social network have some influence on information diffusion. In addition, information reaches elements that are not directly related to the element that produces it. Therefore, this paper proposed two models (maximum flow-based model and Erdös number-based model) capable of measuring the influence that elements of a social network have on each other, even if they are not directly related.

The Erdös number-based model is a new approach for calculating resistance distance, which allows weights to be applied to resistance, considerably increasing the applicability of resistance distances in real-world applications. The maximum flow-based model is another way of calculating the relationship weight, which considers the maximum amount of information to be transmitted between two elements of the social network.

Both models were applied to scientific social networks built using the DBLP database, and a clustering algorithm (*k*-medoids) was used to identify scientific communities. These communities are composed of researchers who have great potential for communication among themselves. The obtained results allowed us to carry out a detailed analysis about the behavior of these models when identifying the scientific communities analyzed to assess whether the Erdös model also produces a good measure to set the relationship weight compared to the maximum flow model. This analysis shows that the results were satisfactory for both models.

By means of a case study, both the analyses and the obtained results confirmed the effectiveness of our approach using weighted resistance distance calculation in scientific social network analysis. However, additional experiments are needed so as to carry out a qualitative analysis.

As future work, the authors intend to improve the Erdös number-based model so that the elements connected by intermediate nodes can reduce their weight in the relationship, as proposed by maximum flow-based model. Thus, the distribution of elements in the groups can be enhanced, producing a smaller number of unit groups.

In this study, we considered only the Laplacian matrix for weighted graphs in the definition of the RENM. However, the spectral graph theory studies the structural properties derived from the matrices that represent graphs. The latter lead to the spectral properties of the representation matrices, which are the central element of the spectral theory of graphs. In this sense, as the spectral graph theory deals with matrices of weighted graphs similar to Laplacian, as future work, we intend to improve the RENM by considering other matrix representations.

Moreover, considering the complexity to set the parameters of *k*-medoids algorithm, the authors intend to explore other clustering algorithms that do not require these settings, such as density-based clustering algorithms.

Footnotes

1. http://dblp.uni-trier.de/

AUTHORS' CONTRIBUTIONS

VS contributed to the data extraction, model development, results analysis, and paper writing. RC contributed to the model development, results analysis, and paper writing. GZ contributed to the students' guideline, model development, and results analysis. JMS contributed to the students' guideline and results analysis, FC, JMND, and RB contributed to the students' guideline, results analysis, and paper writing. All authors read and approved the final manuscript.

REFERENCES

1. Aldenderfer MS, Blashfield RK (1984) Cluster analysis. Sage, Beverly Hills. https://doi.org/10.4135/9781412983648,
2. Bakshy E, Rosenn I, Marlow C, Adamic L (2012) The role of social networks in information diffusion. In: Proceedings of the 21st international conference on World Wide Web SE—WWW '12, pp 519–528
3. Bapat RB (2004) Resistance matrix of a weighted graph. Communications in Mathematical and in Computer Chemistry/ MATCH 50:73–82
4. Bapat RB, Gutman I, Xiao W (2003) A simple method for computing resistance distance. J Phys Sci 58:494–498
5. Bastos MT, Recuero RDC, Zago GDS (2014) Taking tweets to the streets: a spatial analysis of the vinegar protests in Brazil. First Monday. https://doi.org/10.5210/fm.v19i3.5227
6. Bezdek JC, Pal NR (1998) Some new indexes of cluster validity. IEEE Transactions on Systems, Man and Cybernetics, Part B (Cybernetics) 28:301–315. https://doi.org/10.1109/3477.678624
7. Cayley A: A theorem on trees, Quart. J. Pure Appl. Math. 23 (1889); 376–378.
8. Chen J, Song X, Nie L et al (2016) Micro tells macro. In: Proceedings of the 2016 ACM on Multimedia Conference—MM '16. ACM Press, New York, New York, USA, pp 898–907
9. Collected Mathematical Papers Vol. 13, Cambridge University Press 1897, 6–28.
10. Cumpsty NA (2009) Some Lessons Learned. In: Volume 7: Turbomachinery, parts A and B. ASME, Lyon, France, pp 785–794
11. Dias A, Chomutare T, Botsis T (2012) Exploring the community structure of a diabetes forum. Studies in Health Technology and Informatics 180:833–837. https://doi.org/10.3233/978-1-61499-101-4-833
12. Dijkstra EW (1959) A note on two problems in connexion with graphs. Numer Math 1:269–271. https://doi.org/10.1007/BF01386390
13. Edmonds J, Karp RM (1972) Theoretical improvements in algorithmic efficiency for network flow problems. J ACM 19:248–264. https://doi.org/10.1145/321694.321699

14. Evans TS, Lambiotte R, Panzarasa P (2011) Community structure and patterns of scientific collaboration in business and management. Scientometrics 89:381–396. https://doi.org/10.1007/s11192-011-0439-1
15. Ford LR, Fulkerson DR (1956) Maximum flow through a network. Can J Math 8:399–404. https://doi.org/10.4153/CJM-1956-045-5
16. Grabowicz P, Ramasco J, Moro E et al (2012) Social features of online networks: the strength of intermediary ties in online social media. PLoS One 7:e29358. https://doi.org/10.1371/journal.pone.0029358
17. Guille A (2013) Information diffusion in online social networks. Proceedings of the 2013 Sigmod/PODS PhD symposium on PhD symposium 1:31–36. https://doi.org/10.1145/2483574.2483575
18. Han J, Kamber M (2006) Data mining: concepts and techniques. Morgan Kaufmann Publishers, USAzb
19. Hossain L, Zhu D (2009) Social networks and coordination performance of distributed software development teams. J High Technol Management Res 20:52–61. https://doi.org/10.1016/j.hitech.2009.02.007
20. Howard PN, Duffy A, Freelon D et al (2011) Opening closed regimes: what was the role of social media during the Arab Spring? Project on Information Technology and Political Islam:1–30. https://doi.org/10.1007/s13398-014-0173-7.2
21. Huang Y, He P, Li B (2012) IEEE Xplore—Applying centrality measures to the behavior analysis of developers in open source software Communit... In: cloud and green computing (CGC), 2012 second international conference on. IEEE pp 418–423
22. Hughes AL, Palen L (2009) Twitter adoption and use in mass convergence and emergency events. Int J Emerg Manag 6:248. https://doi.org/10.1504/IJEM.2009.031564
23. Ichise R, Takeda H, Ueyama K (2005) Community mining tool using bibliography data. In: Proceedings of the International Conference on Information Visualisation IEEE, pp 953–960
24. Jones JJ, Settle JE, Bond RM et al (2013) Inferring tie strength from online directed behavior. PLoS One 8:e52168. https://doi.org/10.1371/journal.pone.0052168
25. Jung JJ (2010) Integrating social networks for context fusion in mobile service platforms. J Universal Computer Sci 16:2099–2110

26. Kaufman L, Rousseeuw PJ (2005) Finding groups in data: an introduction to cluster analysis. Wiley-Interscience 33:368. https://doi.org/10.1007/s00134-006-0431-z
27. Kempe D, Kleinberg J, Tardos É (2003) Maximizing the spread of influence through a social network. In: Proceedings of the ninth ACM SIGKDD international conference on knowledge discovery and data mining—KDD '03, p 137. https://doi.org/10.1145/956755.956769
28. Kennelly AE (1899) Equivalence of triangles and three-pointed stars in conducting networks. Electrical World and Engineer 34:413–414
29. Kitchenham B, Charters S (2007) Guidelines for performing systematic literature reviews in software engineering version 2.3. Engineering 45:1051. https://doi.org/10.1145/1134285.1134500
30. Kuphaldt TR (2009) Lessons In electric circuits: DC. In: Open Book Project. http://www.ibiblio.org/kuphaldt/electricCircuits/DC/, pp 1–526
31. Michael Barr (2001) Rational Erdős number. https://www.oakland.edu/Assets/upload/docs/Erdos-Number-Project/barr.pdf
32. Newman M (2004a) Who is the best connected scientist?A study of scientific coauthorship networks. In: Complex networks. SFI working paper 00-12-64, Santa Fe, pp 337–370
33. Newman MEJ (2004) Coauthorship networks and patterns of scientific collaboration. In: Proceedings of the National Academy of Sciences 101 (suppl 1) 5200-5205. https://doi.org/10.1073/pnas.0307545100
34. Newman M (2004c) Detecting community structure in networks. Eur Phys J B 38:321–330. https://doi.org/10.1140/epjb/e2004-00124-y
35. Nowell DL, Kleinberg J (2003) The link prediction problem for social networks. In: Proceedings of the twelfth international conference on information and knowledge management, vol 58, pp 556–559. https://doi.org/10.1145/956863.956972
36. Pakhira MK, Bandyopadhyay S, Maulik U (2004) Validity index for crisp and fuzzy clusters. Pattern Recogn 37:487–501. https://doi.org/10.1016/j.patcog.2003.06.005
37. Rice E, Tulbert E, Cederbaum J et al (2012) Mobilizing homeless youth for HIV prevention: a social network analysis of the acceptability of a face-to-face and online social networking intervention. Health Educ Res 27:226–236. https://doi.org/10.1093/her/cyr113

38. Song X, Nie L, Zhang L et al (2015a) Interest inference via structure-constrained multi-source multi-task learning. In: IJCAI International Joint Conference on artificial intelligence, pp 2371–2377
39. Song X, Nie L, Zhang L et al (2015b) Multiple social network learning and its application in volunteerism tendency prediction. In: Proceedings of the 38th International ACM SIGIR Conference on Research and Development in Information Retrieval—SIGIR '15. ACM Press, New York, New York, USA, pp 213–222
40. Strode V, Campos F, Pereira CK et al (2016) Information extraction to improve link prediction in scientific social networks. 2016 IEEE 20th International Conference on Computer Supported Cooperative Work in Design (CSCWD):515–520. https://doi.org/10.1109/CSCWD.2016.7566043
41. Ströele V, Zimbrão G, Souza JM (2013) Group and link analysis of multi-relational scientific social networks. J Syst Softw 86:1819–1830. https://doi.org/10.1016/j.jss.2013.02.024
42. Trajanovski S, Kuipers FA, Ilic A et al (2015) Finding critical regions and region-disjoint paths in a network. IEEE/ACM Trans Networking 23:908–921. https://doi.org/10.1109/TNET.2014.2309253
43. Varlamis I, Eirinaki M, Louta M (2010) A study on social network metrics and their application in trust networks. In: Proceedings - 2010 International Conference on Advances in Social Network Analysis and Mining, ASONAM 2010, pp 168–175
44. Wang D, Pedreschi D, Song C et al (2011) Human mobility, social ties, and link prediction. In: Acm. ACM, New York, NY, USA, pp 1100–1108
45. Wasserman S, Faust K (1994) Social network analysis: methods and applications. https://doi.org/10.1525/ae.1997.24.1.219
46. Watts J (2013) Brazil protests: president to hold emergency meeting. In: The Guardian. http://www.guardian.co.uk/world/2013/jun/21/brazil-protests-president-emergency-meeting. Accessed 27 Mar 2018
47. Xiao W, Gutman I (2003) Resistance distance and Laplacian spectrum. Theoretical Chemistry Accounts: Theory, Computation, and Modeling (Theoretica Chimica Acta) 110:284–289. https://doi.org/10.1007/s00214-003-0460-4

48. Yan E, Guns R (2014) Predicting and recommending collaborations: an author-, institution-, and country-level analysis. Journal of Informetrics 8:295–309. https://doi.org/10.1016/j.joi.2014.01.008
49. Yang C, Ma J, Silva T et al (2014) A multilevel information mining approach for expert recommendation in online scientific communities. Comput J 58:1921–1936. https://doi.org/10.1093/comjnl/bxu033
50. Yang J, Leskovec J (2012) Defining and evaluating network communities based on ground-truth. ACM SIGKDD Work Min Data Semant 42:745–754
51. Yin RK (2009) Case study research: design and methods, fifth edn. Sage Publications, Beverly Hills
52. Yuan G, Murukannaiah PK, Zhang Z, Singh MP (2014) Exploiting sentiment homophily for link prediction. In: Proceedings of the 8th ACM Conference on Recommender systems—RecSys '14, pp 17–24

CHAPTER 4

IDENTIFYING NETWORK PUBLIC OPINION LEADERS BASED ON MARKOV LOGIC NETWORKS

Weizhe Zhang, Xiaoqiang Li, Hui He, and Xing Wang

School of Computer Science and Technology, Harbin Institute of Technology, Harbin 150001, China

ABSTRACT

Public opinion emergencies have important effect on social activities. Recognition of special communities like opinion leaders can contribute to a comprehensive understanding of the development trend of public opinion. In this paper, a network opinion leader recognition method based on relational data was put forward, and an opinion leader recognition system integrating public opinion data acquisition module, data characteristic selection, and fusion module as well as opinion leader discovery module based on Markov

CITATION(APA): Zhang, W., Li, X., He, H., & Wang, X. (2014). Identifying network public opinion leaders based on markov logic networks. The scientific world journal, 2014. (8 pages).

Logic Networks was designed. The designed opinion leader recognition system not only can overcome the incomplete data acquisition and isolated task of traditional methods, but also can recognize opinion leaders comprehensively with considerations to multiple problems by using the relational model. Experimental results demonstrated that, compared with the traditional methods, the proposed method can provide a more accurate opinion leader recognition and has good noise immunity.

INTRODUCTION

As the Internet enters into the We-media era, every individual can be a message sender. However, public opinion emergencies may affect the social activities significantly due to the mixed netizen qualities. To avoid adverse effect, it is necessary to have a comprehensive understanding of the development trend of public opinion and recognize special communities like opinion leaders.

The earliest domestic and foreign researches on the discovery of network opinion leaders focused on using the opinion leader theory in the traditional social sciences and research method transplantation directly to recognize the internet leaders but failed to achieve ideal results. They often determined community opinion leaders based on the quantitative data analysis. These methods [1], regardless of the characteristics of objective fact, only depend on logic reasoning and could not represent the public opinion transmission characteristics on the new media of network. Recently, scholars began to explore the difference between network environment and offline practical society through quantitative measurement. The accuracy of the leader discovery model [2–5] witnessed a continuous improvement from the clustering analysis of statistical properties [6] to social network analysis based on complex network [7–9] and to the semantic analysis of text content.

Attributes of Public Opinion Participants

Behaviors of every netizen involved in the public opinion transmission are described by inherent attribute, content attribute, and social network attribute.

Inherent attribute refers to the independent attributes of participants from the concerning public opinion events such as career, position, internet age, logins, community credits, fans population, and concerns.

Content attribute refers to the behavioral description of the participants in a certain public opinion event including posts, replies, comments received, reposts, number of mentions, number of words, and emotional tendency.

Social network attribute refers to the mutual relationship of participants in the network mainly including fans and concerns of the participants.

Shortcomings of Existing Research Methods

Existing network opinion leader discovery is based on the recognition model involving only one or two attributes of participants. No network opinion leader recognition involving all three attributes of participants has been reported yet. This will affect the accuracy of the opinion leader recognition method.

Existing network opinion leader discovery views the attributes of participants as independent and identically distributed (IID) data. In the theory of probability statistics, mutually independent sequence of variables or other random variables that have the same probability distribution is called IID. However, attribute data of participants are relational data. Different attributes of participants are mutually correlated instead of being independent from each other. For example, fans' population of inherent attribute often is proportional to comments received of content attribute; participants enjoying high attentions from opinion leader are more likely to be the opinion leader. The incomprehensive understanding of such relations will make some opinion leaders remain unidentified. Furthermore, existing network opinion leader discovery has no modeling solution to the relational data.

As a result, simultaneous application of all three attributes and the exploration of relationships in public opinion data can improve the performance of network opinion leader recognition method.

MARKOV LOGIC NETWORKS

Markov Logic Networks refer to a learning method of statistical relation that is combining the Markov network and first-order logic together. It was proposed by Singla and Domingos [10] in 2004 and then improved by Domingos and his students.

Markov network [11] is a joint distribution model of random variable set ($X = (x_1, x_2, \ldots,)$). It is composed of an undirected graph (G) and a potential function (Φ_k) set. Every random variable occupies a node of the graph and each group has a potential function in the model. Potential function is a nonnegative real function, which represents the state of the corresponding group. The joint distribution of Markov network is as follows:

$$P(X = x) = \frac{1}{Z}\prod_{k}\Phi_k\left(x_{\{k\}}\right), \tag{1}$$

where $x_{\{k\}}$ is the state of random variables in the group and Z is partition function (state sum) that is defined as $\sum_{x \in X} \prod_k \Phi_k(x_{\{k\}})$. Weight all characteristic values of potential-use states of each group in the Markov network. Then sum them and calculate the exponentiation. Finally, a log-linear model can be gained as follows:

$$P(X = x) = \frac{1}{Z}\exp\left(\sum_j \omega_j f_j(x)\right). \tag{2}$$

Characteristic function can be any real function of state. In this paper, characteristic function refers to the dual characteristics value ($\{0, 1\}$). Equation (1) is the most direct expression of potentials, in which every possible state of each group has a corresponding characteristic value and a weight. Equation (1) is related to the power of groups. However, the amount of characteristic values can be reduced freely by some methods (e.g., logic function of state), thus enabling the characteristic function to provide a simpler expression under large number of groups compared with the potential functions. This is the principle of the Markov Logic Networks.

Markov Logic Networks are a first-order logic knowledge base where every code has a weight. This first-order logic knowledge base can be viewed as the template of Markov Logic Networks. Viewed from probability, the Markov Logic Networks provide a simple language to define large Markov network as well as a flexible and modularized integration with abundant knowledge. Viewed from the first-order logic, the Markov Logic Networks provide a sound processing to knowledge base with uncertainties, defects, and even contradictories, thus decreasing the vulnerability.

Take the data set of Skyline for example. The simplest situation is as follows: suppose the knowledge base only contains the formula F_1 (weight = 1.5), and the corresponding MLN of the knowledge base is $\{(F_1, 1.5)\}$:

$$F_1 : \text{PostHasWord}(y, x) \longrightarrow (\text{BadWord}(x) \longrightarrow \text{BadPost}(y)), \tag{3}$$

where x and y are individual variables; BadWord(x), BadPost(y), and

PostHasWord(y, x) are predicates, representing whether x has bad word, y is a bad post, and y contains x. F_1 means that if y has the bad word x, y can be deduced as a bad post.

Given an individual constant set $C = \{A, B, D\}$, a closed Markov Logic Network can be generated (Figure 1).

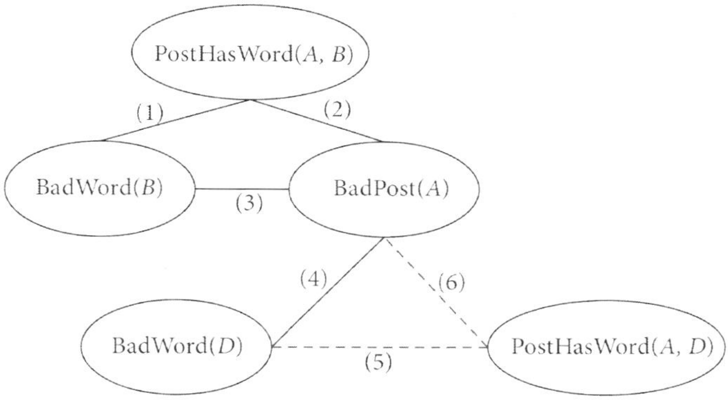

Figure 1: Closed Markov Logic Network.

Markov Logical Reasoning Method

Markov logical reasoning is equal to the probabilistic reasoning of the complex relationship. The basic task of reasoning is to inference the most possible state of world y according to the given evidence x (word set).

There are two basic types of reasoning: First, we search the most possible state satisfying some evidence and probability of computer random condition. Next, lazy reasoning and relieving reasoning are compared in improving the performance of above two types of reasoning in processing more complicated relationships. The lazy reasoning only requires adopting instantiation "default" value to the deviated basic values, while relieving reasoning divides indistinctive atoms into one group and views them as an independent unit.

Markov Logic Networks Learning

The Markov Logic Networks learning includes structure learning and parameter learning. Structure learning is to learn the model structure (network structure of the Markov Logic Networks) from data. In structure learning, it is more difficult to learn the rules. There are two structure learning methods based on the inductive logic programming (ILP): (1) learn model structure by using CLAUDIEN (an LP system) directly; (2) learn structure of Markov Logic Networks from the relational database by combining LP and feature introduction in Markov network together. Parameter learning is committed to find the satisfying rule weight with maximum likelihood or maximum conditional likelihood of associated database. Parameter learning includes production parameter learning and discriminant parameter learning. Parameter learning is needed under known structure of Markov network to acquire weight of each rule. However, when the structure of Markov network is unknown, structure learning is needed before the parameter learning to get the network structure. Markov Logic Networks learning includes parameter learning and structure learning. Parameter learning has production parameter learning and discriminant parameter learning as well as their corresponding formulas and approximate algorithms. Structure learning also includes top-down structure learning and bottom-up structure learning.

SYSTEM FRAMEWORK

The overall structure of the network opinion leader recognition system based on the Markov Logic Networks is shown in Figure 2.

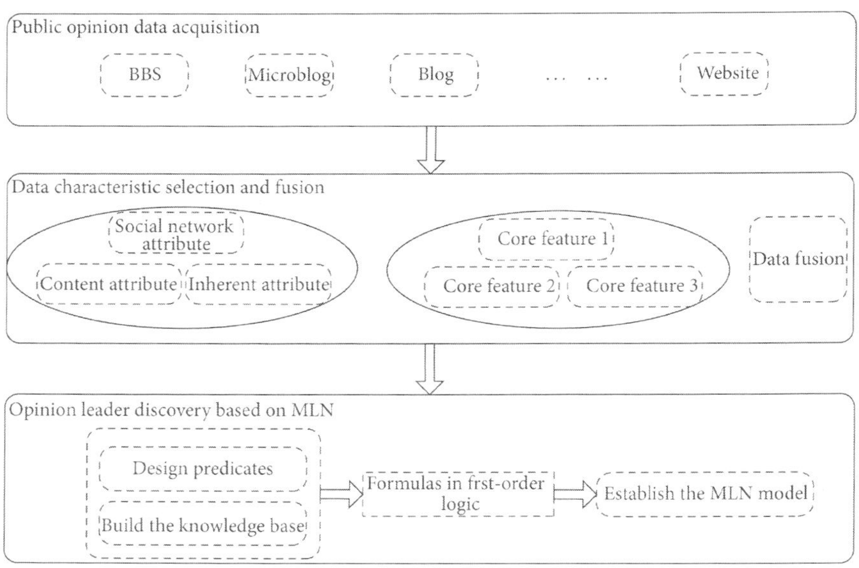

Figure 2: Overall structure of the network opinion leader recognition system based on the Markov Logic Networks.

The designed network opinion leader recognition system includes three modules: public opinion data acquisition module, data characteristic selection and fusion module, and opinion leader discovery module based on Markov Logic Networks. The public opinion data acquisition module is for data collection concerning specific public opinion event. Data characteristic selection and fusion module is for processing and analyzing collected data to disclose the relationship between core characteristics and attributes. The opinion leader discovery module based on Markov Logic Networks is to design predicates, build the knowledge base, and establish the MLN model according to the relationship between core characteristics and attributes.

Opinion Leader Discovery Module Based on Markov Logic Networks

The technical route of opinion leader recognition based on Markov Logic Networks is presented in Figure 3. The top data input, initialization module, and structure learning module are the training module of the model, whereas the bottom data (Leader) verification, verification module, and data output are the verification module of the model.

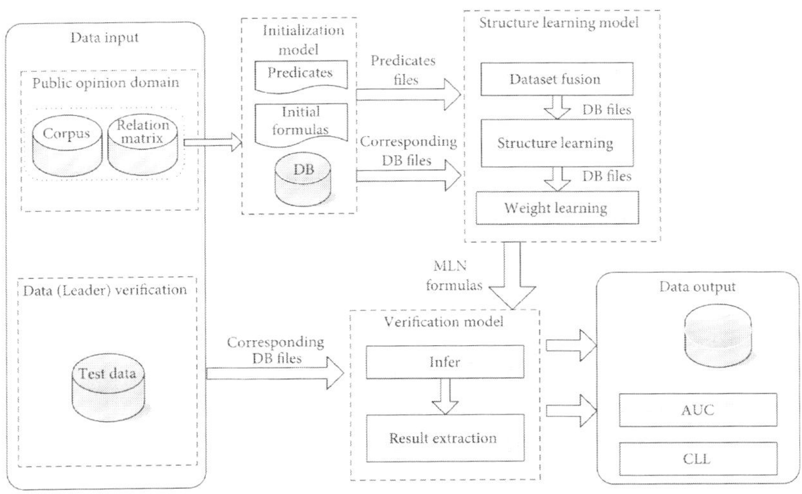

Figure 3: Technical route of opinion leader recognition based on Markov Logic Networks.

The primary task of the training module is to design predicate. Predicate design has two stages: (1) original design of initial predicate set according to the characteristic matrix and relational knowledge gained by the data characteristic extraction and fusion module as well as personal priori knowledge; (2) repetitive reasonable adjustment of predicate design according to the learning results of the model and follow-up experimental analysis of learning results until getting satisfying experimental result.

The initialization module will convert contents in the corpus into DB according to existing predicate design. The structure learning module can conduct the structure learning through the available learnstruct program of Alchemy. Beam search is the default structure learning algorithm. Weight learning can be implemented by the available learning program. Weighted MLN clause can be gained through the structure learning and weight learning, which will be used for reasoning. In the designed system, these weighted MLN clauses are used to deduce user's identity.

In the verification module of the model, both verification data and training data are converted into DB according to the predicate design. The verification module is mainly used to deduce user's identity.

Reasoning results contain all possible predicates and their possibilities. Take the predicate of teachby(course, teacher) for instance. Course represents course and teacher represents teacher, meaning that the course is taught by the teacher. If the course is Chinese, possible teacher set is {Tom, Jack, Lily}. Then, the reasoning results may be as follows:

$$\begin{aligned}&\text{teacherby}(\text{Chinese}, \text{Tom})\ \ 0.9275\\&\text{teacherby}(\text{Chinese}, \text{Jack})\ \ 0.7465\\&\text{teacherby}(\text{Chinese}, \text{Lily})\ \ 0.3556.\end{aligned} \tag{4}$$

Result extraction means to select the reasoning result with highest probability as the final result. In the above case, Tom will be selected as the teacher of Chinese. The data output module will calculate AUC and CLL.

EXPERIMENTAL VERIFICATION

Recognition of Nonrelational Data Model

Netizens were classified according to the nonrelational data model provided by Weka. First of all, the original data have to be converted into ARFF. ARFF mainly includes attribute assertion and data [12, 13].

We chose SVM to classify netizens involved in the "Xu-Ting Event" on the legal forum of Skyline Gossip as Leader, Normal, Passer, and

Waterarmy. In our experiment, every netizen was recognized independently. Experimental results are presented in Figure 4.

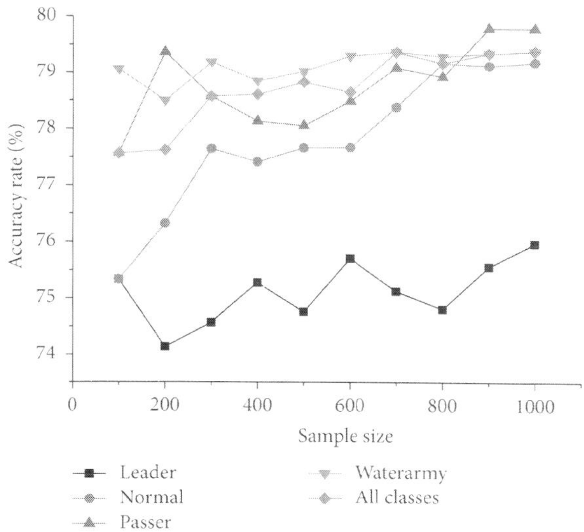

Figure 4: Group recognition of nonrelational data model.

It can be seen from Figure 4 that the nonrelational data model achieved a recognition accuracy of about 77%, which presented a continuous growth as the sample size increases.

Recognition of Relational Data Model

Firstly, we have to design predicates. Our designed predicates according to the characteristic selection and personal priori knowledge are listed in Tables 1, 2, and 3, including three classes (social network attribute, content attribute, and inherent attribute). These three classes of attribute are used to describe different network behaviors and individual characteristics of netizens.

Table 1: Designed predicates of content attribute

Predicate	Meaning
Post(person_id, post_id)	User who is represented by person_id published a post which is represented by post_id.
ReplyNumOfPost(post_id, level_replynumof-post)	The reply number of the post which is represented by post_id is level_replynumofpost.

ClickNumOfPost(post_id, level_clicknumofpost)	The click number of the post which is represented by post_id is level_clicknumofpost.
TotalPostNum(person_id,level_totalpostnum)	The post number of the user who is represented by person_id is level_totalpostnum.
TotalReplyNum(person_id,level_totalreplynum)	The reply number of the user who is represented by person_id is level_totalreplynum.
TotalBeReplyNum(person_id, level_totalbere-plynum)	The number of replies to the user who is represent-ed by person_id is level_totalbereplynum.
Correlation(person_id, post_id, level_correlation)	The correlation level between the user who is represented by person_id published content in the post which is represented by post_id and post topics is level_correlation.
Sentiment(person_id,level_sentiment)	The degree of the emotional tendencies bases on the content published by the user who is repre-sented by person_id is level_sentiment.

Table 2: Designed predicates of social network attribute

Predicates	Meaning
FansNum(person_id, level_fansnum)	The fans number of the user who is represent-ed by person_id is level_fansnum.
FollowNum(person_id, level_follownum)	The follow number of the user who is repre-sented by person_id is level_follownum.
Follow(person_id, person_id)	A user who is represented by the first person_ id follows another user who is represented by the second person_id.
Reply(person_id, person_id, post_id, num)	In the post which is represented by the post_id, the number of a user who is represented by the first person_id reply to another user who is represented by the second person_id is num.

Table 3: Designed predicates of inherent attribute

Predicate	Meaning
Gender(person_id, gender)	The gender of the user who is represented by person_ id is gender.
Age(person_id, level_age)	The age of the user who is represented by person_ id is level_age.
NetworkAge(people, level_networkage)	The network age of the user who is represented by per-son_id is level_networkage.
LogNum(person_id, level_lognum)	The login number of the user who is represented by per-son_id is level_lognum.
CommunityCredits(person_id, level_com-munitycredits)	The community credits of the user who is represented by person_id islevel_communitycredits.

HasPosition(person_id)	The user who is represented by person_id has a communities position.
Role(person_id, role)	The role of the user who is represented by person_ id is role.

After the predicate design, we have to convert original data into DB.

Next, we have to implement structure learning and weight learning. The input MLN file (predicate statements) for structure learning is shown as Box 1.

follow(person_id, person_id)
reply(person_id, person_id)
post(person_id, post_id)
replynumofpost(post_id, level_replynumofpost)
clicknumofpost(post_id, level_clicknumofpost)
totalpostnum(person_id, level_totalpostnum)
totalreplynum(person_id, level_totalreplynum)
totalbereplynum(person_id, level_totalbereply-num)
act(person_id, role)
repeat(person_id, level_repeatnum)

Box 1

And the structure learning results are shown as Box 2.

10.4475 totalbereplynum(a1,a2)
10.7128 clicknumofpost(a1,a2)
10.4475 totalreplynum(a1,a2)
−6.70724 replynumofpost(a1,a2) v !replynumofpost(a1,a3)
5.77628 !repeat(a1,a2) v !repeat(a1,a3)
5.77628 !totalbereplynum(a1,a2) v !totalbereplynum(a1,a3)
5.83318 !totalpostnum(a1,a2) v !totalpostnum(a1,a3)
5.77628 !totalreplynum(a1,a2) v !totalreplynum(a1,a3)
5.83318 !clicknumofpost(a1,a2) v !clicknumofpost(a1,a3)
4.7222 !post(a1,a2) v !post(a3,a2)
5.83318 !act(a1,a2) v !act(a1,a3)

Box 2

The input MLN file (predicate statements and design statements) for weight learning is shown as Box 3.

Waterarmy
repeat(a1,+a2)=>act(a1,+a3)
Leader
totalreplynum(a1,+a2)^totalpostnum(a1,+a3)^totalbereplynum(a1,+a4)=>act(a1,+a5)
post(a1,a2)^replynumofpost(a2,+a3)^clicknumofpost(a2,+a4)=>act(a1,+a5)
Try to take advantage of ralational data
act(a1,+a2)^follow(a1,a3)=>act(a3,+a4)
act(a1,+a2)^reply(a1,a3)=>act(a3,+a4)
Passer
totalreplynum(a1,+a2)^totalpostnum(a1,+a3)^totalbereplynum(a1,+a4)=>act(a1,+a5)

Box 3

The design statements are recognition results of four groups.

The weight learning results are represented as Box 4.

5.8819 act(a1,Waterarmy) v !repeat(a1,Level_repeatnum_20To49)
5.28195 act(a1,Waterarmy) v !repeat(a1,Level_repeatnum_Morethan50)
4.26896 !totalpostnum(a1,Level_totalpostnum_2To2)
v !totalreplynum(a1,Level_totalreplynum_10To49)
v !totalbereplynum(a1,Level_totalbereplynum_2To9)
v act(a1,Normal)
4.93641 !totalpostnum(a1,Level_totalpostnum_Lessthan2)
v !totalreplynum(a1,Level_totalreplynum_Lessthan2)
v !totalbereplynum(a1,Level_totalbereplynum_Lessthan2)
v act(a1,Passer)
3.76921 !totalpostnum(a1,Level_totalpostnum_Lessthan2)
v !totalreplynum(a1,Level_totalreplynum_Lessthan2)
v !totalbereplynum(a1,Level_totalbereplynum_Morethan200)
v act(a1,Leader)
3.74401 !reply(a1,a2) v act(a2,Normal) v !act(a1,Passer)
4.34903 !reply(a1,a2) v act(a2,Leader) v !act(a1,Leader)

Box 4

Table 4 lists some learned valuable clauses such as the following: 4.27395 !reply(a1,a2) v act(a2,Leader) v !act(a1,Leader).

Table 4: Valuable clauses learned from the "Xu-Ting Event"

Weight	Formula
2.76133	!fansnum(a1,Level_fansnum_10To49) v !follownum(a1,Level_follownum_Lessthan10) v gender(a1,Male) v !networkage(a1,Level_networkage) v !lognum(a1,Level_log_num_1000To4999) v !communitycredits(a1,Level_communitycredits_1000To9999)
2.63516	gender(a1,Female) v !lognum(a1,Level_log_num_1000To4999)
3.21897	!fansnum(a1,Level_fansnum_10To49) v !follownum(a1,Level_follownum_Lessthan10) v age(a1,Level_realage_Morethan35) v !networkage(a1,Level_networkage) v !lognum(a1,Level_log_num_1000To4999) v !communitycredits(a1,Level_communitycredits_1000To9999)
3.77246	gender(a1,Female) v !lognum(a1,Level_log_num_Lessthan1000)
4.27395	!reply(a1,a2) v act(a2,Leader) v !act(a1,Leader)
5.65332	gender(a1,a2) v !age(a1,a3) v !age(a1,a4) v lognum(a1,a5) v lognum(a1,a6)
6.06442	!communitycredits(a1,a2) v !communitycredits(a1,a3) v !totalreplynum(a1,a4) v !totalreplynum(a1,a5)
6.06101	!networkage(a1,a2) v !networkage(a1,a3)

This clause means that if a1 is the leader and a1 replies a2, then a2 is a leader too. This is true in real life.

The learned valuable clauses were selected for reasoning of the test set. According to the experimental results (shown as Box 5), Person15, Person30, Person21, Person55, and Person51 were identified as Passer, Normal, Passer, Leader, and Passer, respectively. Four were recognized accurately and 1 was recognized wrongly.

act(Person15,Waterarmy) 0.505	act(Person15,Normal) 0.813969
act(Person15,Leader) 0.495	act(Person15,Passer) 0.99995
act(Person30,Waterarmy) 0.99995	act(Person30,Normal) 0.99995
act(Person30,Leader) 0.99995	act(Person30,Passer) 0.894961
act(Person21,Waterarmy) 0.495	act(Person21,Normal) 0.823968
act(Person21,Leader) 0.504	act(Person21,Passer) 0.99995
act(Person55,Waterarmy) 0.99995	act(Person55,Normal) 0.99995
act(Person55,Leader) 0.99995	act(Person55,Passer) 0.893961

act(Person51,Waterarmy) 0.490001	act(Person51,Normal) 0.794971
act(Person51,Leader) 0.513999	act(Person51,Passer) 0.99995

Box 5

The recognition accuracy comparison results of relational data model and nonrelational data model to different events are listed in Table 5, finding that the relational data model has higher recognition accuracy compared with the nonrelational data model.

Table 5: Recognition accuracy comparison of relational data model and non-relational data model

Event name	Forum	The accuracy of IID model (%)	The accuracy of relation model (%)
Xu-Ting Event	Legal Forum	79.5	82.5
Xu-Ting Event	Tianya By-talk	77.4	80.6
Three years of great Chinese famine	Discussion about the history	77.8	81.8
Three years of great Chinese famine	Tianya By-talk	76.9	80.8

CONCLUSIONS

This paper firstly summarizes and evaluates the shortcomings of existing opinion leader recognition method, describes the advantages of Markov Logic Networks in opinion leader recognition, and summarizes the associated theories of Markov Logic Networks including basic concepts as well as theoretical models (reasoning and learning). The Markov Logic Networks combine the probability theory and first-order logic perfectly, integrating logic/relation expressions, uncertainty processing, and learning. Secondly, this paper designs and implements a network opinion leader recognition system based on previous theories. This designed system firstly collects some public opinion data as the training set for structure learning of Markov Logic Networks and then uses the learning results to reasoning the control results of corresponding public opinion domain of the test data. The experimental results are compared and analyzed to evaluate their validity. Thirdly, this paper carries out an experimental verification, which verifies the superiority of the designed network opinion leader recognition system.

ACKNOWLEDGMENTS

This work was supported by the National Natural Science Foundation of China (NSFC) under Grant no. 61173145, the National Basic Research Program of China under Grant no. G2011CB302605, and the National High Technology Research and Development Program of China under Grant no. 2011AA010705.

REFERENCES

1. Y. Li, S. Ma, Y. Zhang, R. Huang, and Kinshuk, "An improved mix framework for opinion leader identification in online learning communities," Knowledge-Based Systems, vol. 43, pp. 43–51, 2013. View at Publisher · View at Google Scholar
2. R. Goldsborough, "The influence of active online users," Black Issues in Higher Education, vol. 19, no. 5, pp. 30–31, 2002. View at Google Scholar
3. N. Matsumura, Y. Ohsawa, and M. Ishizuka, "Mining and characterizing opinion leaders from threaded online discussions," in Proceedings of the 6th International Conference on Knowledge-Based Intelligent Engineering Systems & Allied Technologies, 2002.
4. N. Matsumura, "Topic diffusion in a community," in Chance Discovery, pp. 84–97, Springer, Berlin, Germany, 2003. View at Google Scholar
5. N. Matsumura, "Collaborative communication strategies in online community," in The 4th International Workshop on Social Intelligence Design (SID '05), 2005.
6. Z. Zhai, H. Xu, and P. Jia, "Identifying opinion leaders in BBS," in Proceedings of the IEEE/WIC/ACM International Conference on Web Intelligence and Intelligent Agent Technology (WI-IAT '08), vol. 3, pp. 398–401, IEEE, December 2008. View at Publisher · View at Google Scholar · View at Scopus
7. L. C. Freeman, "Centrality in social networks conceptual clarification," Social Networks, vol. 1, no. 3, pp. 215–239, 1978. View at Google Scholar · View at Scopus
8. L. Page, S. Brin, R. Motwani, and T. Winograd, The PageRank Citation Ranking: Bringing Order to the Web, 1999.
9. J. M. Kleinberg, "Authoritative sources in a hyperlinked environment," Journal of the ACM, vol. 46, no. 5, pp. 604–632, 1999. View at Google Scholar · View at Scopus
10. P. Singla and P. Domingos, "Markov logic ininfinite domains," In press, http://arxiv.org/abs/1206.5292.
11. M. Richardson and P. Domingos, "Markov logic networks," Machine Learning, vol. 62, no. 1-2, pp. 107–136, 2006. View at Publisher · View at Google Scholar · View at Scopus

12. W. Zhang, H. He, and J. Ye, "A two-level cache for distributed information retrieval in search engines," The Scientific World Journal, vol. 2013, Article ID 596724, 6 pages, 2013. View at Publisher · View at Google Scholar
13. W. Zhang, B. Wang, H. He, and Z. Tan, "Public opinion leader community mining based on the heterogeneous network," Acta Electronica Sinica, vol. 40, no. 10, pp. 1927–1932, 2012.

CHAPTER 5

CONSTRUCTING, CONDUCTING AND INTERPRETING ANIMAL SOCIAL NETWORK ANALYSIS

Damien R. Farine[1,2,3] **and Hal Whitehead**[4]

[1] Department of Zoology, Edward Grey Institute of Field Ornithology, University of Oxford, South Parks Road, Oxford OX1 3PS, UK

[2] Department of Anthropology (Evolutionary), University of California Davis, 1 Shields Avenue, Davis, CA 95616, USA

[3] Smithsonian Tropical Research Institute, Ancon, Panama

[4] Department of Biology, Dalhousie University, 1355 Oxford St, Halifax, NS, Canada B3H 4J1

CITATION (APA): Farine, D. R., & Whitehead, H. (2015). Constructing, conducting and interpreting animal social network analysis. Journal of Animal Ecology, 84(5), 1144-1163. (20 pages).

SUMMARY

- Animal social networks are descriptions of social structure which, aside from their intrinsic interest for understanding sociality, can have significant bearing across many fields of biology.
- Network analysis provides a flexible toolbox for testing a broad range of hypotheses, and for describing the social system of species or populations in a quantitative and comparable manner. However, it requires careful consideration of underlying assumptions, in particular differentiating real from observed networks and controlling for inherent biases that are common in social data.
- We provide a practical guide for using this framework to analyse animal social systems and test hypotheses. First, we discuss key considerations when defining nodes and edges, and when designing methods for collecting data. We discuss different approaches for inferring social networks from these data and displaying them. We then provide an overview of methods for quantifying properties of nodes and networks, as well as for testing hypotheses concerning network structure and network processes. Finally, we provide information about assessing the power and accuracy of an observed network.
- Alongside this manuscript, we provide appendices containing background information on common programming routines and worked examples of how to perform network analysis using the *r* programming language.
- We conclude by discussing some of the major current challenges in social network analysis and interesting future directions. In particular, we highlight the under-exploited potential of experimental manipulations on social networks to address research questions.

INTRODUCTION

Social network analysis is a commonly used toolbox for biologists investigating the causes and consequences of complex social and ecological interactions in animal communities. It is a key technique in sociology (Wasserman & Faust 1994), where it originated in the 1930s, to investigate the link between local patterns of human relationships and social processes, such as the impact of social groups on the likelihood of being obese (Christakis & Fowler 2007). Social network analysis provides a flexible framework for analysing association or interaction data to address a broad set of biological questions (Croft, James & Krause 2008). Most fundamentally, it provides a description of social structure. Network data also forms the substrate for a range of analyses including affiliation or avoidance between conspecifics (Lusseau 2003; Croft, Krause & James 2004), interspecific interactions (Farine, Garroway & Sheldon 2012), mating behaviour (McDonald *et al.* 2013), genetic networks (Gardner *et al.* 2003) and community-level linkages (Montoya, Pimm & Sole 2006; Olesen *et al.* 2008). The strength of social network analysis is that it provides an understanding of how local processes drive group-level properties by taking into account the different social environments experienced by each individual (Strandburg-Peshkin *et al.* 2013), how individual variation in social behaviour can drive population structure (Aplin *et al.* 2013; Jacoby *et al.* 2014; Snijders *et al.* 2014) and how socially transmitted quantities, such as information or disease, flow through individuals in a population (Boogert *et al.* 2008; Hamede *et al.* 2009; Kendal *et al.* 2010). The properties of individuals captured by social network analysis can then be linked to fitness (McDonald 2007; Silk *et al.* 2010; Formica *et al.* 2012; Wey *et al.* 2013; Farine & Sheldon 2015), thus framing sociality in an evolutionary context. Because it provides a means of linking social behaviour across all levels of organization, network analysis is increasingly central to many fields of biology and is quickly becoming the most commonly used approach for describing the structure of social relationships in a broad range of taxa.

In this paper, we focus on the application of social network analysis to non-human animal data (see Fig. 1 & Table 1). Sections 2 through 4 outline the motivations for performing network analyses and give advice on defining what the network represents and collecting data. Sections 5 and 6 provide the information required to decide how to infer edges, calculate the value of each edge and display the resulting network. Section 7 introduces different network metrics and what to consider when interpreting their values and distributions. Section 8 explains how null models and permutation tests can be used to conduct robust statistical testing. Section 9 details a number of current and future approaches for hypothesis testing in animal social network analysis. Section 10 provides guidelines for estimating the power and precision of an observed network and background on different sources of bias. Section 11 discusses future directions that we believe will provide new insight into animal social networks. This paper is accompanied by appendices containing worked empirical and simulation examples using *R*. Because animal social network analysis has become a broad field of research, interpretation of network data is dependent on the definitions and assumptions used in each individual case. Our aim is to introduce the process of studying social behaviour using social network analysis by providing a synthesis of the approaches and considerations that are common to most studies (see Box 1). This topic is considered at greater length in the books by Croft, James & Krause (2008) and Whitehead (2008), but there have been recent advances, especially in the automated collection of very large data sets and in the development of analytical tools to tackle the novel challenges associated with generating networks from tracking data. The application of social network analysis across different domains of animal biology is discussed with greater detail in the chapters of the book edited by Krause *et al*. (2015).

- Match data collection, and node and edge definitions to the biological questions.
- Determine whether edges will be weighted or binary, and directed or undirected.
- Calculate significance by comparing observed results to a distribution of test statistics generated using permutation tests.
- Avoid thresholding the network.
- Ensure that the scale of data collection appropriately captures the biological process.
- Use appropriate sampling period for each analysis
- Maximize the number of observations per individual (and consider the potential trade-off between sampling rate and number of individuals in the study).
- Avoid using network-level metrics unless a high proportion of individuals in the population are identified.
- Consider and control for potential confounding effects generated by other social processes and the method of data collection.
- Keep individuals in the same order in all data used for analysis (attribute data are ordered the same as the rows/columns in the adjacency matrix).

Box 1: Key Points

Table 1: Overview of key considerations in each step of network analysis (see also Fig. 1). In addition to the key references, both Whitehead (2008) and Croft, James & Krause (2008) cover these topics in detail

Step	Important consideration	Key references
Collecting data	1. What is being observed? 1a. Associations/Gambit of the group? 1b. Interactions between individuals? 2. What data sampling method to use: focal follows, group follows, all-event? 3. Is there observation error or bias?	General methods: Whitehead (1997), Whitehead & Dufault (1999), James, Croft & Krause (2009). Sampling methods: Altmann (1974) Gambit of the group: Franks, Ruxton & James (2010)
Building the network	1. What is the biological definition of an edge in the network? 2. What is the sampling period? 3. Is there any observation bias that can be controlled with the association index? 4. Are edges directed or undirected?	Edge inference: Psorakis *et al.* (2012). Association indices: Cairns & Schwager (1987). Affiliation indices (Whitehead & James 2015) Dynamic networks: Blonder *et al.* (2012), Hobson, Avery & Wright (2013),Pinter-Wollman *et al.* (2013).
Hypothesis testing	1. What is the question? 2. What is the null hypothesis? 3. Is it potentially true? 4. What needs to be controlled for? 5. What test statistic(s) should be used? 6. Does the network need to be compared to randomized networks? 6a. What type of randomization is required? 7. Present effect size statistics.	General considerations: Croft *et al.* (2011b). Randomizations: Manly (1997), Bejder, Fletcher & Brager (1998), Whitehead (1999) Whitehead, Bejder & Ottensmeyer (2005), Sundaresan, Fischhoff & Dushoff (2009).

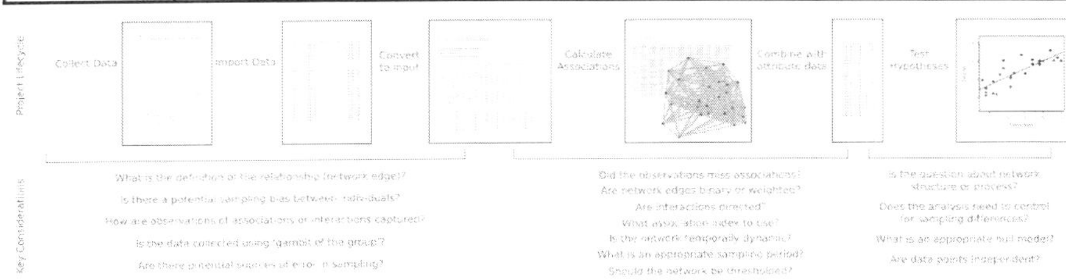

Figure 1: Primary steps and key considerations in the collection and analysis of animal social networks (see also Table 1).

OVERVIEW OF NETWORK ANALYSIS

What is a network?

The term 'network' can mean different things to different people. It can refer to the 'real' set of interactions between animals that integrate to form community dynamics. Hinde (1976) defined social structure in terms of the nature, quality and patterning of the relationships among its members, where a relationship between two individuals integrates the content, quality and patterning of their interactions. For example, animals might have genetic, affiliative, agonistic, cooperative, dominant and many other types of relationships that combine to form the 'true' social system (Barrat *et al.* 2004). We can call this the *real* network.

Most commonly, the networks that biologists create are analytical representations of a combined set (or subset) of measures of the true relationships. We can call these the *observed* networks. To make a simple analogy, consider that for many decades, biologists could only infer genetic structure from observed parent–offspring relationships. Researchers could only obtain the true underlying genetic relationships of a population (or pedigree which is a type of network), once DNA paternity testing was developed.

In this paper, we generally use *network* as a description for the *observed patterns of associations or interactions*. We use *network analysis* as the framework of analytical approaches that use observed networks to try and describe and visualize the real network, as well as to test hypotheses about biological phenomena related to social behaviour. For a broader perspective on this question, we point readers to Borgatti & Halgin (2011).

Network Structure

Networks consist of nodes connected by edges. *Nodes* can represent individuals, groups, classes of individuals or other entities. Each node can possess attributes, such as the identity and phenotypic traits of the individual it represents. *Edges* generally represent how two nodes relate to one another, and can be used to describe how frequently they associate or interact, or to describe other relationships (such as genetic relatedness, see Section 4). They often have numeric values (*weighted edges*), describing the strength of the relationship (e.g. rates or numbers of interactions), although it can sometimes be useful to conceive of edges as *binary* (either 0 or 1) indicating the presence or absence of a relationship (e.g. whether a male–female pair have copulated or not, McDonald *et al.* 2013). Nodes and/or edges may also vary temporally, allowing network structure to change dynamically over time (these are termed *dynamic networks*, see Section 5).

Why quantify networks?

Early applications of social network methods, such as those in dolphins (Lusseau 2003) and guppies (Croft, Krause & James 2004), were largely descriptive. These often investigated variation in social roles (Lusseau & Newman 2004) and phenotypic structure in populations (Croft *et al.* 2005). More recently, the power of the network approach has been clearly demonstrated by two types of studies: those describing potential fitness consequences related to network position, and those investigating the spread of information and disease through animal networks. Network studies can thus be placed into four broad categories: (i) description of social structures (e.g. Connor, Heithaus & Barre 2001; Wittemyer, Douglas-Hamilton & Getz 2005); (ii) studies of the causes and consequences of individual variation in network position – where 'network position' refers to the structural properties that arise as a consequence of an individual's phenotype or patterns of sociality (e.g. McDonald 2007; Pike *et al.* 2008; Oh & Badyaev 2010; Aplin *et al.* 2013); (iii) studies of social processes and the implications of network structure for dynamics of information (e.g. Boogert *et al.* 2008; Aplin *et al.* 2012, 2015; Allen *et al.* 2013) and disease or parasite spread across networks (e.g. Godfrey *et al.* 2009; Hamede *et al.* 2009; Fenner, Godfrey & Bull 2011; Bull, Godfrey & Gordon 2012; MacIntosh *et al.* 2012; Brockmann & Helbing 2013; VanderWaal *et al.* 2013a,b); and (iv) relationship between environment and network structure (Edenbrow *et al.* 2011; Webster *et al.* 2013; Firth & Sheldon 2015; Pinter-Wollman 2015) and vice-versa (Oh & Badyaev 2010; Formica *et al.* 2011; Shizuka & McDonald 2012; Farine & Sheldon 2015). The first category of studies deals with visualizing and describing the natural world (network diagrams and statistics are more easily assimilated by the human brain than raw matrices of association measures). The second category deals with individual traits and provides a framework for quantifying individual position within a broader social structure. The third focuses on the interplay between social dynamics, interaction patterns and socially mediated flows, where networks provide a unique ability to integrate the dyadic nature and non-uniform structure of social relationships when modelling social processes. The last focuses on the external factors that shape network structure, how network structure can alter individuals' environments and to what extent individuals can shape aspects of their environment.

Network approaches also have the potential to characterize emergent properties of social organization. A classic paper by Granovetter (1973) highlighted the potential importance of weak ties (connections that are rarely made) for connecting otherwise disparate groups or communities, in turn shaping their higher-level structure. The idea that societies can have properties that are a function of their overall structure (i.e. emergent properties), rather than of individuals themselves, has stimulated a substantial body of theoretical research into the link between network structure and social processes, in particular the spread of disease or information (e.g. Newman 2002; Keeling 2005; Shirley & Rushton 2005; May 2006; Bansal, Grenfell & Meyers 2007; Perkins *et al.* 2009; Volz *et al.* 2011; Hock & Fefferman 2012; Whitehead & Lusseau 2012; Ashby & Gupta 2013). Many of these studies used analytical mathematical models, and this approach that has made a large contribution to our understanding of the emergent properties of networks.

Network analysis provides a common framework for studying emergent processes in animal societies. For example, many animal populations exhibit fission–fusion dynamics, which can emerge under a range of ecological pressures (Aureli *et al.* 2008). In such populations, the patterns of associations can often appear random. However, social network analysis can elucidate whether patterns of connections between phenotypes (i.e. assortment in the social network) are non-random (Farine *et al.* 2015b). Because such processes may play a role in shaping how selection drives the evolution of social species, future efforts will benefit from a focus on the following: (i) the process of self-organization and orderliness in social networks, (ii) what interactions (including affiliative, genetic, mating and spatial networks) underlie dynamic processes in social networks (such as transmission properties) and (iii) whether network structure can evolve in response to selection pressure (both ecological and social).

DEFINING NETWORK EDGES

Defining Relationships (Edges)

The first consideration when setting out to collect or analyse network data is to define the relationships represented by edges in the network. Castles *et al.* (2014) highlight the importance of carefully considering the edge definition, as networks based on different edge definitions may not be equivalent. Ideally, the definition of network edges will be motivated by the biological process being investigated and edge definitions should be based upon how the animals interact and communicate. For example, studying the spread of a sexually transmitted disease requires edges that represent sexual contact.

However, the technical limitations associated with collecting the data (i.e. the challenges of collecting sufficient data to generate robust estimates for a network given a specific edge definition or of studying replicate networks) must also be considered. Researchers should strive to measure the complete set of interactions (context-specific events, such as grooming or fighting) for each dyad (two individuals that could potentially be connected by an edge). These may need to be captured using more than one network, for example one network to capture affiliative interactions and a separate network to capture agonistic interactions. However, many studies are limited to capturing information about the spatial and temporal co-occurrences of pairs of individuals, which are called associations. If interactions are rare or difficult to observe (e.g. they occur underwater), then association networks could provide a more precise estimate of interaction rates by avoiding having many false negatives: interactions that occurred but were not or could not be observed (Farine 2015). Avoiding false negatives is important because the absence of even a few edges can significantly impact the global structure of the social network (Fig. 2 demonstrates the effects of removing weak edges). Furthermore, associations (such as co-membership in groups) can sometimes better represent animal relationships than dyadic interactions. For example, they capture a broader range of interactions which better reflect the social environment, and therefore more appropriate when investigating broad-scale patterns (such as measuring social selection, Farine & Sheldon 2015). Thus, when defining edges and how edge weights are calculated (Section 5), we recommend addressing the following questions. (i) How can the edges be made most relevant to biological process being studied? (ii) Can enough data be collected to accurately capture different relationships and the overall structure of the population (Section 10)?

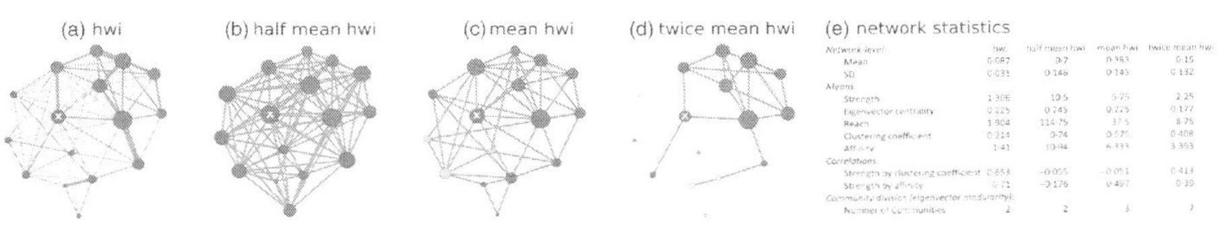

Figure 2: Thresholding networks can have significant impact on the structure and statistical properties of a network. (a) This network of bottlenose whales (*Hyperoodon ampullatus*), based on observations of groups of animals made at sea off Nova Scotia from 1988 to 2003 (see Gowans, Whitehead & Hooker 2001 for methodology), was calculated using the half-weight index (hwi), to account for potentially missed observations of individuals in groups. This network

was then thresholded at (b) half the mean hwi, (c) at the mean hwi (d) and at twice the mean hwi. Nodes are coloured by community (detected using leading eigenvector communities; Newman 2006) and sized by their degree (strength in the original network, binary degree in the others). This figure highlights how thresholding can lead to unpredictable results, such as individual 'x' changing communities, and varying relationships between node properties (such as correlations between node measures, e). For example, the correlation of individuals' rank in terms of strength between the networks (a) and (b) is only 0·57, and the relationship has an R^2 of only 0·28.

Edge Weights and Directionality

Edge values represent either the presence or absence (a binary network of 0s or 1s), or a numeric value describing the strength of the relationship or the number of interactions (a weighted network). Associations are generally symmetrical, but in many cases, interactions are not (i.e. individual A groomed individual B 10 times, but B only groomed A twice). Thus, edges can be either undirected (these are represented by at most one edge between each pair of nodes) or directed (represented by at most two edges – one in each direction). How to calculate edge weights is discussed in Section 5.

DATA COLLECTION METHODS

Collecting Interaction Data

Numerous types of data can be used to construct social networks (see Chapter 3 of Whitehead 2008). A first requirement is that a substantial portion of the individuals (or each unit representing a node) in the population are uniquely identifiable. Methods for identifying individuals include using naturally occurring individual variation in coloration or morphology (Würsig & Jefferson 1990), marking individuals to make them individually identifiable (e.g. with colour-bands McDonald 2007; Farine & Milburn 2013), or electronic tags that provide information about the location or relative distance between individuals. In addition to having individually identifiable study subjects, network analysis requires data on interactions or associations. Altmann (1974) outlined protocols for assembling raw data from direct behavioural observations, particularly focal observations, that are very relevant for animal social networks (see also Chapter 3 of Whitehead 2008).

One method frequently used to capture associations is the 'gambit of the group' (see Franks, Ruxton & James 2010). Gambit of the group defines all individuals within a group of animals observed at a point in time as being associated. Thus, association rates (see Section 5) represent the propensity for each pair of individuals to co-occur in the same group. The fundamental assumptions of the gambit of the group are that all, or almost all, interactions of some kind take place within groups and that interactions of this type occur at a similar rate among all animals when they form groups (Whitehead & Dufault 1999; Farine *et al.* 2015, or that the group itself is meaningful to the animals. This method is particularly useful when groups of animals can be easily observed, and group membership changes over time (Silk *et al.* 2015).

Automated Techniques for Collecting Data

Technological advances in animal tracking are rapidly increasing the amount of data collected, both in the laboratory and in the wild (Krause *et al.* 2013). Methods include using videos to record the position of individuals (e.g. Strandburg-Peshkin *et al.* 2013), fitting tags (e.g. Passive Integrated Transponders or PIT tags) to individuals to make them detectable when they come in proximity to logging stations fitted with antennae (e.g. Farine & Lang 2013), fitting global positioning system (GPS) devices to many individuals to capture their movements relative to one another (Godfrey *et al.* 2014; Strandburg-Peshkin *et al.* 2015) and fitting devices that record when individuals are within a certain proximity of one another (Hamede *et al.* 2009; Rutz *et al.* 2012). With electronic tags and automated data collection, complete or nearly complete records of the associations among members of a population are sometimes available (e.g. Boogert, Farine & Spencer 2014). These technologies are typically limited to creating association (or proximity) networks, thus inferring interactions rather than capturing them explicitly. However, the amount and completeness of the data available can result in a high degree of certainty around the estimates of when individuals could have associated, despite providing little or no data on any one potential interaction.

Attribute Data

The attributes of individuals, typically consisting of phenotypic traits or details about individual state, are a vital part of social network analysis. One aim of social network analysis is to determine how sociality, or the relationship between sociality and fitness, is mediated by individual traits.

Multiple traits can be captured as attributes of each node in the network, such as age, sex, size or breeding status. Some attributes may be central to the research questions, while others may be recorded because they may have an impact on an individual's network position that needs to be controlled for in subsequent analyses (see example in *Thresholding edges or individuals* in the following section).

INFERRING ASSOCIATIONS FROM DATA

The Adjacency Matrix

The most fundamental data structure in animal social network analysis is the association or adjacency matrix. This is an N x N matrix, where N is the number of individuals in the study, and each cell contains the value of an edge in the network that represents associations or interactions. Typically, the matrix is read as the 'actors' along the rows associating or interacting with the 'receivers' along the columns, so that the presence/weight of the edge between individual A and individual C can be found on the first row and third column. If the network is undirected, then the matrix will be symmetrical (edge A to C is equal to edge C to A). The diagonal of the matrix contains the 'self-edges', or the number of associations/interactions an individual has with itself. These are rarely used in animal networks where nodes are individuals, but can be used in networks where nodes represent other properties such as location or the agglomeration of a number of individuals, such as species (Mokross *et al.* 2014).

Calculating edge values from observational data

Once data have been collected on the interactions or associations in the study population, pair-wise observations can be converted into edges to populate the adjacency matrix. For interaction data, the relationship measure is often the total number of interactions observed between each pair of individuals. If the number of observations (or total time) differs between individuals, then the relationship should reflect the rates at which interactions occurred (dividing by observation time, see Farine 2015).

Association indices can be used to define edges in the network (Cairns & Schwager 1987). They typically estimate the proportion of time individuals that are associated and range between 0 and 1, where 0 indicates that they never associate, and 1 indicates that the dyad was always together. If observations are rarely missed, then the simple ratio index can be used. Here, the edge weight is calculated using: $E_{AB} = \frac{x}{x+y_{AB}+y_A+y_B}$ where the undirected edge weight between individuals *A* and *B* is the number of samples or sampling periods where they co-occurred (x) divided by the number in which one or both were identified (y_{AB}) is the number of times both *A* and *B* were observed in the same sample but not together, y_A is the number of samples where only individual *A* was seen, and y_B is the number of samples where only *B* was seen). If individuals are frequently missed (when they should have been observed), then the half-weight index $\left(E_{AB} = \frac{x}{x+y_{AB}+\frac{1}{2}(y_A+y_B)}\right)$ can provide a less biased estimate of the real rate of association. Whitehead (2008) provides a comprehensive list of association indices (table 4·5, page 98) with extensive discussion and examples of their usefulness when dealing with different types of sampling bias (table 4·6, page 99). Table 2provides an overview of some useful software packages to help generate networks, and Supporting Information Appendix S1 provides an overview of different formats to store social data.

Table 2: Key software packages for creating and analysing social networks

Name	Pros	Cons	Key references
ucinet/Netdraw	Fully integrated point-and-click analyses Wide range of hypothesis testing tools Extensive documentation and help files Easily implemented network diagrams in NetDraw	Requires adjacency matrix Only node randomizations Graphs are not vectorized Only available in Windows	Borgatti, Everett & Freeman (2002), Borgatti, Everett & Johnson (2013)
socprog	Fully customized for animal social networks Generates adjacency and generalized affiliation matrix from data Extensive range of tests tailored to biological questions Calculates and fits models to lagged association rates Available as stand-alone package or as functions in matlab Free on all platforms (but see Cons) Output formats compatible with ucinet and NetDraw	Detailed plotting functions available only with matlab. Compiled version only available in Windows Non-compiled version requires matlab and Statistics Toolbox	Whitehead (2008, 2009)

r	Extremely flexible Packages provide most possible tests and network measures Animal network-specific routines available Simple support for dynamic networks Integration of almost every possible statistical test High-quality plotting (including GIS functionality) Free on all platforms	Steep learning curve Requires careful management of variables Requires some knowledge of software coding (see appendices of this paper)	*Animal networks*asnipe: Farine (2013a) *Network metrics*: igraph :Csardi & Nepusz (2006), sna: Butts (2008) *Dynamic networks*: time ordered: Blonder *et al.* (2012)

Inferring Edge Values with Automated Data Collection

As with observation data, edges should ideally be derived from automated data based on how the species interacts and communicates. If animals reliably interact with one another at ranges less than a particular cut-off, this can be used to define association. However, the scale at which associations are maintained can vary over short periods of time as a consequence of different social or environmental contexts. For example, imagine a system with two pairs of birds holding neighbouring territories. When the two groups are well separated from each other (say 200 m), individuals in each group could be a relatively large distance apart from each other (say up to 20 m) and still remain in contact (e.g. acoustically). If the two groups come into territorial conflict, individuals within each group might be very close to each other (say within 1 m), but also very close to the individuals from the other group (say within 5 m). In this example, a fixed threshold of 10 m would introduce both false positives and false negatives into edge data representing spatial associations. This same issue arises when recording visits by PIT-tagged individuals at feeders. In particular, the amount of time a flock spends at a feeder can vary greatly based on the flock size. In both temporal data-stream and spatial proximity networks, we should not impose arbitrary thresholds if these are not clearly defined by the biology of the system (a good example is Hamede *et al.* 2009). Instead, if groups are defined as individuals being closer to each other than to others, then clustering algorithms are a useful approach.

In behavioural observations, we often implicitly define groups as sets of individuals that are closer or interact more within themselves than they do with others. Clustering algorithms, such as Gaussian mixture models (Psorakis *et al.* 2012), can extract such patterns from spatiotemporal data. These algorithms statistically infer the best-fitting temporal boundaries of groups based on the data rather than relying on a fixed inter-group or inter-individual visit interval. The results are often more robust than traditional methods at capturing biological interactions (Psorakis *et al*. 2015).

Associations and Affiliations

Using association indices as edge weights produces a network that represents the pattern of association among individuals. Sometimes we are more interested in affiliations (individuals actively associating with other individuals) and wish to remove other causes of association patterns, such as spatial and temporal overlaps. Generalized affiliation indices are the residuals after regressing association indices or other measures of association (or interaction), on potential structural predictors of association, thus isolating true affiliations using a generalized linear model (Whitehead & James 2015). These generalized affiliation indices then form the adjacency matrix for a network analysis aimed at true affiliations.

Thresholding Edges or Individuals

Thresholding involves either removing individuals with few observations or converting a weighted network to binary by only counting values above or below a certain value. An example of the latter is Croft *et al*. (2009), who used both high and low thresholds to investigate assortment by behavioural type. However, choosing an appropriate threshold may be problematic, and different threshold values can lead to different conclusions in the same network (see Fig. 2). Thresholding networks has been found to generate high rates of both type I and type II errors (Butts 2009; Langer, Pedroni & Jancke 2013; Farine 2014) than using the original weighted network, and should always be applied with caution. Thus, applying a threshold to an association index or interaction rate is usually not recommended other than for displaying networks (Franks, Ruxton & James 2010).

For some purposes, it is useful to remove individuals for whom there are few data. Edges connected (or not connected) to rarely observed individuals may be very inaccurate. As a result, network metrics measured for that node are unlikely to represent the true state or behaviour of the individual. This could, at worst, impact the value of every other node in the network when using global network measures. Thresholding nodes' data should be considered on a case-by-case basis, but the impact of removing individuals may be smaller than the impact of having spurious edge values. A good example of thresholding individuals based on properties of the data is Aplin *et al.* (2013) who removed individuals with fewer than 100 observations as these exhibited a clear relationship between number of observations and the binary degree. In contrast, no such relationship existed for individuals with more than 100 observations.

Creating Temporal Networks

There are two principal types of temporal networks, time-ordered and time-aggregated networks (Blonder *et al.* 2012). In a time-ordered network, each edge is encoded with start and end times, capturing the complete set of information about when edges (i.e. interactions/associations in social analysis) occurred and their duration, and so what edges co-occurred. Such time-ordered data can be used to map the topological flow of information or pathogens through networks when the ordering of interactions or associations are considered to be important (Blonder & Dornhaus 2011). Blonder *et al.* (2012) provide a useful *R* package *time ordered* to perform some analyses on these types of networks.

Time-aggregated networks, in contrast, maintain the same form as regular aggregated networks (an adjacency matrix). Here, a new network is calculated for each time slice, such as for every sampling event, week, month, season or year. Creating these networks is relatively simple as they only require the input data to be subset for each period.

Finally, a useful measure of temporal stability is the lagged association rate (Whitehead 1995). This measure calculates the probability that a given dyad is re-observed after a given time period. This can be estimated at the network level, for different classes of individuals, or at the dyadic level. The lagged association rate is useful for describing and modelling the temporal scales over which social behaviour processes operate, or for comparing how these differ between different classes of individuals (e.g. Aplin *et al.* 2013).

DISPLAYING SOCIAL STRUCTURES

Network Diagrams

Network diagrams, formally called graphs, allow us to visualize social connections and the overall structure of the network. Nodes, usually individuals, can be represented by shapes (circles, squares) of different colours used to represent attributes, such as sex, class or gregariousness. In an undirected network, with a symmetric adjacency matrix, edges are usually drawn as a single line between each node. Relationships in a network that are directed are represented by edges with an arrow pointing in the direction of the interaction. In binary networks, edges are displayed (1), or not (0), whereas the thickness of the line is typically made proportional to the strength of a dyad's association in weighted networks. Weak edges can be omitted from the diagram to aid clarity (see Section 5 on thresholding). In Table 3, we provide information on some of the available software packages for visualizing social networks.

Table 3: Other useful software

Type	Name	References
Visualizing networks	Gephi	Bastian & Heymann (2009)
	GraphViz	Gansner & North (2000)
	Cytoscape	Shannon *et al.* (2003)
	SoNIA	Moody, McFarland & Blender-deMoll (2005)
	Tulip	Auber (2004)
	Netdraw	Borgatti (2002)
	r package *igraph*	Csardi & Nepusz (2006)
Collecting data	J-Watcher	Blumstein & Daniel (2007)
Calculating dominance hierarchies	matman	de Vries, Netto & Hanegraaf (1993)
	CodaReader	Adams (2005)

Community Delineation

Some social networks are highly modular – the nodes form communities such that most edges (in a binary network) or a high proportion of the edge weight (in a weighted network) is within rather than between communities. Identifying such communities – sometimes called clusters or social units – is not trivial. Many techniques have been developed by both statisticians and network analysts, but the Newman (2006) eigenvector modularity technique is often used with animal social networks and usually works well. Many others are also available, and most of these are implemented in the R packages listed in Table 2. Also popular is hierarchical cluster analysis in which communities are embedded within one another, and the results are displayed using a tree-like dendrogram (Whitehead & Dufault 1999). This can be an excellent representation of the social network if the society does actually consist of a hierarchically embedded set of social tiers (e.g. Wittemyer, Douglas-Hamilton & Getz 2005), which community detection algorithms typically fail to detect. If not, dendrograms may be highly misleading (Whitehead 2008).

INTERPRETING NETWORK METRICS

Types of Network Metrics

Network metrics are statistical measures used to characterize properties of individuals (nodes) or the whole network. Most measures are node based; they calculate a separate value for each node. Others, such as density, measure global network properties. A final type of metric are edge metrics (e.g. bridges, edge betweenness), but these are rarely applied to animal social networks and not covered further. Almost all network metrics can be expressed/defined either as a weighted measure or as a binary measure. For example, the most common nodal measures used in network analysis are binary degree and strength, which are the sum of the number of edges each node has or of all the edge weights connected to it, respectively. The average or distribution of the measures can be used to describe more general properties of the social structure. For example, one of the fundamental properties of networks is the degree distribution (see pp. 243–260 in Newman 2010).

We give definitions of some of the most commonly used network metrics in Box 2, Newman (2010) provides detailed accounts of many more, and Borgatti (2005) provides detailed a detailed account of their appropriate use. Network metrics can easily be calculated in most network analysis programs, including *socprog* (Whitehead 2009) and *ucinet* (Borgatti, Everett & Freeman 2002). Further, there are some excellent libraries in *r* (R Development Core Team 2013), in particular *igraph* (Csardi & Nepusz 2006) and *sna* (Butts 2008) which calculate almost all common network algorithms (see Table 2).

B = binary only, W = weighted only, BW = weighted and binary networks.

NODE-LEVEL METRICS

Degree (B): the binary degree is the count of the number of edges connected to the node. If a network is directed, degree can be the partitioned into in-degree and out-degree, representing the number of incoming and outgoing edges, respectively. This measure captures the gregariousness of individuals, in terms of the number of associates or interaction partners.

Strength (W): the weighted equivalent of binary degree is the sum of all edge weights connected to the node (if all edges have a weight of 1, then the strength will equal the binary degree). Strength can also be partitioned into in-strength and out-strength for directed networks. This measure typically represents the expected total interaction or association rate per sample. For example, a node with a strength of 2 would be expected to be observed with approximately two other individuals on average (if using most association indices).

Betweenness Centrality (BW): a count of the number of shortest paths that flow through the node. This measures how important a node is for connecting disparate parts of a social network. Individuals with a high betweenness centrality are likely to connect largely independent communities. This often highlights individuals that have a greater tendency to change groups than others.

Eigenvector Centrality (BW): the sum of the centralities of an individual's neighbours. High centrality can be achieved by having either a large degree or being connected to associates with a high degree (or both). This is a useful property as it captures the potential 'importance' of individuals in the network, as social hubs, or for the propagation of information or diseases through animal populations.

Page Rank (BW): a robust measure of centrality for directed networks that divides the centrality gained through associates by the associate's out-degree. This means that very central nodes only pass on a small amount of centrality to each node that is connected to them, thereby controlling the measure of eigenvector centrality for long tails in the degree distribution. Individuals with a large page rank are disproportionately important for connecting different components of the network, and this measure is likely to be important when investigating flows through networks.

Reach (BW): Measure of what proportion of all other nodes can be reached in one step, two steps and so on. This is the equivalent of calculating how much of the network is in 'n degrees of separation'. This measure has not been explored much in animal social networks, but could be useful for investigating differences in social structure, or implications of changes in social structure, when conducting experiments (such as removals of key individuals). It is also likely to be interesting in models of disease or information transmission in order to estimate how quickly most individuals in a population can become infected/knowledgeable.

NETWORK-LEVEL METRICS

Density (BW): the number of edges in a network divided by the total possible edges (B), or the sum of edge weights divided by the number of possible edges (W). A potentially important measure for normalizing observed degree distributions as larger networks tend towards very low densities.

Homophily/Assortativity (BW): the correlation in the phenotype of connected individuals. Positive assortment suggests that nodes are more connected than expected, whereas negative assortment suggests avoidance of alike nodes. This can now be measured on weighted networks and is a powerful approach for identifying phenotypic structure in social networks. For example, positive assortment by degree (gregariousness) has been linked with rapid spread of information or disease through social networks.

Transitivity (BW): the proportion of triads (trios of nodes) that have three edges divided by the number of triads that have two edges. When compared to null models, this identifies whether trios have a tendency to be more or less connected than expected. This is potentially an important measure, particularly when measuring interactions, as it captures the level of clustering in the network. For example, grooming networks may have low transitivity if grooming is directed up or down a linear hierarchy. Transitivity can be measured for nodes as well as at a network level. However, care should be taken if using the gambit of the group approach to capture associations, as this automatically closes triads (but the impact of this remains unexplored).

Box 2: Common network metrics

Interpretation of Network Metrics

Studies use network metrics to estimate differences between individuals in their placement within a social network. In animal networks, the most common are based on measures of centrality. This generally refers to individuals that are broadly more (or more strongly) connected than others. We recommend caution when interpreting network metrics, as these depend on both the measure used (e.g. degree vs. betweenness) and the edge definitions (e.g. rates vs. number of interactions), and on how the population is structured (e.g. if a population is structured into communities, metrics calculated within a single community may be very different from those calculated for the entire population). For example, the strength (weighted degree) of individuals in a network with edges defined as association indices defined using group membership is roughly proportional to their average typical group size (Jarman 1982). In contrast, the strength of a node in a network using counts as edge values simply represents the total number of interactions observed from that individual. Box 3 provides an example of how the network structure can impact the interpretation of sets of centrality metrics. Because the metric can have different interpretations on different social networks, we recommend visualizing the structure of the network and the correlation structures between different metrics. We also discuss why comparing metrics across different networks is also problematic in Section 11.

The structure of the network can impact the correlation between different measures of centrality. While in many cases, centrality measures might capture the same biological processes, in some other cases, these might differ. In Box 3 (Figure), we present two toy networks. The first (a) contains two clusters joined by a single individual. In this network, node 1 provides a bridge between the two clusters, and it has a high betweenness (b). However, node 1 has the lowest degree. Thus, if we were to investigate dynamics of spread, betweenness might provide a better estimate of relative node importance. The second network (c) contains individuals that are more uniformly connected. Individual 1 in this network has both the highest degree and, by far, the highest betweenness. VanderWaal *et al.* (2014) developed a metric called cut point potential (the potential for individuals to disconnect parts of the network if they are removed) to disentangle the effects of degree from betweenness, and this may warrant more widespread use. In both networks, eigenvector centrality and page rank are both strongly correlated with degree (a rank correlation close to 1), whereas clustering coefficient is not (and rarely is). Thus, when interpreting the relative importance of nodes in a network, the relationships between different network metrics may be informative. The node-level ratio of network metrics could also be useful (such as the ratio of degree to betweenness), though to our knowledge, this has not been explored.

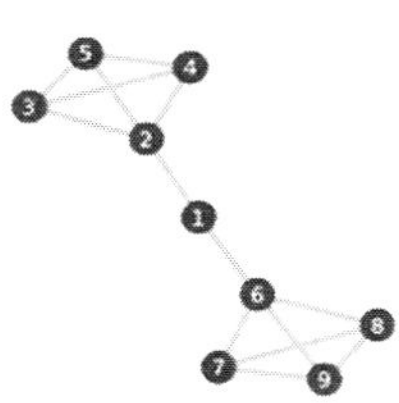

	1	2	3	4	5	6	7	8	9
Degree	2·0	4·0	3·0	3·0	3·0	4·0	3·0	3·0	3·0
Betweenness	16·0	15·0	0·0	0·0	0·0	15·0	0·0	0·0	0·0
Eigen. Cen.	0·24	0·38	0·33	0·33	0·33	0·38	0·33	0·33	0·33
Clust. Coef.	0·0	0·5	1·0	1·0	1·0	0·5	1·0	1·0	1·0
Page Rank	0·08	0·14	0·11	0·11	0·11	0·14	0·11	0·11	0·11

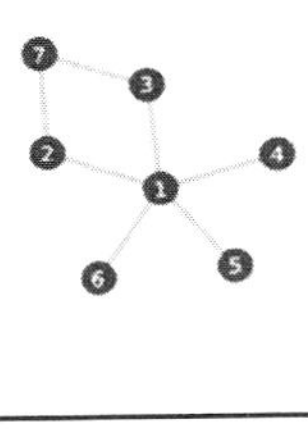

	1	2	3	4	5	6	7
Degree	5·0	2·0	2·0	1·0	1·0	1·0	2·0
Betweenness	12·5	2·0	2·0	0·0	0·0	0·0	0·0
Eigen. Cen.	0·47	0·47	0·47	0·31	0·31	0·31	0·23
Clust. Coef.	0·0	0·0	0·0	0·0	0·0	0·0	0·0
Page Rank	0·34	0·14	0·14	0·08	0·08	0·08	0·14

Box 3. The effect of network structure on network metric correlations

CONSTRUCTING NULL MODELS

What are null models?

Null models are data sets that are based on observed data, but generated in a way that allows some aspects of the data to be random. This can involve generated new data using simulations inspired by the observed data (e.g. creating random networks with the same degree distributions), but more commonly involve shuffling existing data to create expectations of random given certain constraints. Many potentially realistic null models can be tested using social network analysis. For example, one null model might assume that individuals have no preferred affiliates given their spatial and temporal use of the study area, or another that there is no tendency of males to preferentially associate with other males given the overall gregariousness profiles of members of each sex. Note the 'given ...' clauses – these are what make hypothesis testing on social networks so challenging. Hypothesis testing on networks generally relies on null models constructed by randomizing the data. The key issue is whether the null model actually represents the biological null hypothesis being tested.

Using Null Models Accounts for Non-independence

A key consideration when applying a statistical test to a social network is non-independence in the data. Network measures are inherently non-independent and thus violate the assumptions underlying most parametric statistical tests (Croft *et al.* 2011b). In particular, each association index is shared by two individuals. This results in over-inflation of the degrees of freedom used to calculate significance (see Appendix S2 for an example). Randomizations are the most widely used approach to control for non-independence (Croft, James & Krause 2008; Whitehead 2008; Croft *et al.* 2011b). One strength of randomizations is that they can be integrated with almost any statistical test.

Another source of non-independence is temporal structure. For many analyses, the assumption is that different samples of data (i.e. observations) are independent. Because of the persistence of associations and the autocorrelation of interactions, this is often not the case. For example, data collected by following a focal animal for a defined period of time will be pseudo-replicated. Using sampling periods, where data is aggregated to generate independent samples, is one way of reducing the effects of temporal partitioning (e.g. by placing all data from one focal follow into one sampling period). Lagged association rate analyses (Whitehead 1995) can help determine a suitable sampling period for which subsequent samples can be considered independent. Null models are useful for dealing with this issue by randomizing blocks of pseudo-replicated data, thus generating a realistic null distribution when estimating significance with statistical tests such as regressions (see next section).

Randomization Techniques

Data randomizations use the observed data to generate replicated data sets, where each replicate is a shuffled version of the original data (Manly 1997). In each step in the randomization, a new network is created with the same nodes as the original, but with randomized edges based on assumptions about relationship under investigation. Repeating this process many times creates a distribution of values that represents the expected relationship under the assumption of the null hypothesis. The process of defining a null model (how the data will be shuffled) involves keeping certain aspects of the data the same during each randomization step. This has two functions: it controls for particular factors that could influence the data (such as the number of observations of each individual or the spatial distribution of individuals), and it provides a framework that facilitates the comparison of different hypotheses (Gotelli & Graves 1996). A common null model is to randomize who associates with whom, but to restrict swaps between pairs of individuals observed in the same location at the same time. Significance is then estimated by comparing the observed statistic to the distribution of same test statistics measured on the randomized networks. We provide an overview of how to integrate randomizations into hierarchical models in Box 4, and worked examples in Appendix S2.

The application of randomizations is not limited to traditional permutation tests. They can also be used to control for confounding factors in more traditional tests and calculate appropriate *P* values. To demonstrate how to calculate *P* values in a GLMM using randomizations, we generated a simulated network consisting of 40 individuals from two different sampling areas (the code is contained in Appendix S3). In this simulation, we started by generating a Poisson-distributed gregariousness score for each individual, with individuals in area 2 having slightly higher average gregariousness. We then allocated a sex to each individual, with the probability that an individual is male that is proportional to the individual's gregariousness value. Using the *rgraph* function in the *R*package *sna*, we then simulated 100 sampling periods. In each sampling period, the probability that two individuals interacted (had an edge) was proportional to their combined gregariousness scores (but individuals in different areas did not interact). We then generated the networks (Figure Box 4a; males red, females blue, node size is proportional to strength) using the simple ratio index, finding that the degree was higher in males than in females (Figure Box 4b). Fitting the GLMM model, *Strength ~ Sex + (1|Area)* using the *lmer* function in the *lme4* package generated a positive coefficient value (coefficient ± SE = 2·598 ± 0·787, t = 3·302, see Box 4 Table 1). To calculate the *P*value, we then performed 1000 datastream permutations (swapping observations in the sampling periods), controlling for time (sample) and space (area). After each swap, we re-fit the same model but with the strength value of each node calculated from the newly created (and increasingly random) network: $Strength_i \sim Sex + (1|Area)$, where *i* is the current randomization. Comparing the observed coefficient value (2·598, Figure Box 4c red line) with the distribution of the 1000 coefficient values from the permutations (Figure Box 4c, black histogram) captured this significance of this difference (P_{rand} = 0·035).

Fixed effects	**Coefficient**	**Standard error**	***t* statistic**
Intercept	4·597	0·909	5·062
Sex (male)	2·598	0·787	3·302
Random effect	**Variance**	**Standard error**	
Area	0·901	0·953	
Residual	5·895	2·428	

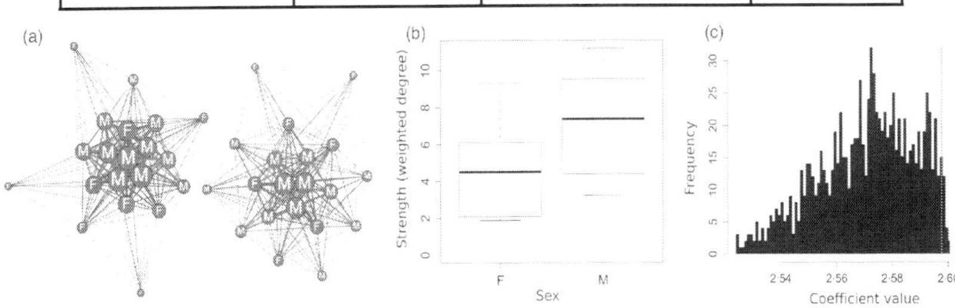

Box 4. Incorporating null models into hierarchical models

There are two main approaches to building null models for animal networks: node-based randomizations and data stream-based randomizations. Node-based randomizations involve entirely re-distributing the attributes of the nodes in the network, while maintaining the same number of each type, when creating each random network (Croft, James & Krause 2008; Whitehead 2008; Croft *et al.* 2011b). This randomization is often used to test for differences in network position between nodes with different attributes (do males have more associates?). It is simple to implement as it relies only on the adjacency matrix. However, this cannot control for parameters other than the number of individuals of each type. Further, it relies on the assumption that the observed network is a strong representation of the true network (Croft *et al.* 2011b), and has the potential for much higher rates of type I and type II errors than randomizations based on shuffling the data stream (Farine 2014).

Permutations of the data stream involve sequentially swapping observations between individuals (Bejder, Fletcher & Brager 1998). Data-stream permutations can be used to test the same hypotheses as node-based randomizations. They can also be used to test for preferred and avoided relationships, either overall or between specific dyads. Swaps can occur at the group level (A is moved from group 10 to group 15, and B is moved from 15 to 10), or at the dyadic level (groomer A is swapped from grooming C to grooming D, B is swapped from D to C). This method is very powerful as it can control for a number of different possible confounding effects, such as controlling for spatial or temporal variation in the presence of individuals (Whitehead 1999; Whitehead, Bejder & Ottensmeyer 2005; Sundaresan, Fischhoff & Dushoff 2009), and for the sampling method by keeping the number of observations of each individual fixed. We recommend always using data-stream permutations when possible.

HYPOTHESIS TESTING IN ANIMAL SOCIAL NETWORKS

Determining if the Network is 'Non-random'

Often, we want to test whether a network is more structured than expected from random. To do so, a test statistic needs to be chosen that will be used to compare a property of the observed data to the same property measured on a set of randomized networks. This statistic should represent an aspect of the structure that is expected to differ between non-random and random networks.

For example, if individuals have preferred and avoided relationships, then the standard deviation of their association strengths should be higher in the observed network than in a network where individuals associate with others at random (and thus equally). Whitehead, Bejder & Ottensmeyer (2005) propose different test statistics for different null hypotheses, such as the lack of short (within sampling periods) and long (between sampling periods) preferred relationships among dyads, or uniform gregariousness among individuals. The most easily interpretable test statistic is the coefficient of variation (CV) of the association indices. This measure has the added benefit of solving the problem of presenting effect size, as its values on the observed network and the mean of the randomized ones indicate how strong the effect is. To test if the network contains more preferred/avoided relationships than expected at random, the CV of the observed network is compared to the CV measured in 1000 or more randomized version of the network (see previous section). The *P* value is then calculated by taking the number of times the CV value of the observed network is smaller than a randomized network, divided by the number of randomizations (see Ramos-Fernandez *et al.* (2009) and Mourier, Vercelloni & Planes (2012) for good examples of this method being applied).

Using Network Data in Linear and Hierarchical Models

Linear models (in particular generalized linear models, GLMs, and generalized linear mixed models, GLMMs) are an attractive framework for investigating relationships between an individual's attributes (phenotype) and its network position (social network metric). This approach can easily control for confounding effects arising from data sampling (e.g. including the number of observations as a fixed effect or using a binomial model that includes both the numerator and denominator of the association index). While GLMMs deal well with repeated observations made on individuals, they cannot readily control for non-independence in the network measures themselves (Snijders 2011). In Box 4, we provide an example of combining GLMM with network randomizations. In Box 5, we show how this approach deals with biases that can easily arise when collecting social data (see *Is there a bias in the data sampling?* in Section 10). When using this method, a table with the coefficient values, standard error, test statistic (e.g. *t*-value) and the *P* value calculated from the randomization test can all be reported. The study by Boogert, Farine & Spencer (2014) is a good example of this framework applied to testing hypotheses on social networks.

A major potential source of error in social networks is sampling bias. If there are differences in how observable individuals are as a function of their class, this can easily lead to spurious results. We demonstrate this by repeating the simulation from Box 4, but this time males and females do not vary in gregariousness (we maintain the same distribution, but now allocate sex randomly). After generating sampling periods, we now allocate a probability that each female is observed with only 70% reliability, whereas bright conspicuous males are observed with 100% reliability (the code for this simulation is found in Appendix S4). Thus, even though both classes had the same average degree measured on the full set of associations (Figure Box 5a), reducing the observation probability of females introduced a difference in their means (Figure Box 5b). Fitting the same GLMM model as in Box 4 suggested that this difference was highly significant (Table Box 5, note the *t*-value is larger than in the Box 4 example). However, when calculating the *P* value using randomizations (see details in Box 4), the observed coefficient value (2·143, Figure Box 5c red line) did not fall outside the distribution of randomized coefficient values (Figure Box 5c histogram), resulting in a non-significant effect (P_{rand} = 0·226). This example highlights how using randomizations improves our ability to capture biological patterns in the *real* rather than in the *observed* network.

Fixed effects	**Coefficient**	**Standard error**	***t* statistic**
Intercept	4·315	0·621	6·949
Sex (male)	2·143	0·578	3·707
Random effect	**Variance**	**Standard error**	
Area	0·371	0·609	
Residual	3·186	1·785	

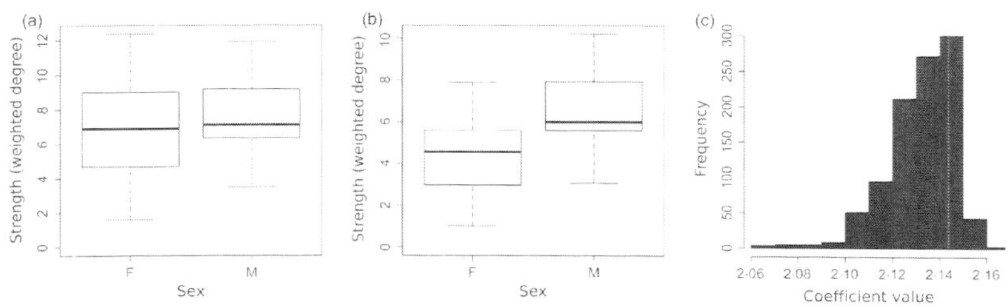

Box 5: Demonstrating how permutations deal with biases

Mantel Tests and MRQAP

Sometimes we wish to test the hypothesis that association strength, or interaction rate, is related/unrelated to some other dyadic measure, such as genetic relatedness or gender similarity. The Mantel test (Mantel 1967) makes such a test, usually by means of node-based permutations. There are a number of nonparametric versions of the Mantel test which may be appropriate especially with interaction numbers or rates, which can have high skew (e.g. Hemelrijk 1990). The matrix correlation coefficient (i.e. the correlation between the corresponding non-diagonal elements of the two matrices) is a suitable effect size measure for Mantel tests.

Building on the Mantel test, we can test for a relationship between a dependent dyadic variable, such as association strength, and an independent variable, such as genetic relatedness, while controlling for one or more further independent variables, such as gender similarity or home range overlap. In this case, regressions can be performed on the matrix data using a procedure called multiple regression quadratic assignment procedure (MRQAP). There are several ways that the MRQAP permutations can be performed, but both theory and practice favour a recently developed double-semi-partialling technique, which randomizes the residuals of a regression model rather than the independent or dependent matrices themselves (Dekker, Krackhardt & Snijders 2007). Partial correlation coefficients are suitable effect size statistics for MRQAP tests. The study by VanderWaal *et al.* (2013b) is a good example of this method being used to test the structural similarities between association and pathogen transmission networks.

Unfortunately, all implementations to date rely on node-based permutations of the data, and the validity of using data-stream permutations with MRQAP has not, to our knowledge, been investigated. An alternative, proposed by Rushmore *et al.* (2013), is to fit edge weights into a logistic mixed-effect model, and this could then be combined with an appropriate null model (see previous section).

Network-based Diffusion Analysis

An increasing number of studies are investigating the role of social networks in mediating population-level processes, such as the spread of information. Network-based diffusion analysis (NBDA) is a powerful tool for inferring the relative rates of social transmission of information in a social network and rates of individual-based learning (Franz & Nunn 2009). NBDA uses the order or times when individuals were observed to have acquired some information to fit parameters of social transmission and non-social learning based on their network links to individuals that already have the information (Hoppitt, Boogert & Laland 2010). Hoppitt & Laland (2011) provide a useful manual and *R* package for using NBDA. They include details on how to incorporate confounding effects into the analysis. Farine *et al.* (2015a) and Nightingale *et al.* (2015) also provide important extensions to the NBDA framework. These deal with the need to compare competing networks that facilitate diffusion (i.e. different hypothesized relationships), and with estimating the uncertainty surrounding the estimation of the independent network. Because NBDA has an inherent null hypothesis (no social transmission), it does not need to be combined with randomizations, although the potential for integrating alternative null models (such as data-stream permutations) into this method has yet to be explored. The *s*, or rate of social transmission, parameter can be used as an effect size and reported with 95% confidence intervals. NBDA has been widely used to investigate transmission of behaviours in animals, for example in fish (Atton *et al.* 2012), cetaceans (Allen *et al.* 2013), primates (Kendal *et al.* 2010; Hobaiter *et al.* 2014) and birds (Boogert *et al.* 2008; Aplin *et al.* 2012, 2015; Farine *et al.* 2015).

Multiple Hypothesis Testing using Null Models

In addition to using randomizations for null hypothesis testing, multiple null models can also be used to evaluate competing hypotheses (Gotelli & Graves 1996). This involves keeping different aspects of the data constant in each model to identify whether they affect social structure (Farine 2013b). For example, to test whether an observed phenotypic structure in a social network is driven by spatial distribution of individuals or by social avoidance, we can build one null model that controls for space and one that does not. If both null models are rejected, this suggests that the observed patterns are driven by social attraction/avoidance because both null models create more random networks than observed. In contrast, if the spatially controlled null model is not rejected, but the non-spatial model is, then individuals are non-randomly distributed in space, but we cannot reject the hypothesis that they associate randomly with respect to phenotype within their location. This approach is demonstrated in Farine *et al.* (2015b).

Simulations

Simulations can be used to infer the mechanisms that underpin network formation or to explore the consequences of network structure for population processes. There are two types of simulations that are useful in network studies: data-driven simulations and fully simulated networks. Data-driven simulations involve building an agent-based model (Railsback & Grimm 2012) in which patterns of interactions are determined by the observed network. This can be used to investigate how social structure mediates population processes, such as the spread of a disease in that population. Alternatively, individuals (nodes in the network) can be removed to determine the impact on social connectivity and network robustness (e.g. Manno 2008; Wey *et al.* 2008). To our knowledge, no study has performed simulations on increasingly random networks (using data-stream permutation methods) to investigate whether some aspects of population processes are conserved under different null models, or how fast these processes are expected to degrade.

The second approach stochastically simulates behaviours of individual actors based on a minimal set of rules, or mechanisms. Comparing the result of simulations to empirical data is a powerful way of inferring candidate mechanisms generating social structure (Cross *et al.* 2012; Sumpter, Mann & Perna 2012; Farine, Downing & Downing 2014). For example, simulations could be used to test whether females preferentially copulated with successful males by comparing the degree distribution from an observed network with the degree distributions from simulated networks that implement preferential mating.

Reporting Statistical Results

A criticism of null hypothesis significance testing is the presentation of 'naked *P* values' (Anderson, Burnham & Thompson 2000) without any indication of the size of the effect being tested. The results of hypotheses tested on social networks are often presented without effect statistics, but these should be reported. In permutation tests, if the test statistic is directly interpretable (e.g. the standard deviation of the association indices or the coefficient values of a model), the effect size can be reported by giving the value measured on the observed network as well as the mean or 95% range of the values measured on the permuted networks. This provides information both on the size of the effect and on the uncertainty associated with the conclusions drawn from the result.

ESTIMATING POWER AND PRECISION

Have enough data been collected?

Clearly, the more the data available, the more the observed network will mirror the real social network of the animals, and the more powerful will be tests against null hypotheses. Whitehead (2008) provides a guideline for estimating the sampling effort required to achieve a reliable social network (defined as a correlation between the edges of the real and the observed network of at least 0·8). A network that is moderately socially differentiated, where the coefficient of variation (CV) of edge weights of the real network is approximately 0·2, requires a mean of about 50 identifications per dyad. This decreases as the network becomes more strongly differentiated, for example as relationships become less mixed and start to resemble pairs forming territories. A highly differentiated population (with a CV of approximately 0·6) requires an average of five identifications per dyad, whereas an extreme population (CV of around 10) requires only 0·02 identifications per dyad (i.e. there is high certainty that a single observation of an edge is accurate). The power of permutation tests to reject null hypotheses is highly dependent on the strength of the pattern being tested and the amount of data collected (Whitehead 2008). In most cases, large data sets are required, a further reason why automated data collection systems have become very useful.

Realistically, studies are often limited either by resources or by logistics. Given this, it can be useful to identify strategies that will maximize the quality of the data that are collected (see 10.4). When sampling effort is limited, the simulation study by Franks, Ruxton & James (2010) suggests that increasing the number of censuses, rather than increasing the proportion of individuals sampled in each census, generates a more robust network when sampling fission–fusion groups. This may also be relevant when sampling populations with stable social groups, in which case collecting more samples from fewer groups may yield better results than sampling more groups.

Dealing with Missing Individuals

The issue of missing individuals need to be assessed in terms of the biological question under investigation. The simulation study by Silk *et al.* (2015) suggests that the correlation between node-level metrics measured on partial and full networks has a relatively linear relationship with the proportion identified. However, sampling only a small proportion of the population may have a greater negative impact on network-level measures, such as phenotypic assortment (see Farine 2014), than on means of pair-wise measures. The issue of how to deal with missing individuals in a social network is an outstanding question. To our knowledge, there is no way, other than using simulations, to estimate how well a network based on a subset of the population captures the properties of the real network.

Is there a bias in the data sampling?

Perhaps a more important issue than missing individuals is biased sampling. There are many ways that sampling may bias observations towards some individuals which become disproportionately represented in the data. For example, brightly coloured or more active individuals may be easier to find, and these individuals may be observed more often. In many cases, bias is inevitable, and these biases need to be considered either in quantitative analyses or when discussing the results. In the case of the bright individuals, we may expect that if these are observed more often than others, then if there was no difference in their real gregariousness, these individuals could still have more associates (edges) in the social network and therefore greater network centrality. Currently, the best ways to deal with these situations are to use generalized affiliation indices (Whitehead & James 2015) or to build null models using permutations that account for these sampling differences (see Box 5).

Estimating the Quality of an Observed Network

How to estimate whether an observed network is robust and precise remains an outstanding question. Above, we have provided some guidelines to estimate how well a population should be sampled, and what to prioritize during data collection. If data have already been collected, Lusseau, Whitehead & Gero (2008) suggest using bootstrapping (resampling data at random with replacement) or jackkniffing (removing a certain percentage of data) as a way to estimate the confidence intervals around network measures. Wey *et al.* (2008) adapted jackkniffing, combined with anovas, to calculate the percentage of data that could be removed without causing a significant change in the different network measures. In their case, they found that network measures were robust up to 75% of data removed. Finally, Cross *et al.* (2012) provide a useful framework using hierarchical models to evaluate the relative importance of difference factors (individual, dyadic and environmental) driving variation in association rates in an observed network. However, more research is required (using mathematical models, simulated networks or very well-sampled networks) to determine the following: (i) whether such approaches provide a good estimate of the potential error in an incompletely sampled network; (ii) how confident we are that the network we observed is a good estimation of the real patterns.

Controlling for Multiple Comparisons

One criticism of many social network studies is that they often examine the relationship of multiple metrics with multiple individual-level attributes (e.g. traits or pathogen status). With many significance tests, type I errors multiply, an issue that taxes statistical analyses generally. Bonferroni and other corrections can be used to reduce the type I error rate, but at the expense of type II errors. In network analysis, we recommend concentrating on effect sizes rather than *P* values whenever possible, thus avoiding the multiple comparison problem.

REMAINING CHALLENGES AND FUTURE DIRECTIONS

Comparing Networks and Comparative Studies

Comparing networks across contexts (e.g. between populations or species) remains one of the main challenges in network analysis (see Chapter 7 in Croft, James & Krause 2008). The lack of measures that can be used to make robust comparisons among networks prevents their use in comparative studies to investigate broad questions in social behaviour. Comparing networks is challenging both because measures are influenced by network topology (e.g. degree distributions by the number of nodes in a network; how discrete or modular is the population being studied) and the lack of standardization in data collection. If data collection is completely standardized, for example all individuals in two populations of similar size are sampled at a high-resolution and in the same context, then network comparison may be possible. This is because differences arising from the network structures are attributable to biological processes. Thus, we suggest that issues with comparing networks should not frighten biologists from collecting data in properly replicated networks. In contrast, if data are collected differently, such as by using focal follows vs. gambit of the group, then the methods themselves may generate fundamental differences in network structure that are not related to the biology of the organisms under investigation.

Dynamic Network Analysis

Temporal dynamics represents a significant analytical challenge in social network analysis. First, the data must be collected and analysed at a scale that is appropriate for the biological questions (Pinter-Wollman *et al.* 2013). This includes collecting sufficient data to generate a representative network at each point in time. Another major obstacle is the development of robust statistical tests for dynamic networks, including appropriate null hypotheses and associated randomization-based null models. For example, to test whether the property of a network (such as mean degree) is increasing more than expected over time requires quantitative null hypotheses that are based on previous knowledge of the system. Hobson, Avery & Wright (2013) provide a useful framework highlighting different scales at which temporal information can be incorporated into network analysis. Computer scientists are also rapidly developing tools for analysing temporal networks, such as dynamic community analysis (Tantipathananandh & Berger-Wolf 2011), though these may take some time to filter down to biological users and in their current formulations, they are not realistic for many animal social networks.

Repeatability of Network Position

Whether individuals have consistent network positions across different samples (i.e. repeatability) may be a critical consideration when exploring social factors that mediate individual fitness (Wilson *et al.* 2013). An increasing number of studies are demonstrating that network position is repeatable across different samples (Boogert, Farine & Spencer 2014; Jacoby *et al.* 2014). Attempts to understand the determinants – ecological, behavioural or genetic – of individual network properties should begin with an exploration of the stability of these properties across repeated observations of individuals. Nakagawa & Schielzeth (2010) describe a framework for measuring repeatability based on repeated measurements made on the same individuals, which with repeatedly sampled networks can determine whether individual network positions are repeatable. This framework uses GLMMs to calculate the proportion of variance in the distribution of network measures that is attributable to the individual identities (as a random effect) in the network. Repeatability may also be a useful measure of robustness. For example, data could be partitioned into two samples and the rank correlations of individual network metrics measured. Both these methods estimate the consistency of individual positions and network structure. This may be an important assumption to examine when testing hypotheses based on relationships such as the link between network centrality and reproductive success.

Using Networks in an Experimental Framework

Most published social network studies still remain largely descriptive. This may potentially result in over-interpretation of the strength attributed to published findings in a given field (James, Croft & Krause 2009). For example, different studies may observe similar patterns arising from different mechanisms (or worse, from having similar biases in their observation data). Thus, there is a pressing need for studies that combine network analyses with experimental manipulation of animal groups that explicitly confirm the results of observational studies and identify underlying mechanisms. This can be done by experimentally removing individuals, where the network is used to identify key mediators of group structure (such as in Flack *et al.* 2006), or by altering ecological conditions experienced by groups. The latter has generally been restricted to natural experiments, such as measuring community structure in groups across habitat remnants of different size (e.g. Mokross *et al.* 2014). An exception is Firth & Sheldon (2015) who experimentally controlled individual's access to different food resources to quantify the impact of spatially breaking-up flocks of birds. They found that new network connections formed under this regime were carried over into other contexts. Another experiment introduced novel behaviours into replicated animal populations to track the spread of information through networks (Aplin *et al.* 2015). Finally, Croft *et al.* (2011a) experimentally manipulated parasite loads and found that infected fish were actively avoided by conspecifics. However, controlled manipulation experiments remain rare. Yet, they are critical for understanding the role of individuals and social structure in social dynamics, or what causes individuals to have different network positions. Only through experiments will social networks be able to provide definitive causative evidence for socially mediated mechanisms underpinning evolutionary processes.

CONCLUSION

Studies of animal social networks can break new ground across a broad range of disciplines. This may require an increased reliance on experimental manipulations, repeated sampling of individuals across individuals' lifetimes and network analyses that move beyond dyadic measures. Combining observed data with simulation models is a promising avenue to quantitatively assess competing mechanisms. For example, this approach could be used to discover how the rules that govern processes such as group joining and leaving drive social structure. In turn, this may, in the future, help to inform management or conservation action. In all cases, network approaches will be improved by defining biologically appropriate network edges, ensuring well-sampled networks, including robust null models in statistical testing, and evaluating the uncertainty surrounding the results.

ACKNOWLEDGEMENTS

We thank Ben Sheldon, Lucy Aplin, Neeltje Boogert and Meg Crofoot for extensively commenting on drafts of the manuscript. We also thank three anonymous referees for their comments. DRF was funded by grants from the BBSRC (BB/L006081/1) and ERC (AdG 250164) awarded to Ben C. Sheldon and the NSF (NSF-IOS 1250895) awarded to Margaret C. Crofoot.

REFERENCES

1. Adams, E.S. (2005) Bayesian analysis of linear dominance hierarchies. *Animal Behaviour*, 69, 1191–1201.
2. Allen, J., Weinrich, M., Hoppitt, W. & Rendell, L. (2013) Network-based diffusion analysis reveals cultural transmission of lobtail feeding in humpback whales. *Science*, 340, 485–488.
3. Altmann, J. (1974) Observational study of behavior – sampling methods. *Behaviour*, 49, 227–267.
4. Anderson, D.R., Burnham, K.P. & Thompson, W.L. (2000) Null hypothesis testing: problems, prevalence, and an alternative. *Journal of Wildlife Management*, 64, 912–923.
5. Aplin, L.M., Farine, D.R., Morand-Ferron, J. & Sheldon, B.C. (2012) Social networks predict patch discovery in a wild population of songbirds. *Proceedings of the Royal Society B*, 279, 4199–4205.
6. Aplin, L.M., Farine, D.R., Morand-Ferron, J., Cole, E.F., Cockburn, A. & Sheldon, B.C. (2013) Individual personalities predict social behaviour in wild networks of great tits (*Parus major*). *Ecology Letters*, 16, 1365–1372.
7. Aplin, L.M., Farine, D.R., Morand-Ferron, J., Cockburn, A., Thornton, A. & Sheldon, B.C. (2015) Experimentally induced innovations lead to persistent culture via conformity in wild birds. *Nature*, 518, 538–541.
8. Ashby, B. & Gupta, S. (2013) Sexually transmitted infections in polygamous mating systems. *Philosophical Transactions of the Royal Society of London. Series B, Biological Sciences*, 368, 20120048.
9. Atton, N., Hoppitt, W., Webster, M.M., Galef, B.G. & Laland, K.N. (2012) Information flow through threespine stickleback networks without social transmission. *Proceedings of the Royal Society B*, 279, 4272–4278.
10. Auber, D. (2004) Tulip—a huge graph visualization framework. *Graph Drawing Software* (eds M. Junger & P. Mutzel), pp. 105–126. Springer, Berlin Heidelberg.
11. Aureli, F., Schaffner, C.M., Boesch, C., Bearder, S.K., Call, J., Chapman, C.A. *et al.* (2008) Fission–fusion dynamics new research frameworks. *Current Anthropology*, 49, 627–654.
12. Bansal, S., Grenfell, B.T. & Meyers, L.A. (2007) When individual behaviour matters: homogeneous and network models in epidemiology. *Journal of the Royal Society Interface*, 4, 879–891.

13. Barrat, A., Barthelemy, M., Pastor-Satorras, R. & Vespignani, A. (2004) The architecture of complex weighted networks. *Proceedings of the National Academy of Sciences of the United States of America*, 101, 3747–3752.
14. Bastian, M., Heymann, S. & Jacomy, M. (2009) *Gephi: an open source software for exploring and manipulating networks*. International AAAI Conference on Weblogs and Social Media.
15. Bejder, L., Fletcher, D. & Brager, S. (1998) A method for testing association patterns of social animals. *Animal Behaviour*, 56, 719–725.
16. Blonder, B. & Dornhaus, A. (2011) Time-ordered networks reveal limitations to information flow in ant colonies. *PLoS ONE*, 6, e20298.
17. Blonder, B., Wey, T.W., Dornhaus, A., James, R. & Sih, A. (2012) Temporal dynamics and network analysis. *Methods in Ecology and Evolution*, 3, 958–972.
18. Blumstein, D.T. & Daniel, J.C. (2007) *Quantifying Behavior the JWatcher Way*. Sinauer Associates Inc, Sunderland, MA.
19. Boogert, N.J., Farine, D.R. & Spencer, K.A. (2014) Developmental stress predicts social network position. *Biology Letters*, 10, 20140561.
20. Boogert, N.J., Reader, S.M., Hoppitt, W. & Laland, K.N. (2008) The origin and spread of innovations in starlings. *Animal Behaviour*, 75, 1509–1518.
21. Borgatti, S.P. (2002) *NetDraw Software for Network Visualization*. Analytic Technologies, Lexington, KY.
22. Borgatti, S.P. (2005) Centrality and network flow. *Social Networks*, 27, 55–71.
23. Borgatti, S.P., Everett, M.G. & Freeman, L.C. (2002) *Ucinet for Windows: Software for Social Network Analysis*. Analytic Technologies, Harvard, MA.
24. Borgatti, S.P., Everett, M.G. & Johnson, J.C. (2013) *Analyzing Social Networks*. SAGE Publications Ltd, London, UK.
25. Borgatti, S.P. & Halgin, D.S. (2011) On network theory. *Organization Science*, 22, 1168–1181.
26. Brockmann, D. & Helbing, D. (2013) The hidden geometry of complex, network-driven contagion phenomena. *Science*, 342, 1337–1342.
27. Bull, C.M., Godfrey, S.S. & Gordon, D.M. (2012) Social networks and the spread of Salmonella in a sleepy lizard population. *Molecular*

Ecology, 21, 4386–4392.

28. Butts, C.T. (2008) Social network analysis with SNA. *Journal of Statistical Software*, 24, 1–50.
29. Butts, C.T. (2009) Revisiting the foundations of network analysis. *Science*, 325, 414–416.
30. Cairns, S.J. & Schwager, S.J. (1987) A comparison of association indexes. *Animal Behaviour*, 35, 1454–1469.
31. Castles, M., Heinsohn, R., Marshall, H.H., Lee, A.E.G., Cowlishaw, G. & Carter, A.J. (2014) Social networks created with different techniques are not comparable. *Animal Behaviour*, 96, 59–67.
32. Christakis, N.A. & Fowler, J.H. (2007) The spread of obesity in a large social network over 32 years. *New England Journal of Medicine*, 357, 370–379.
33. Connor, R.C., Heithaus, M.R. & Barre, L.M. (2001) Complex social structure, alliance stability and mating access in a bottlenose dolphin 'super-alliance'. *Proceedings of the Royal Society B*, 268, 263–267.
34. Croft, D.P., James, R. & Krause, J. (2008) *Exploring Animal Social Networks*. Princeton University Press, Princeton, NJ.
35. Croft, D.P., Krause, J. & James, R. (2004) Social networks in the guppy (Poecilia reticulata). *Proceedings of the Royal Society B*, 271, S516–S519.
36. Croft, D.P., James, R., Ward, A.J.W., Botham, M.S., Mawdsley, D. & Krause, J. (2005) Assortative interactions and social networks in fish. *Oecologia*, 143, 211–219.
37. Croft, D.P., Krause, J., Darden, S.K., Ramnarine, I.W., Faria, J.J. & James, R. (2009) Behavioural trait assortment in a social network: patterns and implications. *Behavioral Ecology and Sociobiology*, 63, 1495–1503.
38. Croft, D.P., Edenbrow, M., Darden, S.K., Ramnarine, I.W., van Oosterhout, C. & Cable, J. (2011a) Effect of gyrodactylid ectoparasites on host behaviour and social network structure in guppies *Poecilia reticulata. Behavioral Ecology and Sociobiology*, 65, 2219–2227.
39. Croft, D.P., Madden, J.R., Franks, D.W. & James, R. (2011b) Hypothesis testing in animal social networks. *Trends in Ecology and Evolution*, 26, 502–507.
40. Cross, P.C., Creech, T.G., Ebinger, M.R., Heisey, D.M., Irvine, K.M. & Creel, S. (2012) Wildlife contact analysis: emerging methods,

questions, and challenges. *Behavioral Ecology and Sociobiology*, 66, 1437–1447.

41. Csardi, G. & Nepusz, T. (2006) The igraph software package for complex network research. *InterJournal, Complex Systems*, 1695.
42. Dekker, D., Krackhardt, D. & Snijders, T. (2007) Sensitivity of MRQAP tests to collinearity and autocorrelation conditions. *Psychometrika*, 72, 563–581.
43. Edenbrow, M., Darden, S.K., Ramnarine, I.W., Evans, J.P., James, R. & Croft, D.P. (2011) Environmental effects on social interaction networks and male reproductive behaviour in guppies, *Poecilia reticulata*. *Animal Behaviour*, 81, 551–558.
44. Farine, D.R. (2013a) Animal social network inference and permutations for ecologists in R using asnipe. *Methods in Ecology and Evolution*, 4, 1187–1194.
45. Farine, D.R. (2013b) Emergent social structure and collective behaviour from individual decision-making in wild birds. DPhil Thesis, University of Oxford, Oxford.
46. Farine, D.R. (2014) Measuring phenotypic assortment in animal social networks: weighted associations are more robust than binary edges. *Animal Behaviour*, 89, 141–153.
47. Farine, D.R. (2015) Proximity as a proxy for interactions: issues of scale in social network analysis. *Animal Behaviour*, 104, e1–e5.
48. Farine, D.R., Downing, C.P. & Downing, P.A. (2014) Mixed-species associations can arise without heterospecific attraction. *Behavioural Ecology*, 25, 574–581.
49. Farine, D.R., Garroway, C.J. & Sheldon, B.C. (2012) Social network analysis of mixed-species flocks: exploring the structure and evolution of interspecific social behaviour. *Animal Behaviour*, 84, 1271–1277.
50. Farine, D.R. & Lang, S.D.J. (2013) The early bird gets the worm: foraging strategies of wild songbirds lead to the early discovery of food sources. *Biology Letters*, 9, 20130578.
51. Farine, D.R. & Milburn, P.J. (2013) Social organisation of thornbill-dominated mixed-species flocks using social network analysis. *Behavioral Ecology and Sociobiology*, 67, 321–330.
52. Farine, D.R. & Sheldon, B.C. (2015) Selection for territory acquisition is modulated by social network structure in a wild songbird. *Journal of Evolutionary Biology*, 28, 547–556.

53. Farine, D.R., Spencer, K.A. & Boogert, N.J. (2015) Early-Life Stress Triggers Juvenile Zebra Finches to Switch Social Learning Strategies. *Current Biology*, http://dx.doi.org/10.1016/j.cub.2015.06.071
54. Farine, D.R., Aplin, L.M., Sheldon, B.C. & Hoppitt, W. (2015a) Interspecific social networks promote information transmission in wild songbirds. *Proceedings of the Royal Society B*, 282, 20142804.
55. Farine, D.R., Firth, J.A., Aplin, L.M., Crates, R.A., Culina, A., Garroway, C.J. *et al.* (2015b) The role of social and ecological processes in structuring animal populations: a case study from automated tracking of wild birds. *Royal Society Open Science*, 2, 150057.
56. Fenner, A.L., Godfrey, S.S. & Bull, C.M. (2011) Using social networks to deduce whether residents or dispersers spread parasites in a lizard population. *Journal of Animal Ecology*, 80, 835–843.
57. Firth, J.A. & Sheldon, B.C. (2015) Experimental manipulation of avian social structure reveals segregation is carried over across contexts. *Proceedings of the Royal Society B*, 282, 20142350.
58. Flack, J.C., Girvan, M., de Waal, F.B.M. & Krakauer, D.C. (2006) Policing stabilizes construction of social niches in primates. *Nature*, 439, 426–429.
59. Formica, V.A., McGlothlin, J.W., Wood, C.W., Augat, M.E., Butterfield, R.E., Barnard, M.E. *et al.* (2011) Phenotypic assortment mediates the effect of social selection in a wild beetle population. *Evolution*, 65, 2771–2781.
60. Formica, V.A., Wood, C.W., Larsen, W.B., Butterfield, R.E., Augat, M.E., Hougen, H.Y. *et al.* (2012) Fitness consequences of social network position in a wild population of forked fungus beetles (Bolitotherus cornutus). *Journal of Evolutionary Biology*, 25, 130–137.
61. Franks, D.W., Ruxton, G.D. & James, R. (2010) Sampling animal association networks with the gambit of the group. *Behavioral Ecology and Sociobiology*, 64, 493–503.
62. Franz, M. & Nunn, C.L. (2009) Network-based diffusion analysis: a new method for detecting social learning. *Proceedings of the Royal Society B*, 276, 1829–1836.
63. Gansner, E.R. & North, S.C. (2000) An open graph visualization system and its applications to software engineering. *Software: Practice & Experience*, 30, 1203–1233.

64. Gardner, T.S., di Bernardo, D., Lorenz, D. & Collins, J.J. (2003) Inferring genetic networks and identifying compound mode of action via expression profiling. *Science*, 301, 102–105.

65. Godfrey, S.S., Bull, C.M., James, R. & Murray, K. (2009) Network structure and parasite transmission in a group living lizard, the gidgee skink, Egernia stokesii. *Behavioral Ecology and Sociobiology*, 63, 1045–1056.

66. Godfrey, S.S., Anasri, T.H., Gardner, M.G., Farine, D.R. & Bull, C.M. (2014) A contact based social network of lizards is defined by low genetic relatedness among strongly-connected individuals. *Animal Behaviour*, 97, 35–43.

67. Gotelli, N.J. & Graves, G.R. (1996) *Null Models in Ecology*. Smithsonian Institution Press, Washington, DC, USA.

68. Gowans, S., Whitehead, H. & Hooker, S.K. (2001) Social organization in northern bottlenose whales (*Hyperoodon ampullatus*): not driven by deep water foraging? *Animal Behaviour*, 62, 369–377.

69. Granovetter, M. (1973) The strength of weak ties. *American Journal of Sociology*, 78, 1360–1380.

70. Hamede, R.K., Bashford, J., McCallum, H. & Jones, M. (2009) Contact networks in a wild Tasmanian devil (*Sarcophilus harrisii*) population: using social network analysis to reveal seasonal variability in social behaviour and its implications for transmission of devil facial tumour disease. *Ecology Letters*, 12, 1147–1157.

71. Hemelrijk, C.K. (1990) A matrix partial correlation test used in investigations of reciprocity and other social-interaction patterns at group level. *Journal of Theoretical Biology*, 143, 405–420.

72. Hinde, R.A. (1976) Interactions, relationships and social-structure. *Man*, 11, 1–17.

73. Hobaiter, C., Poisot, T., Zuberbuhler, K., Hoppitt, W. & Gruber, T. (2014) Social network analysis shows direct evidence for social transmission of tool use in wild chimpanzees. *PLoS Biology*, 12, e1001960.

74. Hobson, E.A., Avery, M.L. & Wright, T.F. (2013) An analytical framework for quantifying and testing patterns of temporal dynamics in social networks. *Animal Behaviour*, 85, 83–96.

75. Hock, K. & Fefferman, N.H. (2012) Social organization patterns can lower disease risk without associated disease avoidance or immunity. *Ecological Complexity*, 12, 34–42.

76. Hoppitt, W., Boogert, N.J. & Laland, K.N. (2010) Detecting social transmission in networks. *Journal of Theoretical Biology*, 263, 544–555.

77. Hoppitt, W. & Laland, K.N. (2011) Detecting social learning using networks: a users guide. *American Journal of Primatology*, 73, 834–844.

78. Jacoby, D.M.P., Fear, L.N., Sims, D.W. & Croft, D.P. (2014) Shark personalities? Repeatability of social network traits in a widely distributed predatory fish. *Behavioral Ecology and Sociobiology*, 68, 1995–2003.

79. James, R., Croft, D.P. & Krause, J. (2009) Potential banana skins in animal social network analysis. *Behavioral Ecology and Sociobiology*, 63, 989–997.

80. Jarman, P.J. (1982) Prospects for interspecific comparisons in sociobiology. *Current Problems in Sociobiology* (ed. K.S.C.S. Group), pp. 323–342. Cambridge University Press, Cambridge, UK.

81. Keeling, M.J. (2005) The implications of network structure for epidemic dynamics. *Theoretical Population Biology*, 67, 1–8.

82. Kendal, R.L., Custance, D.M., Kendal, J.R., Vale, G., Stoinski, T.S., Rakotomalala, N.L. *et al.* (2010) Evidence for social learning in wild lemurs (Lemur catta). *Learning & Behavior*, 38, 220–234.

83. Krause, J., Krause, S., Arlinghaus, R., Psorakis, I., Roberts, S. & Rutz, C. (2013) Reality mining of animal social systems. *Trends in Ecology & Evolution*, 28, 541–551.

84. Krause, J., James, R., Franks, D.W. & Croft, D.P. (2015) *Animal Social Networks*. Oxford University Press, Oxford, UK.

85. Langer, N., Pedroni, A. & Jancke, L. (2013) The problem of thresholding in small-world network analysis. *PLoS ONE*, 8, e53199.

86. Lusseau, D. (2003) The emergent properties of a dolphin social network. *Proceedings of the Royal Society B*, 270, S186–S188.

87. Lusseau, D. & Newman, M.E.J. (2004) Identifying the role that animals play in their social networks. *Proceedings of the Royal Society B*, 271, S477–S481.

88. Lusseau, D., Whitehead, H. & Gero, S. (2008) Incorporating uncertainty into the study of animal social networks. *Animal Behaviour*, 75, 1809–1815.

89. MacIntosh, A.J.J., Jacobs, A., Garcia, C., Shimizu, K., Mouri, K., Huffman, M.A. *et al.* (2012) Monkeys in the middle: parasite transmission through the social network of a wild primate. *PLoS ONE*, 7, e51144.

90. Manly, B.F.J. (1997) *Randomization, Bootstrap and Monte Carlo Methods in Biology*, 2nd edn. Chapman and Hall, London, UK.

91. Manno, T.G. (2008) Social networking in the Columbian ground squirrel, Spermophilus columbianus. *Animal Behaviour*, 75, 1221–1228.

92. Mantel, N. (1967) The detection of disease clustering and a generalized regression approach. *Cancer Research*, 27, 209–220.

93. May, R.M. (2006) Network structure and the biology of populations. *Trends in Ecology & Evolution*, 21, 394–399.

94. McDonald, D.B. (2007) Predicting fate from early connectivity in a social network. *Proceedings of the National Academy of Sciences of the United States of America*, 104, 10910–10914.

95. McDonald, G.C., James, R., Krause, J. & Pizzari, T. (2013) Sexual networks: measuring sexual selection in structured, polyandrous populations. *Philosophical Transactions of the Royal Society of London. Series B, Biological Sciences*, 368, 20120356.

96. Mokross, K., Ryder, T.B., Cortes, M.C., Wolfe, J.D. & Stouffer, P.C. (2014) Decay of interspecific avian flock networks along a disturbance gradient in Amazonia. *Proceedings of the Royal Society B*, 281, 20132599.

97. Montoya, J.M., Pimm, S.L. & Sole, R.V. (2006) Ecological networks and their fragility. *Nature*, 442, 259–264.

98. Moody, J., McFarland, D.A. & Blender-deMoll, S. (2005) Visualizing network dynamics. *American Journal of Sociology*, 110, 1206–1241.

99. Mourier, J., Vercelloni, J. & Planes, S. (2012) Evidence of social communities in a spatially structured network of a free-ranging shark species. *Animal Behaviour*, 83, 389–401.

100. Nakagawa, S. & Schielzeth, H. (2010) Repeatability for Gaussian and non-Gaussian data: a practical guide for biologists. *Biological Reviews*, 85, 935–956.

101. Newman, M.E.J. (2002) Spread of epidemic disease on networks. *Physical Review E*, 66, 016128.

102. Newman, M.E.J. (2006) Finding community structure in networks using the eigenvectors of matrices. *Physical Review E*, 74, 036104.

103. Newman, M.E.J. (2010) *Networks: An Introduction*. Oxford University Press, Oxford.

104. Nightingale, G.F., Boogert, N.J., Laland, K.N. & Hoppitt, W. (2015) Quantifying diffusion on social networks: a Bayesian Approach. *Animal Social Networks* (eds J. Krause, R. James, D.W. Franks & D.P. Croft), pp. 38–52. Oxford University Press, Oxford, UK.

105. Oh, K.P. & Badyaev, A.V. (2010) Structure of social networks in a passerine bird: consequences for sexual selection and the evolution of mating strategies. *American Naturalist*, 176, E80–E89.

106. Olesen, J.M., Bascompte, J., Elberling, H. & Jordano, P. (2008) Temporal dynamics in a pollination network. *Ecology*, 89, 1573–1582.

107. Perkins, S.E., Cagnacci, F., Stradiotto, A., Arnoldi, D. & Hudson, P.J. (2009) Comparison of social networks derived from ecological data: implications for inferring infectious disease dynamics. *Journal of Animal Ecology*, 78, 1015–1022.

108. Pike, T.W., Samanta, M., Lindstrom, J. & Royle, N.J. (2008) Behavioural phenotype affects social interactions in an animal network. *Proceedings of the Royal Society B*, 275, 2515–2520.

109. Pinter-Wollman, N. (2015) Persistent variation in spatial behavior affects the structure and function of interaction networks. *Current Zoology*, 61, 98–106.

110. Pinter-Wollman, N., Hobson, E.A., Smith, J.E., Edelman, A.J., Shizuka, D., de Silva, S. *et al.* (2013) The dynamics of animal social networks: analytical, conceptual, and theoretical advances. *Behavioral Ecology*, 25, 242–255.

111. Psorakis, I., Roberts, S.J., Rezek, I. & Sheldon, B.C. (2012) Inferring social network structure in ecological systems from spatio-temporal data streams. *Journal of the Royal Society Interface*, 9, 3055–3066.

112. Psorakis, I., Voelkl, B., Garroway, C.J., Radersma, R., Aplin, L.M., Crates, R.A. *et al.* (2015) Inferring social structure from temporal data. *Behavioral Ecology and Sociobiology*, 69, 857–866.

113. R Development Core Team (2013) *R: A Language and Environment for Statistical Computing*. R Foundation for Statistical Computing, Vienna, Austria.

114. Railsback, S.F. & Grimm, V. (2012) *Agent-based and Individual-based Modeling: A Practical Introduction*. Princeton University Press, Princeton, NJ.

115. Ramos-Fernandez, G., Boyer, D., Aureli, F. & Vick, L.G. (2009) Association networks in spider monkeys (*Ateles geoffroyi*). *Behavioral Ecology and Sociobiology*, 63, 999–1013.

116. Rushmore, J., Caillaud, D., Matamba, L., Stumpf, R.M., Borgatti, S.P. & Altizer, S. (2013) Social network analysis of wild chimpanzees provides insights for predicting infectious disease risk. *Journal of Animal Ecology*, 82, 976–986.

117. Rutz, C., Burns, Z.T., James, R., Ismar, S.M.H., Burt, J., Otis, B. *et al.* (2012) Automated mapping of social networks in wild birds. *Current Biology*, 22, R669–R671.

118. Shannon, P., Markiel, A., Ozier, O., Baliga, N., Wang, J., Ramage, D. *et al.* (2003) Cytoscape: a software environment for integrated models of biomolecular interaction networks. *Genome Research*, 13, 2498–2504.

119. Shirley, M.D.F. & Rushton, S.P. (2005) The impacts of network topology on disease spread. *Ecological Complexity*, 2, 287–299.

120. Shizuka, D. & McDonald, D.B. (2012) A social network perspective on measurements of dominance hierarchies. *Animal Behaviour*, 83, 925–934.

121. Silk, J.B., Beehner, J.C., Bergman, T.J., Crockford, C., Engh, A.L., Moscovice, L.R. *et al.* (2010) Strong and consistent social bonds enhance the longevity of female baboons. *Current Biology*, 20, 1359–1361.

122. Silk, M.J., Jackson, A.L., Croft, D.P., Colhoun, K. & Bearhop, S. (2015) The consequences of unidentifiable individuals for the analysis of an animal social network. *Animal Behaviour*, 104, 1–11.

123. Snijders, T.A.B. (2011) Statistical models for social networks. *Annual Review of Sociology*, 37, 131–153.

124. Snijders, L., van Rooij, E.P., Burt, J.M., Hinde, C.A., van Oers, K. & Naguib, M. (2014) Social networking in territorial great tits: slow explorers have the least central social network positions. *Animal Behaviour*, 98, 95–102.

125. Strandburg-Peshkin, A., Twomey, C.R., Bode, N.W.F., Kao, A.B., Katz, Y., Ioannou, C.C. *et al.* (2013) Visual sensory networks and

effective information transfer in animal groups. *Current Biology*, 23, R709–R711.

126. Strandburg-Peshkin, A., Farine, D.R., Couzin, I.D. & Crofoot, M.C. (2015) Shared decision-making drives collective movement in wild baboons. *Science*, 348, 1358–1361.
127. Sumpter, D.J.T., Mann, R.P. & Perna, A. (2012) The modelling cycle for collective animal behaviour. *Interface Focus*, 2, 764–773.
128. Sundaresan, S.R., Fischhoff, I.R. & Dushoff, J. (2009) Avoiding spurious findings of nonrandom social structure in association data. *Animal Behaviour*, 77, 1381–1385.
129. Tantipathananandh, C. & Berger-Wolf, T. (2011) Finding Communities in Dynamic Social Networks. IEEE International Conference on Data Mining (ICDM). Conference Publishing Services, Vancouver, Canada.
130. VanderWaal, K.L., Atwill, E.R., Hooper, S., Buckle, K. & McCowan, B. (2013a) Network structure and prevalence of Cryptosporidium in Belding's ground squirrels. *Behavioral Ecology and Sociobiology*, 67, 1951–1959.
131. VanderWaal, K.L., Atwill, E.R., Isbell, L.A. & McCowan, B. (2013b) Linking social and pathogen transmission networks using microbial genetics in giraffe (*Giraffa camelopardalis*). *Journal of Animal Ecology*, 83, 406–414.
132. VanderWaal, K.L., Atwill, E.R., Isbell, L.A. & McCowan, B. (2014) Quantifying microbe transmission networks for wild and domestic ungulates in Kenya. *Biological Conservation*, 169, 136–146.
133. Volz, E.M., Miller, J.C., Galvani, A. & Meyers, L.A. (2011) Effects of heterogeneous and clustered contact patterns on infectious disease dynamics. *PLoS Computational Biology*, 7, e1002042.
134. de Vries, H., Netto, W.J. & Hanegraaf, P.L.H. (1993) Matman – a program for the analysis of sociometric matrices and behavioral transition matrices. *Behaviour*, 125, 157–175.
135. Wasserman, S. & Faust, K. (1994) *Social Networks Analysis: Methods and Applications*. Cambridge University Press, Cambridge, UK.
136. Webster, M.M., Atton, N., Hoppitt, W.J.E. & Laland, K.N. (2013) Environmental complexity influences association network structure and network-based diffusion of foraging information in fish shoals. *American Naturalist*, 181, 235–244.

137. Wey, T., Blumstein, D.T., Shen, W. & Jordan, F. (2008) Social network analysis of animal behaviour: a promising tool for the study of sociality. *Animal Behaviour*, 75, 333–344.

138. Wey, T.W., Burger, J.R., Ebensperger, L.A. & Hayes, L.D. (2013) Reproductive correlates of social network variation in plurally breeding degus (*Octodon degus*). *Animal Behaviour*, 85, 1407–1414.

139. Whitehead, H. (1995) Investigating structure and temporal scale in social organizations using identified individuals. *Behavioral Ecology*, 6, 199–208.

140. Whitehead, H. (1997) Analysing animal social structure. *Animal Behaviour*, 53, 1053–1067.

141. Whitehead, H. (1999) Testing association patterns of social animals. *Animal Behaviour*, 57, F26–F29.

142. Whitehead, H. (2008) *Analyzing Animal Societies*. University of Chicago Press, Chicago, IL, USA.

143. Whitehead, H. (2009) SOCPROG programs: analysing animal social structures. *Behavioral Ecology and Sociobiology*, 63, 765–778.

144. Whitehead, H., Bejder, L. & Ottensmeyer, C.A. (2005) Testing association patterns: issues arising and extensions. *Animal Behaviour*, 69, e1–e6.

145. Whitehead, H. & Dufault, S. (1999) Techniques for analyzing vertebrate social structure using identified individuals: review and recommendations. *Advances in the Study of Behavior*, 28, 33–74.

146. Whitehead, H. & James, R. (2015) Generalized affiliation indices extract affiliations from social network data. *Methods in Ecology and Evolution*, DOI: 10.1111/2041-210X.12383.

147. Whitehead, H. & Lusseau, D. (2012) Animal social networks as substrate for cultural behavioural diversity. *Journal of Theoretical Biology*, 294, 19–28.

148. Wilson, A.D.M., Krause, S., Dingemanse, N.J. & Krause, J. (2013) Network position: a key component in the characterization of social personality types. *Behavioral Ecology and Sociobiology*, 67, 163–173.

149. Wittemyer, G., Douglas-Hamilton, I. & Getz, W.M. (2005) The socioecology of elephants: analysis of the processes creating multitiered social structures. *Animal Behaviour*, 69, 1357–1371.

150. Würsig, B. & Jefferson, T.A. (1990) Methods of photo-identification for small cetaceans. Reports of the International Whaling Commission. Special, 42-43.

SECTION II
NEURAL NETWORKS

CHAPTER 6

CONNECTIVITY INFERENCE FROM NEURAL RECORDING DATA: CHALLENGES, MATHEMATICAL BASES AND RESEARCH DIRECTIONS

IldefonsMagrans de Abril[ac], JunichiroYoshimoto[b], KenjiDoya[a]

[a]Okinawa Institute of Science and Technology, Graduate University, Japan

[b]Nara Advanced Institute of Science and Technology, Japan

[c]ARAYA, Inc., Tokyo, Japan

ABSTRACT

This article presents a review of computational methods for connectivity inference from neural activity data derived from multi-electrode recordings or fluorescence imaging. We first identify biophysical and technical challenges in connectivity inference along the data processing pipeline. We then review connectivity inference methods based on two major

CITATION: de Abril, I. M., Yoshimoto, J., & Doya, K. (2018). Connectivity inference from neural recording data: Challenges, mathematical bases and research directions. Neural Networks. (18 pages, including Table of Contents)

mathematical foundations, namely, descriptive model-free approaches and generative model-based approaches. We investigate representative studies in both categories and clarify which challenges have been addressed by which method. We further identify critical open issues and possible research directions.

Keywords: Connectivity inference, Functional connectivity, Effective connectivity, Multi-electrode recording, Calcium fluorescence imaging

INTRODUCTION

Understanding operational principles of neural circuits is a major goal of recent international brain science programs, such as the BRAIN Initiative in the U.S. Insel et al. (2013), Martin and Chun (2016), the Human Brain Project in the E.U. Amunts et al. (2016), Markram (2012), and the Brain/MINDS program in Japan Okano and Mitra (2015), Okano et al. (2016). A common emphasis in these programs is utilization of high-throughput, systematic data acquisition and advanced computational technologies. The aim of this paper is to present a systematic review of computational methods for inferring neural connectivity from high-dimensional neural activity recording data, such as multiple electrode arrays and calcium fluorescence imaging.

Why do we need to infer neural connectivity? High-dimensional neural recording data tell us a lot about information representation in the brain through correlation or decoding analyses with relevant sensory, motor, or cognitive signals. However, in order to understand operational principles of the brain, it is important to identify the circuit mechanisms that encode and transform information, such as extraction of sensory features and production of motor action patterns (Churchland & Sejnowski, 1992). Knowing the wiring diagram of neuronal circuits is critical to explain how such representations are produced, predicting how the network would behave in a novel situation, and extracting the brain's algorithms for technical applications (Sporns, Tononi, & Kötter, 2005).

The network of the brain can be analyzed at various spatial scales (Gerstner, Kistler, Naud, & Paninski, 2014). At the macroscopic level, there are more than a hundred anatomical brain areas and connection structure across those areas give us an understanding of the overall processing architecture of the brain. At the mesoscopic level, connections of neurons within each local area, as well as their projections to other areas, are characterized in order to understand computational mechanisms of neural circuits. At the microscopic level, locations and features of synapses on dendritic arbors of each neuron are analyzed to understand operational mechanisms of single neurons.

This review focuses on the mesoscopic level, inferring connections between neurons in local circuits on the basis of neural activity recording data from multi-electrode recordings or fluorescence imaging. Connectivity inference from anatomical data, such as diffusion MRI at the macroscopic level, tracer injection at the mesoscopic level, and serial electron microscopy data at the microscopic data, are beyond the scope of this review. Some methods, especially those of model-free approaches, may also be applicable to connectivity inference from functional MRI data at the macroscopic level.

This paper presents an overview of challenges in network inference and different mathematical approaches to address those challenges. We first review the data processing pipeline of connectivity analysis and identify both biophysical and computational difficulties. From mathematical viewpoint, we classify connectivity inference methods broadly into descriptive, model-free approaches and generative, model-based approaches and explain representative methods in each category. We then examine which methods offer solutions for specific challenges and identify open issues and important research directions.

There have been several recent reviews of specific mathematical frameworks in network connectivity inference, such as Bayesian approaches (Chen, 2013) and the maximum entropy method Roudi et al. (2015), Yeh et al. (2010). Some reviews focus on macroscopic connectivity analysis using MRI Friston (2011), Lang et al. (2012), Sporns (2012), Sporns (2013). Reports from the First Neural Connectomics Challenge (http://connectomics.chalearn.org) review top-ranked methods for connectivity inference from calcium imaging data Guyon et al. (2014), Orlandi et al. (2014).

It is important to distinguish several types of connectivity that have been addressed previously Aertsen et al. (1989), Friston (2011), Valdes-Sosa et al. (2011). *Functional* connectivity (FC) is defined as statistical dependence among measurements of neuronal activity. It can be computed using correlation or other model-free methods (see Section 4). *Effective*connectivity (EC) characterizes the direct influence exercised between neuron pairs after discounting any indirect effects. EC is usually computed by optimizing parameters of a model that is assumed to have generated the observed data (see Section 5). *Anatomical* connectivity signifies existence of actual synapses, either excitatory or inhibitory. Even if there is an anatomical connection between neurons, it may not be detected when, for example, the source neuron is inactive or the recipient neuron is unresponsive due to strong inhibition by other neurons. Detection of functional or effective connectivity does not warrant the existence of anatomical connectivity. An example would be a false positive connection inferred between two neurons that receive common inputs from a third neuron.

Neural connectivity can be described at different levels of detail: existence, direction, sign, magnitude, and temporal dynamics. The level of description that is most reliable and useful depends on constraints on instrumentation and the amount of data available. One aim of this paper is to clarify how those constraints affect the choice of inference and validation methods for a given application.

Although it is beyond the scope of this review, graph theoretical analysis of the inferred network can play a key role in understanding the interplay between the brain structure and its function (Bullmore & Sporns, 2009). Such graph theoretic characterization includes clustering and connectivity degree distribution Bonifazi et al. (2009), Shimono and Beggs (2014), Yu et al. (2008). These abstract metrics facilitate comparison of the structure of diverse neural populations. For example, Hu et al., (2016) proposed a method to relate the network statistics of connectivity of linear point processes or Hawkes models (see Section 5.1.6) to its function.

DATA PROCESSING PIPELINE

This section presents a generic data processing pipeline, starting with data acquisition and continuing with data pre-processing, network inference, post-processing, and validation of results. Fig. 1 shows the overall view of the experimental setup.

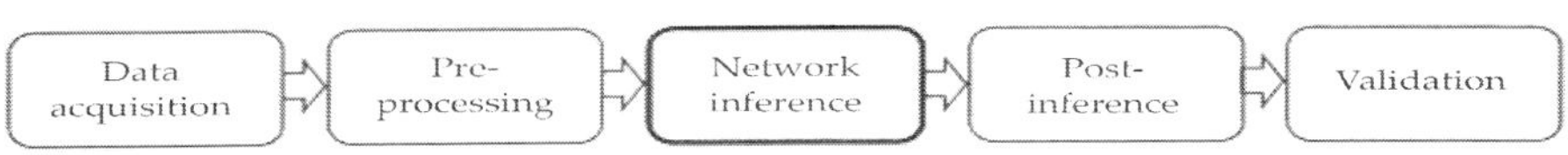

Figure 1: Generic data processing pipeline, starting with data acquisition and continuing with data pre-processing, network inference, post-processing, and validation of results.

Data Acquisition

Multiple electrodes and fluorescence imaging are two major methods for making high-dimensional measurements of single neuronal activities.

Multiple Electrode Recording

For *in vivo* multiple neural recordings, the most commonly used device is the so-called "tetrode", a bundle of four wire electrodes. By implanting tens of tetrodes and applying a spike sorting method, it is possible to record hundreds of neurons simultaneously Buzsaki (2004), Jog et al. (2002). Linear or matrix arrays of electrodes, often using semiconductor technologies, are also used for recording hundreds of neurons near the cortical surface (Fujishiro, Kaneko, Kawashima, Ishida, & Kawano, 2014). For brain tissue slices and cultures, electrode grids patterned on a plate enable recording of hundreds of neural activities Fujisawa et al. (2008), Shimono and Beggs (2014). In such *in vitro* experiments, it is also possible to use intracellular glass electrodes to record sub-threshold membrane potentials of selected neurons in a population (Brette & Destexhe, 2012, chap. 3).

Optical Imaging

Optical imaging allows activity data gathering from hundreds to thousands of neurons simultaneously, using voltage sensitive dyes (VSDs) or genetically encoded calcium indicators (GECIs). Chemical VSDs offer high temporal resolution, but lack the cell type-specificity that GECIs offer. However, genetically encoded voltage indicators (GEVIs) are in active development (Yang & St-Pierre, 2016). Recently GECIs have become increasingly popular because they can be expressed under the control of cell-type specific promoters (Looger & Griesbeck, 2011). The weaknesses of GECIs have been their slow temporal response and low sensitivity, but recent GECIs have begun to achieve time constants on the order of ten milliseconds, making detection of single spikes feasiible (Pnevmatikakis et al., 2016).

Fast CCD cameras and confocal or two-photon laser scanning microscopes are commonly used with GECIs. CCD cameras enable simultaneous imaging of all neurons in the focal plane, allowing recording of as many as a thousand frames per second, but measurements are limited to neurons near the surface of the tissue. Two-photon microscopes use infrared light, which is less subject to refraction and which excites fluorescent molecules only at the focal point, allowing recording of neurons several hundreds of microns beneath the surface Dombeck et al. (2009), Lütcke and Helmchen (2011). Most recently, head-mount miniature microscopes using gradient index (GRIN) rod lenses have allowed access to deep neural structures, such as the hippocampus of awake behaving animals (Ziv et al., 2013).

Pre-processing

Pre-processing steps depend on recording methods and specific input requirements of the network inference method.

In multiple electrode recording, each electrode can receive signals from multiple neurons and signals from the same neuron can be detected by multiple electrodes. Spike sorting algorithms are used to identify spikes from each neuron by applying principal component analysis, independent component analysis, and any biophysical knowledge of spike shapes and intervals Mahmud et al. (2015), Shimono and Beggs (2014).

The task in optically imaged data pre-processing is to transform an image sequence into a multi-dimensional time-series of neural activities. Pre-processing steps for optical imaging are: (i) image segmentation to identify regions corresponding to each neuron, (ii) extraction of the fluorescence trace for each neuron and (iii) spike inference (Pnevmatikakis et al., 2016). These pre-processing operations have to deal with light scattering, motion artifacts, and slow fluorescence changes with respect to the underlying membrane potential.

Connectivity Inference

Connectivity inference methods can be largely classified into two classes. Descriptive model-free methods are based on descriptive statistics without assuming any particular process that generated the data. On the other hand, generative model-based methods assume a certain mathematical model that generates the data and infer parameters and structure of the model. We will explain those methods in detail in subsequent sections.

Most connectivity inference methods require time-series of spikes from each neuron in the population. In some studies, spike inference and connectivity inference are performed using an integrated optimization algorithm directly from time-series of fluorescence (Mishchenko, Vogelstein, & Paninski, 2011). Some other studies use fluorescence signals directly for model-free analysis of connectivity without an explicit spike inference mechanism (Veeriah, Durvasula, & Qi, 2015).

Post-processing

After the connectivity matrix is obtained by applying any of the connectivity inference methods, it is often useful to perform post-processing to achieve a biologically realistic result. For example, inference methods that consider only simultaneous activities yield only symmetric connectivity matrices. A heuristic method to determine the direction is to use the average temporal order of activations of two neurons (Sutera et al., 2014). Another issue is that inferred connectivity can be a mix of direct causal effects between neuron pairs and indirect effects through other neurons. One way to address this problem is to use matrix deconvolution Barzel and Barabási (2013), Feizi et al. (2013), Magrans and Nowe (2014), as described in Section 6.1.2.

Furthermore, employment of a network inference method depends on the choice of parameters. It is possible to improve both robustness and accuracy of the connectivity matrix by combining several matrices computed using different parameters and/or different inference methods Magrans and Nowe (2014), Sutera et al. (2014).

Validation

In order to validate the connectivity inference method itself, use of synthetic data from a network model with a known connection matrix is the first step. For estimating connectivity of real biological neural networks, validation is more challenging, as anatomical or physiological verification of all pair-wise synaptic connectivity is extremely laborious.

Synthetic Data

The strength of synthetic data for validation is that we have all information about simulated connectivity and other biophysical parameters of the model. Therefore, we can use standard error metrics to measure the similarity between the inferred connectivity matrix and the one used for data generation.

When the final objective is to infer connection weights, one difficulty with many methods is that they deliver estimated connectivity values with different scales. The relative mean squared error

$$relative\,MSE = \frac{min_{\alpha} \sum_{ij} |W_{ij} - \alpha \hat{W}_{ij}|^2}{\sum_{ij} |W_{ij}|^2}. \quad (1)$$

where $\hat{W}_{ij}$ is the estimate of W_{ij}, can alleviate the scaling problem (Fletcher & Rangan, 2014).

If the objective is to infer the graphical structure of the network, what matters is the binary existence/nonexistence of connections. The Area Under the Curve of the Receiver–OperatorCharacteristic (AUROC) is a popular performance metric used in such a case Garofalo et al. (2009), Guyon et al. (2014), Stetter et al. (2012). The ROC Curve describes the relationship between the False Positive (FP) Rate ($\frac{FP}{FP+TN}$) or Fall-out and the True Positive (TP) Rate ($\frac{TP}{TP+FN}$) or Recall at different thresholds. The perfect classifier has an AUROC of 1, while a random classifier has an AUROC of 0.5. However, this metric may overestimate performance in highly biased data-sets (Schrynemackers, Küffner, & Geurts, 2013). The Area under the Precision Recall Curve (AUPRC) was proposed as an alternative measure to improve validation accuracy of sparsely connected neural networks Orlandi et al. (2014), Sutera et al. (2014). Specifically, it is the area below the curve that describes the relationship between the Precision ($\frac{TP}{TP+FP}$) versus the Recall ($\frac{TP}{TP+FN}$) at different thresholds.

In order to find the best trade-off between TPs and FPs, Garofalo et al. (2009) defined the Positive Precision Curve (PPC), which describes the relationship between the True False Sum (TFS=TP+FP) and the True False Ratio ($TFR = \frac{TP-FP}{TFS+TN}$). The peak of this curve can be used to extract a binary connectivity map from a weight matrix inferred by a network inference method. To assess the required recording duration and bin size to achieve a desired reconstruction performance, Ito et al. (2011) proposed the use of the curves of the TPR at a fixed FPR as a function of different recording durations and bin sizes.

Real Data

In real neural recording, although the ground truth of connectivity is not generally available, we can assess quality of inferred connections using statistical significance testing and cross validation. Significance testing is used to accept or reject the null hypothesis (H_0) that a connection between two neurons does not exist (Lizier, 2014). To approximate H_0, we can run our network inference method on many surrogate time-series created by perturbing the training time-series such that it destroys the connectivity information Fujisawa et al. (2008), Lizier et al. (2011), Oba et al. (2016),

Shimono and Beggs (2014). Then, the test itself consists of computing the probability that the inferred connectivity value is generated uniquely by chance. However, sometimes an accept/reject test is not enough, as for instance, when we wish to infer a weighted connectivity matrix or parameters like the synaptic delay or a time constant from a biophysical model.

Model structures and/or parameter sets in a certain class can be compared using relative quality measures like the likelihood ratio and other criteria that penalize larger complexity models like the Akaike Information Criterion (AIC) and the Bayesian Information Criterion (BIC) (Aho, Derryberry, & Peterson, 2014).

An alternative strategy is cross validation, in which the inferred model is tested against a separate test data set that was not used for model inference. A standard method in probabilistic model-based methods is to compute a normalized likelihood of the model for a test data set (Gerwinn, Macke, & Bethge, 2010). In addition, comparison of any statistical features, such as average firing rates and spike-train statistics of the sample data produced from the inferred model and the real data is helpful in validating the inferred model (Pillow, Paninski, Uzzell, Simoncelli, & Chichilnisky, 2005). When a certain graph-theoretic feature of the network is known, for example, by previous anatomical studies, comparison of such features, such as the node degree distribution, can be a helpful way of validation (Bullmore & Sporns, 2009).

CHALLENGES

This section summarizes two types of challenges in connectivity inference: biophysical and technical. We first describe complexities arising from biophysical properties of neurons and synapses, and then from technical difficulties due to constraints in instrumentation and computation.

Biophysical Challenges

Apparent Connectivity

As mentioned in the Introduction, functional connectivity (FC) and effective connectivity (EC) may not be due to direct synaptic connectivity. A typical example is a common input that activates two or more neurons simultaneously. In such a case, functional connectivity can be inferred between the recipient neurons even if there is no direct connection between them. Other cases are a common input with different time delays or an indirect connection through a third neuron, which can activate two neurons in sequence. In this case, effective connectivity can be inferred even if there is no direct synaptic connection between them.

Directionality:

Methods based on simple correlation cannot detect the direction of the causal or temporal relationship. Even when the temporal order of activities is considered, if time resolution of the measurement or analysis is coarse, ordered activation can appear as simultaneous activation, making it difficult to determine directionality.

Cellular Diversity:

Many neurons show intrinsic refractoriness such that spike frequency gradually drops even with a constant level of input. Some neurons also show a burst property such that once excited above a threshold, they keep spiking even without excitatory inputs. In such cases, it is not straightforward to discriminate whether a change in the activity of a neuron is due to some network dynamic or the neuron's intrinsic properties (Gerstner et al., 2014: chap. 1). Such refractoriness or burstiness can be categorized as inhibitory or excitatory self-connection, but whether it actually results from self-connecting synapses (autapse) (Bekkers, 2009) must be carefully interpreted. Even in a local circuit, there are many different types of neurons with a large diversity of biophysical parameters (Kandel, Schwartz, Jessell, et al., 2000a, chap. 4). Thus a simple assumption of uniform cellular properties may not be valid.

Synaptic Diversity:

Diversity also applies to connectivity because underlying synapses can have complex and diverse characteristics (e.g. excitatory/inhibitory, facilitatory/depressive, delays, etc.) (Kandel, Schwartz, Jessell, et al., 2000b, chap. 8). Understanding how a given effective network emerges may require the inference of many additional parameters beyond a simple weight matrix.

Non-stationarity:

Synaptic weights are subject to both short-term (from tens of milliseconds to few minutes) and long-term changes (from minutes to hours) (Gerstner et al., 2014: chap. 19). Physiological states of neurons can also drift over time, especially during experimental manipulations, as in slice preparations, with electrodes inserted, or with exposure to light. *In vitro* cultured neural networks often manifest switching between states of low- and high-firing rates. When using fluorescence imaging, the high firing state can cause erroneous connectivity inference because the low temporal resolution makes it difficult to discriminate between direct and indirect effects (Stetter et al., 2012).

Technical Challenges

Noise:

Every instrument is subject to noise. Electrodes can pick up both biophysical (e.g. thermal noise) and anthropogenic noise (e.g. electromagnetic interference from power lines) (Van Drongelen, 2006). Optical imaging is susceptible to light scattering artifacts, and motion artifacts in awake subjects with *in vivo* imaging. Even though motion correction programs can track shifts of neurons within an image, movement of cells out of the focal plane is difficult to compensate.

Time/space Resolution:

Multi-electrode recording can monitor activities of a few hundred neurons with sampling rates on the order of 10kHz (Shimono & Beggs, 2014). Fluorescence imaging can work with a trade-off of space and time, ranging from few hundred neurons recorded at 100Hz to a hundred thousand neurons, or 80% of a zebra fish brain, at a rate of 0.8Hz (Ahrens, Orger, Robson, Li, & Keller, 2013). The poor temporal resolution of fluorescence imaging has the undesirable feature of mixing indirect causal effects with real direct connectivity effects.

Hidden neurons/External inputs:

Even though recent advances in two-photon imaging have enabled us to capture thousands of neurons at one time (Ahrens et al., 2012), it is still difficult to simultaneously record the activity of every single neuron in a circuit. Neglecting hidden neurons can lead to spurious detection of connectivity between neurons connected via hidden neurons. Even though most experiments take into account sensory stimuli or motor actions in the analysis of neural activities, the brain shows various intrinsic dynamics.

Unknown external inputs, especially those common to multiple neurons, can also result in a detection of spurious connections.

Prior knowledge:

Prior knowledge of the anatomy and physiology of the neural networkunder investigation can be leveraged to enhance connectivity inference. How to incorporate prior knowledge about the composition of neuron types in a population Shimono and Beggs (2014), Sohn et al. (2011) or connection probabilities between different types of neurons (White, Southgate, Thomson, & Brenner, 1986) to constrain the solution space is an important issue in connectivity inference. Additional data, such as the local field potential (Destexhe & Bedard, 2013), can also facilitate understanding of non-stationary behaviors at the microcircuit level.

Scalability:

As the number of neurons measured simultaneously grows from hundreds to thousands, the number of potential connections grows from tens of thousands to millions. Therefore, inference methods need to be designed to maximize computational efficiency and to minimize the cost of parallelized implementations (Gerstner et al., 2014: chap. 10).

MODEL-FREE METHODS

We first review *model-free* methods, which do not assume any mechanism to generate the observed data. These methods tend to be simpler than model-based methods, but they are not able to generate activity data for model validation or prediction. We will review model-free methods in two categories. The first is based on descriptive statistics and the second on information theoretic concepts. Table 1.1, Table 1.2 summarize the major model-free methods and examples that use those methods.

We denote the neural activity data set as

$D = x_i(t) | i = 1, \ldots, P; t = 1, \ldots, T,$

where $x_i(t)$ is the activity of neuron i at the tth time point. $x_i(t)$ may be continuous, for instance, when values are raw data obtained from multiple electrodes or fluorescence imaging, or it may be binary when data are transformed into a spike train by a spike-sorting algorithm.

Table 1.1: Summary of model-free connectivity inference methods, Descriptive Statistics (4.1)

Method	Principle	Examples of application	Features
Correlation (4.1.1)	**Linear relationship**	**Magrans and Nowe (2014): simulation data that assume calcium imaging** **Fujisawa et al. (2008): Multi-electrode recordings from medial prefrontal cortex of rats** **Cohen and Kohn (2011)**	**• Acausal indicator** **• Does not discriminate direct/indirect effects** **• Low computational cost**
Cross-correlation (4.1.2)	Linear relationship with time shift	Garofalo et al. (2009): ROC and PPC curves for CC below JE and TE Ito et al. (2011): ROC curve for CC below TE variants curves Knox (1981) Garofalo et al. (2009): simulation data and multi-electrode recordings from in vitro cultured cortical neurons Ito et al. (2011): simulation data	• Causal indicator • Takes into account the spike train history • Low computational cost
Partial-correlation (4.1.3)	Linear relationship excluding the effect from other neurons	Sutera et al. (2014): simulation data that assume calcium imaging	• Acausal indicator • Discriminates direct/indirect effects • Solution based on PC won First Neural Connectomics Challenge

Table 1.2: Summary of model-free connectivity inference methods, Information Theoretic (4.2) and Supervised Learning (4.3, 6.1.3)

Method	Principle	Examples of application	Features
Mutual information (4.2.1)	**Statistical dependence**	**Garofalo et al. (2009): ROC and PPC curves for MI below TE, JE and CC curves** **Garofalo et al. (2009): simulation data and multi-electrode recordings from in vitro cultured cortical neurons**	**• Acausal indicator** **• Does not discriminate direct/indirect effects** **• Does not discriminate between excitatory and inhibitory connections**

Joint entropy (4.2.2)	Entropy of cross-inter-spike-intervals	Garofalo et al. (2009): ROC and PPC curves for JE below TE and above MI and CC curves Garofalo et al. (2009): simulation data and multi-electrode recordings from in vitro cultured cortical neurons	• Acausal indicator • Does not discriminate direct/indirect effects • Does not discriminate between excitatory and inhibitory connections
Transfer entropy (4.2.3)	Gain in past-future mutual information when the activity of another neuron is considered	Garofalo et al. (2009): ROC and PPC curves for TE above the rest (JE, MI and CC curves) Schreiber (2000) Garofalo et al. (2009): simulation data and multi-electrode recordings from in vitro cultured cortical neurons Shimono and Beggs (2014): multi-electrode recordings from from slice cultures of rodent somatosensory cortex	• Causal indicator • Does not discriminate direct/indirect effects • Does not discriminate between excitatory and inhibitory connections. However, it is possible to perform a subsequent analysis of the correlation or conditional distribution to infer if it is excitatory or inhibitory • Takes into account the spike train history
Delayed transfer entropy (4.2.4)	Transfer entropy with delayed source signal	Ito et al. (2011): ROC curve and TPR at constant FPR better than TE and below HOTE Ito et al. (2011): simulation data	• Same as transfer entropy
High order transfer entropy (4.2.5)	Transfer entropy with multiple time steps of activities	Ito et al. (2011): ROC curve and TPR at constant FPR better than TE and DTE Ito et al. (2011): simulation data	• Same as transfer entropy
Generalized transfer entropy (4.2.6)	Transfer entropy with present source activity, after removing high activity periods	Orlandi, Stetter et al. (2014), Stetter et al. (2012): ROC curve for GTE above TE curve Orlandi, Stetter et al. (2014), Stetter et al. (2012): simulation data that assume calcium imaging and data recorded using calcium imaging from in vitro networks derived from cortical neurons	• The use of information from the same time bin for both neurons enhances reconstruction performance when the data source has a low sampling rate (e.g. calcium imaging) • It copes well with synchronized bursting episodes
Convolutional neural network (6.1.3)	Probability of connection generalized from training data	Romaszko (2014), Veeriah et al. (2015): simulation data that assume calcium imaging	• It assumes the ground truth connectivity labels of a training data set that has similar properties as the network that we want to reconstruct

Descriptive Statistics

This method class utilizes statistical measures to capture the degree of connection between neurons from a sample of activities from a neural population.

Correlation

Correlation indicates the strength of the linear relationship between two random variables that represent two neurons. The most commonly used measure of correlation between activities x_i and x_j of two neurons i and j is the Pearson correlation coefficient defined as:

$$\rho_{ij} = \frac{\Sigma_{ij}}{\sqrt{\Sigma_{ii}\Sigma_{jj}}}, \qquad (2)$$

where $\Sigma_{ij} = \frac{1}{T}\sum_{t=1}^{T}(x_i(t) - \mu_i)(x_j(t) - \mu_j)$ is the covariance and $\mu_i = \frac{1}{T}\sum_{t=1}^{T}x_i(t)$ is the mean activity.

When we use correlation ρ_{ij} to perform network inference of neuronal circuits, we are measuring the rate of co-occurring spikes of neurons i and j (Cohen & Kohn, 2011) and this rate is interpreted as the functional connectivity strength. While Pearson correlation is computationally least costly, it has a number of drawbacks. It is not able to indicate the causal direction and it cannot distinguish direct connections from indirect ones. It is also not suited to deal with external inputs. Despite such limitations, a winning solution of the First Neural Connectomics Challenge used correlation as a key component in a more complex method (see Section 6.1.2).

Cross Correlation

Cross correlation (CC) indicates the strength of the delayed linear relationship between two neurons i and j. Knox (1981) defines CC as "the probability of observing a spike in one train x_j at time t+τ, given that there was a spike in a second train x_i at time t". Ito et al. (2011) remark that, despite the extensive literature, there is no standard definition, and they discuss the two most popular ones:

$$\rho_{i\to j}(\tau) = \frac{1}{T-1}\sum_{t=1}^{T}\frac{(x_i(t) - \mu_i)(x_j(t+\tau) - \mu_j)}{\sigma_i\sigma_j} \qquad (3)$$

where the parameter τ defines the delay of neuron j with respect to neuron i, and μ and σ are the sample average and standard deviation, respectively.

The second CC definition uses the total number of spikes instead of the averages and standard deviations:

$$\rho_{i\to j}(\tau) = \sum_{t=1}^{T} \frac{x_i(t)x_j(t+\tau)}{\sqrt{n_i n_j}}, \tag{4}$$

where n_i and n_j are the total number of spikes from neuron i and j respectively. Both cross-correlation definitions tend to be equivalent when μ_i and μ_j approach 0.

CC is a causal indicator that is able to indicate the direction of the connection. Inference performance of the connection direction depends upon the instrumentation sampling rate and the choice of time delay τ. Despite being able to detect the direction of connectivity, CC has the same limitations as correlation in dealing with indirect connections and external inputs.

We should also note that the optimal parameter τ to detect a connection can be different for each connection. A measure to address this issue is the Coincidence Index (CI), which combines several CCs computed at different τ_s:

$$CI_{i\to j} = \frac{\sum_{\tau=0}^{r} \rho_{i\to j}(\tau)}{\sum_{\tau=0}^{T} \rho_{i\to j}(\tau)}, \tag{5}$$

where r specifies the interval of cross-correlation delays, called the coincidence interval. A large CI indicates a larger reproducibility of correlated spike timing Chiappalone et al. (2007), Shimono and Beggs (2014), Tateno and Jimbo (1999).

Garofalo et al. (2009) evaluated the network inference performance of several model-free methods using data from simulated neuronal networks with only excitatory synapses and including both excitatory and inhibitory connections. According to both ROC and PPC criteria, CC demonstrated a performance just below transfer entropy (TE, see Section 4.2.3) in the fully excitatory setting, and below TE and joint entropy (JE, see Section 4.2.2) when inhibitory connections were also included. Ito et al. (2011) evaluated the inference performance of the two CC variants discussed in this section and several transfer entropy variants (see Sections 4.2.4, 4.2.5 High order transfer entropy (HOTE), 4.2.6 Generalized transfer entropy (GTE)) using data from simulated neuronal networks with both excitatory and inhibitory connections. According to the ROC criterion, both CC variants showed a performance inferior to all variants of transfer entropy.

Partial Correlation

Let $R=[R_{ij}]$ be an inverse of a covariance matrix Σ of which the (i,j) thcomponent is Σ_{ij}. The definition of partial correlation (PC) between activities of neurons i and j is:

$$PC_{ij} = -\frac{R_{ij}}{\sqrt{R_{ii}R_{jj}}} \tag{6}$$

The most salient difference between PC and other methods described in this section is that PC takes into account the activities of all neurons in the population to compute a connectivity indicator between each neuron pair. An important property of PC is that, assuming that $x(t)=(x_1(t),\ldots,x_p(t))$ is normally distributed, then PC_{ij} is 0 if and only if neurons i and j are independent given all the rest. Therefore, PC can be used to distinguish a direct effect from an indirect one. Despite this multivariate feature, PC shares the same limitations as correlation and CC in dealing with external inputs, with the additional computational complexity required to invert the covariance matrix. The first prize solution of the First Neural Connectomics Challenge, discussed in Section 6.1.1, is a good example of how to use PC for network inference (Sutera et al., 2014).

Information Theoretic Methods

Information theory is a mathematical discipline initiated by Shannon to characterize the limits of information management, including transmission, compression and processing (Shannon, 1948). This section presents the application of several information theory measures to the inference of neural microcircuits.

Mutual Information

Mutual information (MI) is a measure of the statistical dependence between stochastic variables. MI of activities x_i and x_j of two neurons is mathematically defined as:

$$MI_{i,j} = \sum_{i,j} P(x_i, x_j) \log \frac{P(x_i, x_j)}{P(x_i)P(x_j)}. \quad (7)$$

This indicator is symmetric; therefore it is unable to identify the direction of the connection. It cannot discriminate direct effects from indirect effects. One way to overcome the directionality limitation is to introduce a delay as in CC, and to consider multiple delays as in CI (Garofalo et al., 2009).

In a comparative study by Garofalo et al. (2009) using data from simulated neuronal networks, MI delivered the performance below TE, JE and CC, according to both ROC and PPC criteria.

Joint Entropy

Joint entropy (JE) is a bi-variate causal measure between the activity of two neurons, testing whether neuron i is a cause of the activity in neuron j. For each reference spike in x_i, the closest future spike on x_j is identified and the cross-inter-spike-interval is computed as cISI=$tx_i - tx_j$, where tx_i and tx_j are the time of successive spikes of neuron i and j, respectively. Then JE is defined as the entropy of the cISI distribution:

$$JE_{i \to j} = -\sum_{k} P(cISI_k) \log_2 P(cISI_k), \quad (8)$$

where k indexes the level of cISI. If neurons i and j are strongly connected, then JE is expected to be close to 0 as the cISI distribution becomes sharp.

In a comparison by Garofalo et al. (2009), JE showed the worst performance below TE, CC and MI in a fully excitatory setting, but the second best performance below TE when inhibitory connections were also included.

Transfer Entropy

Transfer entropy (TE) is a causal indicator defined between the activity of two neurons i and j. The definition of TE was originally formulated as the Kullback–Leibler divergence between the distributions of the neural activity of target neuron j conditioned by its previous activities alone versus conditioned also by previous activities of source neuron i (Schreiber, 2000):

$$TE_{i \to j} = \sum P(x_j(t), x_j(t-1), \ldots, x_j(t-k), x_i(t-1), \ldots, x_i(t-l)) \log_2 \frac{P(x_j(t)|x_j(t-1), \ldots, x_j(t-k), x_i(t-1), \ldots, x_i(t-l))}{P(x_j(t)|x_j(t-1), \ldots, x_j(t-k))} \quad (9)$$

TE can also be defined as the conditional mutual information between the future activity of target neuron j and past activity of source neuron i conditioned by past activity of target neuron j, which can also be expressed as the amount of uncertainty reduced in the future activity of neuron j knowing the past activity of i given past activity of j:

$$\begin{aligned} TE_{i \to j} \\ &= MI(x_j(t), (x_i(t-1), \ldots, x_i(t-l)) | (x_j(t-1), \ldots, x_j(t-k))) \\ &= H(x_j(t) | (x_j(t-1), \ldots, x_j(t-k))) \\ &\quad - H(x_j(t) | (x_j(t-1), \ldots, x_j(t-k), x_i(t-1), \ldots, x_i(t-l))) \end{aligned} \quad (10)$$

From this second definition, we can see that TE is a positive measure that takes a low value when the past activity of neuron i does not convey information about the future activity of neuron j.

TE is equivalent to Granger causality for Gaussian variables (Barnett, Barrett, & Seth, 2009). A nice feature of TE is that it can detect non-linear relationships (Stetter et al., 2012). Although TE does not identify whether the connection is excitatory or inhibitory, the polarity of connection can be tested separately by recording (real data) or simulating the same network with the inhibitory connections pharmacologically blocked Orlandi, Stetter et al. (2014), Stetter et al. (2012). The performance of TE in discriminating the causal direction depends whether the sampling rate is faster than the network dynamics Fujisawa et al. (2008), Shimono and Beggs (2014).

In practice, the parameters k and l in Eq. (9) are set to 1 (e.g., Garofalo et al. (2009)), so that we only have to consider a small number of patterns (i.e. $2^3=8$ for binary data). While increasing k and l can allow detection of delayed effects of a connection, it requires larger amount of data to have reliable estimates of the probability distribution to compute the entropies.

Before the First Neural Connectomics Challenge, TE was arguably the most successful model-free method Garofalo et al. (2009), Stetter et al. (2012). This success encouraged a number of variants as follows Ito et al. (2011), Orlandi, Stetter et al. (2014), Stetter et al. (2012).

Delayed Transfer Entropy (DTE)

A limitation of TE using k=l=1 is that it assumes a constant one-time bin delay between the action potential in neuron i and the post-synaptic action potential of neuron j, which is not a realistic assumption. A suboptimal way to deal with this problem is to create longer time bins at the cost of losing detailed temporal information. Instead, delayed TE (DTE) was proposed to measure TE at a user-defined time delay d (Ito et al., 2011). Compared to the original definition, instead of $x_i(t-1),\ldots,x_i(t-l)$ in Eq. (9), they suggest using a delayed signal $x_i(t-d),\ldots,x_i(t-d-l+1)$. Shimono and Beggs (2014) improve reconstruction performance by identifying the delay parameter for each individual connection using the coincidence index (Eq. (5)).

High Order transfer Entropy (HOTE)

Ito et al. (2011) considered how to increase k and l to take into account longer spike train history, while avoiding negative effects of the increase in the possible patterns (2^{k+l+1}). The additional parameters d in DTE and k and l in HOTE give multiple measures for each neuron pair. They evaluated the reconstruction performance by the peak value and the coincidence index (Eq. (5)) of DTE and HOTE, using a simulated network with both excitatory and inhibitory connections. HOTECI and DTECI had better performance than $HOTE_{pk}$ and DTE_{pk}, which in turn were better than TE with k=l=1.

Generalized Transfer Entropy (GTE)

As described in Section 3, periods of synchronized bursting convey very low connectivity information due to the simultaneous spike of a large percentage of neurons. This phenomenon is especially critical with calcium imaging recordings because of their limited temporal resolution. Generalized TE (Stetter et al., 2012) was proposed to alleviate this problem with two modifications to Eq. (9): (i) to compensate for the slow sampling rate in fluorescence imaging (≥10ms), GTE uses presynaptic activity from the same time bin $x_i(t)$ instead of the previous one $x_i(t-1)$; (ii) to discard the poor connectivity information conveyed by synchronized bursting periods, GTE restricts the computation of TE to those time bins with an overall network activity below a given user-defined threshold.

Just like TE, GTE is unable to differentiate between excitatory and inhibitory connections. Application of GTE to simulated neural networks with excitatory connections shows improved performance with respect to TE, and modified versions of CC and MI implementing the same generalization (Stetter et al., 2012). Orlandi, Stetter et al. (2014) showed that GTE is able to reconstruct inhibitory connections; however, its reconstruction performance is below its reconstruction performance of excitatory connections.

Information Gain

Information gain (IG) is a less known causal indicator between the activity of two neurons i and j. Its definition is:

$$IG_{i \to j} = H(x_j(t)) - H(x_j(t)|x_i(t-1)), \tag{11}$$

where H is the entropy, x_i is the activity vector of the source neuron and x_j is the activity vector of the target neuron. In the case of calcium imaging, it may be convenient to employ the same strategy as with GTE, using the sample from the same time bin $x_i(t)$ instead of the previous one $x_i(t-1)$. This connectivity indicator was used as one of multiple indicators combined by one of the top solutions of the First Neural Connectomics Challenge (Czarnecki & Jozefowicz, 2014).

Supervised Learning Approach

Motivated by recent successes of deep neural networks in challenging pattern classification problems, a new approach has been proposed to apply convolutional neural networks (CNNs) for prediction of the existence of synaptic connectivity Romaszko (2014), Veeriah et al. (2015). In this approach, time-series data of spike trains or fluorescent signals of a pair or more of neurons are given as the input and a CNN is trained to classify if there is a connection between the neurons.

A natural limitation of this approach is that training requires sufficient ground truth data about the existence of connectivity. Although this approach has been successful with simulated data Romaszko (2014), Veeriah et al. (2015), it is hard to apply it directly to real data, for which experimental identification of synaptic connectivity is highly costly.

MODEL-BASED METHODS

Now we review *model-based* methods, in which the connectivity is estimated by explicitly modeling the data generation process. The basic paradigm of the model-based method is: (1) to assume a generative model that generates neural data, and (2) to determine model parameters to fit observed data. Table 2 summarizes the major model-based methods and examples that used those methods. We should keep in mind here that any model for this approach aims to explain the observed data, but does not consider all physiological properties of the real neural networks: The large number of factors affects structure and function of real neural systems, and it is infeasible for all of them to be considered in a single model (See more details about similarities and gaps between the model and biological reality in Ask and Reza (2016). Due to the limitation, the applicability of model-based methods is shown only empirically using simulation studies, but there is no theoretical guarantee that the estimation result are consistent with underlying connectivity. However, by reviewing model-based methods here, we will see approximation techniques to fill the gaps with a feasible number of parameters, as well as practical solutions for the connectivity estimate.

In the following, $x(t)=(x_1(t),\ldots,x_P(t))$ represents a set of P signals observed at the tth time point in a discrete time domain, where $x_i(t)$ denotes the ith neuron's activity at that time. $x_i(t)$ may be continuous when the measurement is raw data collected from calcium imaging and multiple-electrode recording, or binary when the data is transformed into a spike trainby a spike-sorting algorithm. We denote a generative model $p_\theta(x)$, where θ is a parameter vector, including the connection weights.

Table 2: Summary of model-based connectivity inference methods

Method	Principle	References & demonstrations (The asterisk (*) shows inclusion of applications to real data)	Features
Autoregressive models (5.1.1)	**Directed linear interaction**	***Harrison, Penny, and Friston (2003): human fMRI data *Franaszczuk, Blinowska, and Kowalczyk (1985): multichannel EEG time series Smith, Fall, and Sornborger (2011): simulation data that assume calcium imaging**	**• analytical solutions are available • not suitable for spike-train data**
Generalized linear model (5.1.2)	Spike probability based on linear summation of inputs	*Song, Wang, Tu, Marmarelis, Hampson, Deadwyler, and Berger (2013): simulation data and multi-electrode recordings from hippocampal CA3 and CA1 regions of rats *Gerwinn et al. (2010): simulation data and multi-electrode recordings from salamander retinal ganglion cells	• easy but iterative optimization is required
Stochastic leaky integrate-and-fire model (5.1.3)	Stochastic spike with state resetting	Paninski, Pillow, and Simoncelli (2004): simulation data Koyama and Paninski (2010): simulation data Isomura, Ogawa, Kotani, and Jimbo (2015): simulation data	• model for a continuous time domain • a special case of generalized linear models

Network likelihood model (5.1.4)	Continuous-time model with Poisson spikes	*Okatan, Wilson, and Brown (2005): simulation data and multi-electrode recordings from hippocampal CA1 region of a rat *Kim, Putrino, Ghosh, and Brown (2011): simulation data and multi-electrode recordings from the primary motor cortex (MI) of a cat *Stevenson, Rebesco, Hatsopoulos, Haga, Miller, and Kording (2009): simulation data and multi-electrode recordings from the primary motor cortex (MI) and dorsal premotor cortex (PMd) of a monkey	• a special case of generalized linear models • instantaneous firing rates are directly considered
Calcium fluorescence model (5.1.5)		Mishchenko et al. (2011): simulation data	
Hawkes process (5.1.6)	GLM with Poisson observation	Linderman and Adams (2014) Linderman and Adams (2015): simulation data that assume calcium imaging	• The network consists of purely excitatory neurons
Dynamic Bayesian network (5.1.7)	Directed acyclic graph	Eldawlatly, Zhou, Jin, and Oweiss (2010) *Patnaik, Laxman, and Ramakrishnan (2011): simulation data and multi-electrode recordings from dissociated cortical cultures of rats	• Simulated annealing for optimizing both network structures and parameters • The computational efficiency is degraded in case of cyclic graphs • Due to the nature of Bayesian approach, the result depends on hypothetical choices of the prior distribution.
Maximum entropy model (5.1.8)	Maximum entropy under constraint by an energy function	*Tkacik, Mora, Marre, Amodei, Berry, Michael, and Bialek (2014): multi-electrode recordings from retinal ganglion cells	• Inspired by Ising models • Up to the second-order statistics are considered • Heavy computational complexity for parameter estimation

Generative Models

Autoregressive Model

A basic example of a generative model is the *autoregressive* (AR) model mathematically expressed as

$$x_i(t) = A_{i0} + \sum_{j=1}^{P} \sum_{k=1}^{K} A_{ij}(k) x_j(t-k) + \epsilon_i(t), \quad i = 1, \ldots, P.$$

$$\epsilon_i(t) \sim \mathcal{N}(\cdot|0, \sigma_i^2) \tag{12}$$

where K is the degree of the model, $A_{ij}(k)$ is a parameter called an AR coefficient, Ai0 is a bias term, and $N(\cdot|\mu,\sigma^2)$ denotes the Gaussian distribution with mean μ and standard deviation σ (for simplicity, we assume that σ is known in this example).

By integrating two equations in Eq. (12), $x_i(t)$ can be regarded as a sample according to the following conditional distribution:

$$x_i(t) \sim \mathcal{N}\left(\mu_i(t), \sigma_i^2\right), \tag{13}$$

$$\mu_i(t) = A_{i0} + \sum_{j=1}^{P} \sum_{k=1}^{K} A_{ij}(k) x_j(t-k) \tag{14}$$

where $\theta \equiv \{A_{i0}|i=1,\ldots,P\} \cup \{A_{ij}(k)|i,j=1,\ldots,P;k=1,\ldots,K\}$ is a set of parameters in this case.

After determining the parameter θ, if $A_{ij}(k)$ for all k is exactly or very close to zero, it implies that there is no direct interaction from the jth neuron to the ith neuron. In contrast, if $A_{ij}(k)$ for some k deviates enough from zero, the jth neuron x_j can directly affect the ith neuron x_i with a k time-step delay.

Even though a neuron's dynamics are usually nonlinear, for the virtue of simplicity, this method has been used in a wide range of neural data analyses, in addition to calcium imaging and multiple electrode recording Franaszczuk et al. (1985), Harrison et al. (2003), Smith et al. (2011), Valdés-Sosa et al. (2005).

Generalized Linear Model

When $x_i(t)$ is a binary variable indicating whether a spike is generated at the tth time step, the generalized linear models (GLM) (Song et al., 2013) provide a tractable extension of the AR models. A GLM describes spike generation as a point process as

$$x_i(t) \sim \mathrm{Ber}(\cdot|\rho_i(t)) \tag{15a}$$

$$\rho_i(t) \equiv \phi\left(A_{i0} + \sum_{j=1}^{P} \sum_{k=1}^{K} A_{ij}(k) x_j(t-k)\right). \tag{15b}$$

where $\mathrm{Ber}(\cdot|p)$ denotes the Bernoulli distribution with an event probability of p. $\phi(\cdot)$ is a so-called an inverse link function, for which the exponential function $\phi(x)=\exp(x)$ or the sigmoid function $\phi(x)=(1+\exp(-x))^{-1})$ is often used.

Stochastic Leaky Integrate-and-fire Model

A stochastic leaky integrate-and-fire (LIF) model Isomura et al. (2015), Koyama and Paninski (2010), Paninski et al. (2004), is one of the most widely used models for analyzing the behavior of spiking neural networks (Gerstner et al., 2014). The model assumes that the subthreshold membrane potentialof the ith neuron, denoted by V_i, evolves according to the following stochastic differential equation:

$$dV_i(t) = \left(-g_i V_i(t) + \sum_{j=1}^{P} I_{ij}(t)\right) dt + \sigma_i dW_i(t), \tag{16}$$

where g_i is the membrane leak conductance and $dW_i(t)$ is an increment of a Wiener process. I_{ij} is an influence from the jth neuron to the ith neuron, which is often assumed to be

$$I_{ij}(t) = \sum_{\{f:t_j^{(f)}<t\}} \kappa_{ij}(t - t_i^{(f)}), \tag{17}$$

where $\kappa_{ij}(s)$ represents the effect of the jth neuron to the ith neuron after a time delay s and $t_j(f)$ is the fth spike of the jth neuron.

In the hard-threshold version (Paninski et al., 2004), whenever $V_i(t)$ goes over a threshold Vth, the neuron generates a spike. In the soft-threshold case Isomura et al. (2015), Koyama and Paninski (2010), the probability for the neuron to generate a spike in a small time interval dt is given by

$$\Pr\{\text{a spike in } [t, t+dt)\} = f(V(t))dt, \tag{18}$$

where $f(\cdot)$ is a non-negative intensity function.

When $\sigma_i(t)=0$ in Eq. (16), the solution is given by

$$V_i(t) = V_i(0)e^{-g_i t} + \sum_j \int_{-\infty}^{t} \exp(-g_i(t-s)) I_{ij}(t) ds. \tag{19}$$

If we assume that $\kappa_{ij}(\cdot)$ is the Dirac's delta function and that t is much greater than $1/g_i$. Then, Eq. (19) can be approximated by

$$V_i(t) = \sum_j \sum_{\{f:t_j^{(f)}<t\}} \exp(-g_i(t - t_j^{(f)})). \tag{20}$$

By choosing $f(V) \propto \exp(-\beta(V-V_{th}))$ and discretizing the time so that $dt=1$, the combination of Eqs. (20), (18) reduces to the GLM (15) with $A_{i0}=-\beta Vt_h$, $A_{ij}(k)=\beta\exp(-g_i k)$, and exponential inverse link function $\phi(\cdot)=\exp(\cdot)$. For more detailed discussion, Gerstner & Kistler (2002), Gerstner et al. (2014), Paninski, Pillow, & Lewi (2007).

Network Likelihood Models

Network likelihood models (NLMs) Kim et al. (2011), Okatan et al. (2005), Stevenson et al. (2009) are often adopted as a generative model for spike-train data in a continuous-time domain. Let $N_i(t)$ be the total number of the ith neuron's spikes observed before time t. In NLMs, it is assumed that $N_i(t)$follows an inhomogeneous Poisson process $\text{Poisson}(\cdot|\lambda_i(t))$ with the conditional intensity function

$$\lambda_i(t) = \exp\left(A_{i0} + \sum_{j=1}^{P}\sum_{k=1}^{K} A_{ij}(k) I_{jk}(t)\right), \tag{21}$$

where

$$I_{jk}(t) = \int_0^t \xi_k(t-s) N_{j'}(s)ds \tag{22}$$

is a convolution of a linear filter $\xi_k(\cdot)$ and a spiking history of the jth neuron. A typical example of the linear filter is a rectangular window of duration Wdefined as

$$\xi_k(u) = \begin{cases} 1 & (\text{if } u \in [(k-1)W, kW)) \\ 0 & (\text{otherwise}). \end{cases} \tag{23}$$

NLMs can be converted into GLMs if the spiking history $N_i(t)$ is discretized into bins with a fixed window W, and W is so small that any bin has at most a single spike.

Calcium Fluorescence Model

When $x_i(t)$ is a continuous variable indicating the intensity of calcium fluorescence, we have to consider that the fluorescence signal has a fast initial rise upon an action potential followed by a slow decay. A standard model to reflect this feature is as follows (Mishchenko et al., 2011):

$$z_i(t+1) = (1-\alpha_i)z_i(t) + s_i(t) \tag{24}$$

$$x_i(t) = a_i z_i(t) + b_i + \zeta_i(t), \quad \zeta_i(t) \sim \mathcal{N}(0, \tau), \tag{25}$$

where $z_i(t)$ denotes the intracellular calcium concentration and $s_i(t)$ is a spike indicator variable such that $s_i(t)=1$ if the ith neuron fires at time t and $s_i(t)=0$ otherwise. $\theta=\{\alpha_i,a_i,b_i,\tau_i|i=1,\ldots,P\}$ comprises the parameters of the model. Nonlinearities such as saturation can also be modeled as proposed by Vogelstein et al., (2009). In general, we cannot directly observe the variable $s_i(t)$ in this setting. To cope with this issue, Eqs. (24), (25) are often combined with Eqs. (15a), (15b) in which $x_i(t)$ is replaced by $s_i(t)$.

Hawkes Process Model

Linderman and Adams (2014), Linderman and Adams (2015) propose a Hawkes multivariate process to model a network of purely excitatory neurons. The model is considered as a generalized linear model (Section 5.1.2) with Poisson observations and a linear link function. The network model is defined by a binary matrix that indicates the existence of a directed connection between two neurons, a second weight matrix to represent the strength of each connection, and a vector specifying the transmission delay distribution for each directed connection.

Linderman & Adams (2015) enhanced the computational performance of their former approach (Linderman & Adams, 2014) with a new discrete time model that assumes that neurons do not interact faster than an arbitrary time unit and a variational approximation of the former Gibbs sampling solution in order to make the model fully conjugate.

Dynamic Bayesian Network

Dynamic Bayesian Networks (DBN) (Murphy, 2002) extend Bayesian Networks for time-series modeling. A DBN is usually defined as a directed acyclic graph where nodes represent random variables at particular times, and edges indicate a conditional probability dependence $P(x_{i,t}|x_j,t-k)$where $x_{i,t}$ denotes the ith neuron's activity at that time t.

Eldawlatly et al. (2010) demonstrated the feasibility of using DBM to infer the effective connectivity between spiking cortical neurons simulated with a generalized linear model (see Section 5.1.2). They use a simulated annealing algorithm to search the space of network structures and conditional probability parameters. Patnaik et al. (2011) present a significantly faster fitting algorithm that identifies, for each node, parent-sets with mutual information higher than a user-defined threshold. An upper bound of the Kullback–Leibler divergence between the inferred distribution and the true distribution is defined as a function of the user-defined threshold.

Maximum Entropy Model

The maximum-entropy model Roudi et al. (2015), Yeh et al. (2010) assumes that the network state probability distribution is given by an exponential function of the network energy $E[S_i]=E[(x_1,\ldots,x_n)]$ such that entropy is maximized while satisfying any statistical constraints. When first and second-order statistics are given, the state probability distribution is given by

$$\begin{aligned} P(x_1,\ldots,x_n) &= \frac{1}{Z}\exp(-E[(x_1,\ldots,x_n)]) \\ &= \frac{1}{Z}\exp\Big(\sum_i b_i x_i + \sum_{i\neq j} J_{ij}x_i x_j\Big). \end{aligned} \qquad (26)$$

This is an extension of the Ising model (McCoy, 2010) with spatial connections potentially occurring between any neurons as well as temporal correlations.

The main limitation of the maximum entropy model is its computational complexity. Recent studies demonstrated its applicability to a few tens of neurons using only first and second-order statistics without temporal correlations (Yeh et al., 2010).

Estimation of Model Parameters

Maximum Likelihood Method

The standard method to determine model parameters is the maximum likelihood (ML) method. The likelihood of a parameter vector θ given a data set D is defined as $p(D|\theta)$, the probability of reproducing the data. Here we denote the negative log-likelihood function as $J(\theta|D)=-\log p(D|\theta)$. In the ML method, parameters are chosen such that

$$\theta^* = \underset{\theta}{\operatorname{argmin}} J(\theta|D).$$

When the generative model uses a Gaussian distribution for observed data, the ML method reduces to *the least squares method.* For example, in the case of the AR model (12), maximization of the log likelihood is equivalent to minimization of *the sum of squared residuals*

$$J(\theta|D) = \sum_{i=1}^{P}\sum_{t=1}^{T}\epsilon_i^2(t)$$
$$= \sum_{i=1}^{P}\sum_{t=K+1}^{T}\left[x_i(t) - A_{i0} - \sum_{j=1}^{P}\sum_{k=1}^{K}A_{ij}(k)x_j(t-k)\right]^2. \quad (27)$$

where $D \equiv \{x_i(t)|i=1,\ldots,P;t=1,\ldots,T\}$ denotes the observed data set.

This optimization can be achieved analytically in the case of least square problem, or in general by iterative optimization algorithms such as the gradient descent methods. When the model is represented using a set of hidden variables, a standard way is to use the expectation–maximization (EM) algorithms.

Regularization and Bayesian Inference

A disadvantage of the ML method is that it often suffers from overfitting when the number of parameters is large relative to the amount of data. A common way to deal with this issue is to introduce regularization to the parameters. From a Bayesian statistical viewpoint, it assumes a prior distribution for the parameters. This makes explicit the assumption for connection inference, which is necessary due to the ill-posed nature of the inverse problem.

The objective function with a regularization term is given as

$$\theta^* = \underset{\theta}{\operatorname{argmin}}\{J(\theta|D) + \lambda\mathcal{R}(\theta)\},$$

where $\mathcal{R}(\cdot)$ is a non-negative function and λ is a constant that controls the strength of regularization. The most common regularizer is L2-norm regularizer, or the Ridge regularizer,

$$\mathcal{R}(\theta) = \sum_r \theta_r^2, \quad (28)$$

where r indexes the element of the set of parameters. In the Bayesian framework, this is equivalent to assuming a Gaussian prior distribution for the parameters. Another common regularizer is the L1-norm regularizer, or least absolute shrinkage and selection operator (LASSO) regularizer,

$$\mathcal{R}(\theta) = \sum_r |\theta_r|, \tag{29}$$

which favors a sparse solution with many parameters being zero. From a Bayesian viewpoint, this is equivalent to assuming a Laplace (exponential in absolute value) prior distribution.

Besidesintroducing regularization, Bayesian inference has several advantages over the ML method. For example, if the model has hidden variables, Bayesian inference is often used for estimating them as well as the model parameters (Mishchenko et al., 2011). Also, the reliability of each value in the parameter space can be evaluated as the posterior distribution. The high density region of the posterior distribution is called the Bayesian credible region, which can be used as an alternate to the confidence interval in the statistical test (Bowman, Caffo, Bassett, & Kilts, 2008). Another advantage is that the marginal likelihood, defined by the Bayes' theorem, can be used for a criterion to select the best of several possible models (Friston & Penny, 2011) (Section 2.5).

Approximate Bayesian Inference Methods

While Bayesian inference offers favorable features as above, exact computation for the posterior distribution and the marginal likelihood is often analytically intractable. Thus in practice it is important to select an approximation method. In the following, we briefly review approximation methods for the Bayesian inference. Details of their derivations and algorithms are beyond the scope of this paper and can be found in other articles Chen (2013), Murphy (2012).

1. Monte Carlo Sampling approximates the target distribution (or value) as an aggregation (or average) of the finite number of random samples. Gibbs sampling (Casella & George, 1992), Metropolis–Hastings (Hastings, 1970) and sequential importance re/sampling (Liu & Chen, 1998) algorithms are typical examples of this class.
2. Laplace Approximation (Raftery, 1996) is a deterministic approximation applicable to cases in which the MAP estimator can be easily obtained. The central idea is to approximate the target distribution as a multivariate Gaussian distribution using the Hessian matrix (the second-order partial derivatives) of the logarithm of the posterior distribution; thus, the Laplace approximation is not suitable for cases in which the posterior

distribution is multi-modal or asymmetric.

3. Variational Bayes (Attias, 1999) is another deterministic approximation method derived from the variational mean-field theory in statistical mechanics. Let us consider a case in which we want to obtain the posterior distribution p(Z|D), where Z=(z_1,…,z_m) is a set of all unknown variables (i.e. hidden variables and model parameters) such that zi (i=1,…,m) is a partition of Z. Also, consider the probability distribution family of Z that can be factored as q(Z)= $\prod_{i=1}^{M} q(z_i)$. In the variational Bayes method, we approximate p(Z|D) by optimizing q(Z) to minimize the KL divergence between q(Z) and p(Z|D). Optimization is achieved by an algorithm similar to the expectation and maximization (EM) algorithm. Variational Bayes is suitable for cases in which the joint distribution p(D,Z) is an instance of the exponential families (Wainwright & Jordan, 2007).

CASE STUDY

In this section, we review several examples of connectivity inference studies, from both model-free and model-based approaches. We focus on challenges addressed by each study at each stage of the data processing pipeline (Fig. 1).

Model-free Approaches

The First Neural Connectomics Challenge (Guyon et al., 2014) encouraged development of very diverse solutions. Remarkably, the top three solutions were all model-free methods. The organizers provided simulated neural network data sets for training and testing, as well as key references about the problem and sample code to get started. Neural network data sets were simulated using leaky integrate-and-fire neurons with short term synaptic depression implemented in the NEST simulator and spike trains were transformed to fluorescence time-series using a fluorescence response model of calcium markers inside neurons. Each data set consisted of a one-hour time-series of neural activities obtained from fluorescence signals sampled at 20 ms intervals with values normalized in the interval [0,1]. Organizers also provided information about the position of each neuron in an area of 1mm^2 and inter-neuron connectivity labels (i.e. excitatory connection or no connection). A validation data set was provided for performance comparison by the Area Under the ROC Curve (AUC; see Section 2.5). Finally, the score on a separated test data set was used for the final ranking of teams.

The three best solutions and the baseline method provided by challenge organizers all consisted of several components: data pre-processing, main connectivity inference, and post-processing steps as discussed in Section 2(Fig. 2). Below we review these top-ranked methods.

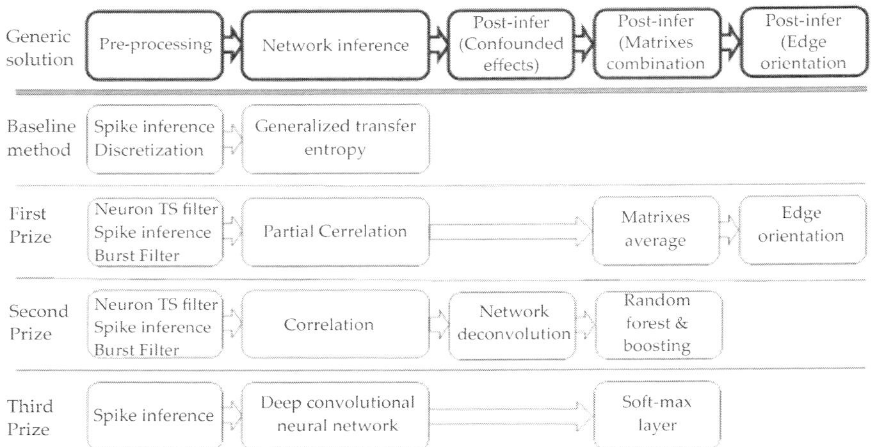

Figure 2: Summary of top solutions of the First Neural Connectomics Challenge. The baseline method and the three winning methods are outlined according to their pre-processing, main connectivity inference, and post-processing steps. Note that not all methods implement all steps.

Partial-correlation and Averaging

The first prize solution (Sutera et al., 2014) achieved a performance of 0.94161 (AUC score). Its pre-processing consisted of four steps: (i) low-pass filtering to remove noise, excessive fluctuations, and light scattering artifacts, (ii) backward differentiation for spike detection, (iii) hard thresholding to eliminate small values below a positive parameter τ, and (iv) weighting spike importance as a function of the inverse of network overall activity. The motivation for this last step was to eliminate effects of synchronized bursting periods, much as generalized transfer entropy does (see Section 4.2).

The main connectivity inference employed the Partial-Correlation (PC) matrix (Section 4.1), using the real valued signal computed in previous pre-processing steps. Post-processing consisted of two steps: (i) Several PC matrices were averaged according to different τ's and different low pass filters to increase the robustness of the solution. (ii) The symmetric average PC matrix was transformed into a directional indicator by multiplying each edge by an orientation weight computed according to the average activation delay between each neuron pair.

The main difficulty in scaling this solution to larger neural networks is the computation burden for calculating the inverse matrices. As an example, computation of the challenge solution connection matrix with 1000 neurons took 30 h on a 3 GHz i7 desktop PC with 7 GB of RAM.

Matrix Deconvolution and Adaptive Averaging

The major features of the second prize solution (Magrans & Nowe, 2014) are that it used matrix deconvolution for eliminating indirect connections and introduced learning to optimally combine several connectivity matrices rather than just using simple averaging. The pre-processing pipeline consisted of three steps: (i) spike train inference based on the OOPSI algorithm (Vogelstein, 2009), (ii) hard thresholding according to a parameter value τ, as in the first prize solution, and (iii) removal of time segments with overall network activity above a given threshold θ.

Connectivity inference employed Pearson correlation between the real valued signal computed from pre-processing. A network deconvolution algorithm (Feizi et al., 2013) was then used to eliminate the combined effect of indirect paths. If the effect of direct connection is given by matrix W_d, under the assumption of linearity, combined effects of all direct and indirect connections follow

$$W_o = W_d + W_d^2 + W_d^3 + \cdots = W_d(I - W_d)^{-1}. \tag{30}$$

From this relationship, the matrix deconvolution method estimates the direct connection matrix W_d from the observed connections matrix W_o by

$$W_d = W_o(I + W_o)^{-1}. \tag{31}$$

Finally, connectivity matrices computed according to different values of τ and θ are combined, but unlike the first prize solution, using a function learned with a non-linear regression method for optimal performance.

Computation of the challenge solution took nearly 48 h on a 3 GHz i7 desktop PC with 32 GB of RAM. The most significant limitation of this solutions is its high computational cost. It does not try to infer self-connections. It does not try to identify the causal direction. It assumes that both training and testing networks have similar statistical properties.

Convolutional Neural Network Approach

The third prize solution (Romaszko, 2014) proposed to automatically extract features from binary spike train pairs using a Convolutional Neural Network (CNN). Pre-processing consisted of three steps: (i) spike train inference using discrete differentiation, (ii) hard thresholding as in the first prize solution, and (iii) spike train normalization in [0,1]. Feature extraction was done with a CNN based on Lenet5 (LeCun, Bottou, Bengio, & Haffner, 1998) followed by a softmax layer that produced binary output in the event of a connection. This method was not designed to evaluate directions of connections.

Input data consisted of binary matrices of 3 by 330 where rows 1 and 2 corresponded to segments of the spike trains of two neurons. Row 3 contained information about the overall network activity, making it possible to identify synchronous bursting episodes. Network training was done with gradient descent.

The computational cost for this solution is very high. For instance, the challenge solution took more than 3 days on 8 server machines working in parallel, where each machine was equipped with 32 GB of RAM and GPU unit with 2496 cores. In addition to the extremely high computational cost, this solution has the same limitations as the second prize solution.

After the Connectomics Challenge, Veeriah et al. (2015) further advanced the CNN approach by integrating both the pre-processing step of spike train inference and connectivity inference into a single neural network architecture to surpass the first prize performance using the same data set. Its architecture consisted of two sub-networks and a final classification layer: (i) a convolutional neural network with max-pooling layers responsible for identifying relevant shapes in the fluorescence time-series with tolerance for minor time translations, (ii) a recurrent neural network (RNN) to model temporal sequences of relevant events. These are duplicated to capture the features of each neuron pair. Finally, a dynamically programmed layer aligned the RNN outputs and computed a connection probability for each neuron pair. Remarkably, this method does not require a separate pre-processing step to handle the synchronized bursting phenomenon.

Although the concept of an end-to-end artificial neural network is appealing, the performance comparison in Veeriah et al. (2015) deserves further consideration. Their reconstructed performances for generalized transfer entropy and partial correlation methods are lower than the performance documented during the First Neural Connectomics Challenge using the same data sets Guyon et al. (2014), Orlandi et al. (2014), Sutera et al. (2014).

While these CNN-based approaches presented good performance with simulated data, it remains to be seen whether such classifiers generalize well to real data, for which ground truth training data are rarely available.

Other Model-free Approaches

Beyond the main results of the First Neural Connectomics Challenge, several *in vivo* and *in vitro* studies rely on model-free methods. Fujisawa et al. (2008) use silicon microelectrode arrays for *in vivo* recording from layers 2, 3 and 5 of the medial prefrontal cortex in rats during a working memory task. Connectivity inference is performed by identifying sharp spikes or troughs in the cross-correlograms. Connectivity significance is assessed by first creating several sets of slightly perturbed spike trains and then computing the statistical significance of the original cross-correlograms with respect to those constructed with the perturbed data.

Cohen & Kohn (2011) discuss a recent surge in cortical processing studies enabled by recording technologies like microelectrode arrays and calcium imaging (see Section 2.5.2). They summarize possible causes of discrepant findings in several correlation-based studies due to technical reasons, i.e., poor spike sorting, experimental factors like different external stimuli, different time bins and/or spike train durations, or confounded effects due to non-stationary internal states. They suggest that multivariate point process models provide a more complete and statistically principled approach to cope with the challenges of inferring the connectivity structure.

Model-based Approaches

Generalized Linear Models

Pillow and colleagues applied GLM frameworks for a stochastic integrate-and-fire neuron model (Paninski et al., 2004) to data of a complete neural recording from ganglion cells in a retina slice (Pillow et al., 2008). They constructed a model including both stimulus response and cross-connections of all neurons and showed through model-based analysis that correlated firing can provide additional sensory information.

Stevenson and colleagues derived a Bayesian inference framework for a spiking neuron model with arbitrary synaptic response kernels (Stevenson et al., 2009). They applied the method to neural recording data from monkey motor cortex and used the log likelihood and correlation coefficient criteria for cross validation. They further used infinite relational model clustering (Kemp, Tenenbaum, Griffiths, Yamada, & Ueda, 2006) to detect cluster structures among recorded neurons.

While Stevenson et al. (2009) used arbitrary synaptic kernel functions with discrete time bins, Song and colleagues (Song et al., 2013) proposed using a set of kernel functions, such as B-splines, with which smoothness of kernel functions is built-in with a relatively small number of parameters.

Oba et al. (2016) propose a GLM framework combined with empirical Bayesian testing. Null samples are created by time-shifting real spike trains for a sufficiently large time lag. They show improved computational performance without decreasing inference performance on a simulated neural network. They also apply this method to real calcium imaging data from a cultured slice of the CA3 region of rat hippocampus.

Calcium Fluorescence Model

In applying a connection estimation method to calcium fluorescent imaging data, a basic way is to estimate spike trains from fluorescence waveforms by spike deconvolution and then to apply a spike-based connectivity inference algorithm. Mishchenko et al. (2011) proposed a framework combining both steps of spike estimation and connection estimation into a unified stochastic generative model. Fletcher & Rangan (2014) proposed a computationally less expensive variant of this approach.

DISCUSSION

Challenges and Solutions

While model-free methods usually address pre- and post-processing separately from connectivity inference, model-based methods often address issues such as noise and connection directionality in the main inference process by incorporating those factors into generative models.

The distinction between apparent and real connections, as well as their directionality, is incorporated into neural dynamics equations in model-based methods. Weights most consistent with data are selected by the maximum likelihood or maximum posterior probability criterion. Among model-free methods, transfer entropy methods address connection directionality directly during the inference, while removal of apparent connections can be addressed by matrix deconvolution in post-processing (Magrans & Nowe, 2014). Partial-Correlation can also remove apparent connections but is not able to identify the causal direction. Sutera et al. (2014) propose a post-processing step to identify the causal direction. Temporal resolution of calcium imaging imposes severe performance limitations to these methods. Better solutions to identify the direction and to infer apparent connections should involve improving both algorithms and instrumentation. Table 3summarizes solutions that were devised to overcome challenges along the data processing pipeline.

Table 3. Summary matrix of the solutions that were devised to overcome each of the challenges. We can observe that each solution was implemented either as a pre/pro-processing step or within the main inference method

Challenges	**Pre-processing**	**Connection inference**		**Post-processing**
		Model-free	**Model-based**	
Apparent connection		**Sutera et al. (2014) and Veeriah et al. (2015)**	**Pillow et al. (2008), Stevenson et al. (2009), Mishchenko et al. (2011) and Fletcher and Rangan (2014)**	**Magrans and Nowe (2014)**
Directionality		Stetter et al. (2012), Garofalo et al. (2009), Ito et al. (2011) and Veeriah et al. (2015)	Pillow et al. (2008), Stevenson et al. (2009), Mishchenko et al. (2011) and Fletcher and Rangan (2014)	Sutera et al. (2014)
Cellular diversity				
Synaptic diversity			Pillow et al. (2008), Stevenson et al. (2009), Mishchenko et al. (2011) and Fletcher and Rangan (2014)	

Non-stationarity	Cohen and Kohn (2011), Sutera et al. (2014), Magrans and Nowe (2014)and Garofalo et al. (2009)	Wollstadt, Martínez-Zarzuela, Vicente, Díaz-Pernas, and Wibral (2014), Stetter et al. (2012) and Veeriah et al. (2015)	Linderman, Stock, and Adams (2014)	
Noise	Cohen and Kohn (2011), Sutera et al. (2014), Magrans and Nowe (2014), Wollstadt et al. (2014), Stetter et al. (2012) and Garofalo et al. (2009)	Veeriah et al. (2015)	Pillow et al. (2008), Stevenson et al. (2009), Mishchenko et al. (2011) and Fletcher and Rangan (2014)	
Time/Space resolution		Veeriah et al. (2015)	Mishchenko et al. (2011) and Fletcher and Rangan (2014)	
Hidden neuron/External input			Kulkarni and Paninski (2007), Vidne, Ahmadian, Shlens, Pillow, Kulkarni, Litke, Chichilnisky, Simoncelli, and Paninski (2012) and Rezende, Wierstra, and Gerstner (2011)	
Prior knowledge			Pillow et al. (2008), Stevenson et al. (2009), Mishchenko et al. (2011) and Fletcher and Rangan (2014)	

Scalability		Sutera et al. (2014) and Magrans and Nowe (2014)	Fletcher and Rangan (2014)	

Non-stationarity

Modeling synaptic plasticity is a key to understand how memory and learning mechanisms are realized in neural circuits. Recent solutions to this challenge incorporate synaptic dynamics into a GLM framework and improved scalability within Bayesian (Linderman et al., 2014) and convex optimization settings (Stevenson & Koerding, 2011). Using a model-free approach to understand evolution of synaptic weights, Wollstadt et al. (2014) proposed a transfer entropy estimator that requires an ensemble of spike trains from equivalent and independent experiments.

Synaptic plasticity is not the sole source of non-stationarity. Time periods when a large percentage of neurons fire at a high rate is an additional non-stationary phenomenon and a source of confusion for most network inference methods. Generalized transfer entropy (Section 4.2.6), screens out the use of non-informative time periods. A similar selection could be developed for other model-free methods.

In Section 6.1 we also discussed how winning solutions of the First Neural Connectomics Challenge chose a modular approach with a dedicated pre-processing step to remove non-informative time periods before applying different connectivity indicators. We further discussed a deep neural networksolution by Veeriah et al. (2015) that surprisingly, despite the lack of any pre-processing step or any specific design feature to remove high-rate time periods, is able to achieve improved performance with respect to the winning solution. A generalization of the above bi-modal switching behavior can also be accommodated with a model-based approach using a hidden Markov model (Linderman, Johnson, Datta, & Adams, 2015).

Architecture

Pre/post-processing methods have the potential to improve network reconstruction performance across different methods. For instance, modular architectures like the top solutions of the connectomics challenge (Section 6.1) implement a number of pre-processing steps to reduce noise and to infer spike trains. However, in applying such modular approaches, care has to be taken not to lose valuable information about the underlying direct connectivity and the directionality. On the other side of the spectrum, the approach using a deep neural network (Veeriah et al., 2015) is unique in providing solutions for multiple pre-processing and post-processing steps within a coherent model-free architecture.

In the model-based approach, Mishchenko et al. (2011) and later Fletcher & Rangan (2014) employed time-series of neural activities obtained from fluorescence signals as inputs, while spike times at an arbitrary sampling rate were additional latent variables. Table 3 summarizes solutions that were devised to overcome each of the challenges. We can observe that each solution was implemented either as a pre/pro-processing step or within the main inference method.

Scalability

A major drawback to model-based methods is their scalability. The more sophisticated the model, the more parameters need to be estimated, which requires more data sampled from a stationary distribution. Prior knowledge, such as sparseness in connections, can be addressed in model-based methods by assuming a prior distribution in the model parameters and by applying Bayesian inference. Introduction of a sparseness prior, or an equivalent regularization term, can make inference from smaller samples more reliable, but this often increases the computational burden.

Hidden Neurons

While neural recording methods are progressing rapidly, allowing whole-brain imaging with cellular resolution in small animals like *Caenorhabditis elegans* (Nguyen et al., 2016) and zebra fish (Ahrens et al., 2013), monitoring all neurons in the brain is still not possible. In typical calcium fluorescent imaging, neurons in only one section of a cortical column are recorded. Neglecting hidden neurons within the network, or unobservable external inputs, can infer erroneous connections. Methods for addressing this hidden input problem are still in early stages of development (Roudi et al., 2015).

The issue of hidden nodes, however, is not the only problem in neural connectivity inference. Similar problems have been addressed in gene regulation and signaling cascade networks, for example, and methods developed in other fields of computational biology and network science may provide helpful intuition and guidance (Su, Lai, Wang, & Do, 2014).

The challenge of hidden nodes groups different application scenarios under the same name, but the common denominator here is a generative model that consists of observed and hidden components. GLM (Section 5.1.2) and maximum entropy settings (Section 5.1.8) are the most common approaches to describe the observed neurons. However, the key differential contribution in all cases is the hidden neurons model. A single, latent process can model hidden inputs as random effects, enhancing network inference among observed neurons by avoiding false connections due to common hidden inputs Kulkarni and Paninski (2007), Vidne et al. (2012). A more sophisticated group of contributions proposes multiple latent processes, for instance to study the feasibility of biophysically plausible hypotheses about how multi-level neural circuits are able to learn and express complex sequences (Rezende et al., 2011). Switching behavior, described in the previous section, can also be seen as a special case of the hidden nodes challenge.

Incorporating Prior Knowledge

Detailed anatomical synaptic maps (Yook, Druckmann, & Kim, 2013), and connectivity maps between cell types across cortical layers (Potjans & Diesmann, 2014) are valuable sources of prior information that should be exploited to improve both inference and computational performance. A straightforward, model-based approach would be to exploit anatomical prior information using more sophisticated regularizers like the adaptive elastic net (Zou & Zhang, 2009), by embedding anatomical information in the adaptive weights instead of computing them using ordinary least squares (Wu et al., 2016). Prior information could also be incorporated as a post-processing method using graph sparsification algorithms to preserve certain graph theoretical properties Ebbes et al. (2008), Lindner et al. (2015).

A possible way to deal with hidden neurons when analyzing cortical microcircuits is to take advantage of the highly replicated structure across layers and cortical columns. A possible framework to implement this idea, proposed by Kim and Leskovec (2011), infers parameters of a Kronecker model. In other words, a low-dimensional connectivity graph that approximately self-replicates across observed and the hidden parts. Within this application, exploitation of prior knowledge could enhance both computational and inferential performance.

The network inference problem could also be considered asa process of selecting a sub-set of pre-synaptic input neurons for each neuron. Therefore, it would be reasonable to explore application of feature selection algorithms

(Liu & Motoda, 2012) to the problem at hand. From this point of view, multi-task feature selection algorithms Obozinski et al. (2006), Wang et al. (2016), Zhou et al. (2010) may be another interesting way to take advantage of the highly replicated structure across layers and cortical columns. These algorithms propose clever ways to jointly learn features across several tasks, maximizing information sharing while minimizing negative transfer. Therefore, application of these methods has the potential to deliver superior inference and computational performance with respect to single-task learning approaches while using the same amount of data.

A common assumption of many inference methods is a simplistic model structure that does not represent the true synaptic and cellular diversity of local neural microcircuits (see Section 3.1). Future research methods should aim for approaches able to fit more biophysically plausible models. Recent wide-field calcium imaging of thousands of neurons over millimeters of brain tissue (Mohammed, Gritton, Tseng, Bucklin, Yao, & Han, 2016), the large number of latent parameters to model hidden neurons and non-stationarity, are all sources of increasing computational complexity and a strong impetus to continue improving the computational efficiency of network inference methods (Lee, Lim, & Ong, 2016).

CONCLUSION

In this paper, we reviewed methods for inference of neural connectivity based on activity recordings from a large number of neurons. We first identified biophysical and technical challenges along the data processing pipeline and then formulated model-free and model-based approaches for the core process of connectivity inference. We further investigated how previous work addressed those challenges using various methods. As a result, we identified favorable methods that deserve further technical developments, most notably coping with hidden neurons.

Connectivity inference itself is an interesting and deep mathematical problem, but the goal of connectivity inference is not only to precisely estimate the connection weight matrix, but also to illustrate how neural circuits realize specific functions, such as sensory inference, motor control, and decision making, and to understand baseline brain dynamics, upon which those function would be based. If we can perfectly estimate network connections from anatomical and activity data, then computer simulation of the network model should be able to reproduce network function as well as resting-state dynamics. But given inevitable uncertainties in connectivity inference, reconstruction of functions in a purely data-driven way might be difficult. How to extract or infer a functional or computational network from a data-driven network, or even to combine known functional constraints as a prior for connectivity inference, is a possible direction of future research.

ACKNOWLEDGMENTS

This research was supported by the program for Brain Mapping by Integrated Neurotechnologies for Disease Studies (Brain/MINDS) from Japan Agency for Medical Research and development (AMED) under the Grant Number JP17dm0207030, KAKENHI Grant 16H06563 from Ministry of Education, Culture, Sports, Science and Technology (MEXT), and internal funding from the Okinawa Institute of Science and Technology Graduate University . We thank Steven D. Aird for editing the manuscript.

REFERENCES

1. Aertsen, A., Gerstein, G., Habib, M., & Palm, G. (1989). Dynamics of neuronal firing correlation: modulation of "effective connectivity". Journal of Neurophysiology, 61(5), 900–917.
2. Aho, K., Derryberry, D., & Peterson, T. (2014). Model selection for ecologists: the worldviews of AIC and BIC. Ecology, 95(3), 631–636.
3. Ahrens, M. B., Li, J. M., Orger, M. B., Robson, D. N., Schier, A. F., Engert, F., et al. (2012). Brain-wide neuronal dynamics during motor adaptation in zebrafish. Nature, 485(7399), 471–477.
4. Ahrens, M. B., Orger, M. B., Robson, D. N., Li, J. M., & Keller, P. J. (2013). Whole-brain functional imaging at cellular resolution using light-sheet microscopy. Nature Methods, 10(5), 413–420.
5. Amunts, K., Ebell, C., Muller, J., Telefont, M., Knoll, A., & Lippert, T. (2016). The human brain project: Creating a european research infrastructure to decode the human brain. Neuron, 92(3), 574–581. URL http://www.sciencedirect.com/ science/article/pii/ S0896627316307966.
6. Ask, M., & Reza, M. (2016). Computational models in neuroscience: How real are they? A critical review of status and suggestions. Austin Neurology & Neurosciences, 1(2), 1008.
7. Attias, H. (1999). A variational Bayesian framework for graphical models. Advances in Neural Information Processing Systems (NIPS), 209–215.
8. Barnett, L., Barrett, A. B., & Seth, A. K. (2009). Granger causality and transfer entropy are equivalent for Gaussian variables. Physical Review Letters, 103(23), 238701.
9. Barzel, B., & Barabási, A.-L. (2013). Network link prediction by global silencing of indirect correlations. Nature biotechnology, 31(8), 720–725.
10. Bekkers, J. M. (2009). Synaptic transmission: excitatory autapses find a function? Current Biology, 19(7), R296–8.
11. Bonifazi, P., Goldin, M., Picardo, M. A., Jorquera, I., Cattani, A., Bianconi, G., et al. (2009). Gabaergic hub neurons orchestrate synchrony in developing hippocampal networks. Science, 326(5958), 1419–1424.

12. Bowman, F. D., Caffo, B., Bassett, S. S., & Kilts, C. (2008). A Bayesian hierarchical framework for spatial modeling of fMRI data. NeuroImage, 39(1), 146–156.
13. Brette, R., & Destexhe, A. (2012). Intracellular recording. In Handbook of neural activity measurement. Cambridge University Press.
14. Bullmore, E., & Sporns, O. (2009). Complex brain networks: graph theoretical analysis of structural and functional systems. Nature Reviews Neuroscience, 10(3), 186–198.
15. Buzsaki, G. (2004). Large-scale recording of neuronal ensembles. Nature Neuroscience, 7(5), 446–451. URL https://www.ncbi.nlm.nih.gov/pubmed/15114356, http://www.nature.com/neuro/journal/v7/n5/pdf/nn1233.pdf.
16. Casella, G., & George, E. I. (1992). Explaining the Gibbs sampler. The American Statistician, 46(3), 167–174.
17. Chen, Z. (2013). An overview of Bayesian methods for neural spike train analysis. Computational Intelligence and Neuroscience, 2013, 1.
18. Chiappalone, M., Vato, A., Berdondini, L., Koudelka-Hep, M., & Martinoia, S. (2007). Network dynamics and synchronous activity in cultured cortical neurons. International Journal of Neural Systems, 17(02), 87–103.
19. Churchland, P. S., & Sejnowski, T. J. (1992). The computational brain. MIT Press. Cohen, M. R., & Kohn, A. (2011). Measuring and interpreting neuronal correlations. Nature Neuroscience, 14(7), 811–819.
20. Czarnecki, W. M., & Jozefowicz, R. (2014). Neural connectivity reconstruction from calcium imaging signal using random forest with topological features. In ECML workshop - "Neural connectomics: From imaging to connectivity". Nancy, France.
21. Destexhe, A., & Bedard, C. (2013). Local field potential. Scholarpedia, 8(8), 10713. Revision number 135811.
22. Dombeck, D. A., Graziano, M. S., & Tank, D. W. (2009). Functional clustering of neurons in motor cortex determined by cellular resolution imaging in awake behaving mice. Journal of Neuroscience, 29(44), 13751–13760. URL http://www. ncbi.nlm.nih.gov/pubmed/19889987.
23. Ebbes, P., Huang, Z., Rangaswamy, A., Thadakamalla, H. P., & Unit, O. (2008). Sampling large-scale social networks: Insights from simulated

networks. In 18th annual workshop on information technologies and systems. Paris, France:

24. Citeseer. Eldawlatly, S., Zhou, Y., Jin, R., & Oweiss, K. G. (2010). On the use of dynamic bayesian networks in reconstructing functional neuronal networks from spike train ensembles. Neural Computation, 22(1), 158–189.
25. Feizi, S., Marbach, D., Médard, M., & Kellis, M. (2013). Network deconvolution as a general method to distinguish direct dependencies in networks. Nature biotechnology, 31(8), 726–733.
26. Fletcher, A. K., & Rangan, S.(2014). Scalable inference for neuronal connectivity from calcium imaging. In Advances in neural information processing systems (Vol. 27) (pp. 2843–2851).
27. Franaszczuk, P. J., Blinowska, K. J., & Kowalczyk, M. (1985). The application of parametric multichannel spectral estimates in the study of electrical brain activity. Biological Cybernetics, 51(4), 239–247.
28. Friston, K. J. (2011). Functional and effective connectivity: a review. Brain Connectivity, 1(1), 13–36.
29. Friston, K., & Penny, W. (2011). Post hoc Bayesian model selection. NeuroImage, 56(4), 2089–2099.
30. Fujisawa, S., Amarasingham, A., Harrison, M. T., & Buzsáki, G. (2008). Behaviordependent short-term assembly dynamics in the medial prefrontal cortex. Nature Neuroscience, 11(7), 823–833.
31. Fujishiro, A., Kaneko, H., Kawashima, T., Ishida, M., & Kawano, T. (2014). In vivo neuronal action potential recordings via three-dimensional microscale needleelectrode arrays. Scientific Reports, 4, 4868. URL http://www.ncbi.nlm.nih.gov/ pubmed/24785307.
32. Garofalo, M., Nieus, T., Massobrio, P., & Martinoia, S. (2009). Evaluation of the performance of information theory-based methods and cross-correlation to estimate the functional connectivity in cortical networks. PloS One, 4(8), e6482.
33. Gerstner, W., & Kistler, W. M. (2002). Spiking neuron models: Single neurons, populations, plasticity. Cambridge University Press.
34. Gerstner, W., Kistler, W. M., Naud, R., & Paninski, L. (2014). Neuronal dynamics: From single neurons to networks and models of cognition. Cambridge, UK: Cambridge Universty Press.

35. Gerwinn, S., Macke, J. H., & Bethge, M. (2010). Bayesian inference for generalized linear models for spiking neurons. Frontiers in Computational Neuroscience, 4.
36. Guyon, I., Battaglia, D., Guyon, A., Lemaire, V., Orlandi, J. G., Ray, B., et al. (2014). Design of the first neuronal connectomics challenge: From imaging to connectivity. In Neural networks, 2014 international joint conference on (pp. 2600– 2607). IEEE.
37. Harrison, L., Penny, W. D., & Friston, K.(2003). Multivariate autoregressive modeling of fMRI time series. NeuroImage, 19(4), 1477–1491.
38. Hastings, W. K. (1970). Monte Carlo sampling methods using Markov chains and their applications. Biometrika, 57(1), 97–109.
39. Hu, Y., Brunton, S. L., Cain, N., Mihalas, S., Kutz, J. N., & Shea-Brown, E. (2016). Feedback through graph motifs relates structure and function in complex networks. ArXiv preprint arXiv:1605.09073.
40. Insel, T. R., Landis, S. C., & Collins, F. S. (2013). The NIH brain initiative. Science, 340(6133), 687–688.
41. Isomura, T., Ogawa, Y., Kotani, K., & Jimbo, Y. (2015). Accurate connection strength estimation based on variational bayes for detecting synaptic plasticity. Neural Computation, 27(4), 819–844.
42. Ito, S., Hansen, M. E., Heiland, R., Lumsdaine, A., Litke, A. M., & Beggs, J. M. (2011). Extending transfer entropy improves identification of effective connectivity in a spiking cortical network model. PloS One, 6(11), e27431.
43. Jog, M., Connolly, C., Kubota, Y., Iyengar, D., Garrido, L., Harlan, R., et al. (2002).
44. Tetrode technology: advances in implantable hardware, neuroimaging, and data analysis techniques. Journal of Neuroscience Methods, 117, 141–152.
45. Kandel, E. R., Schwartz, J. H., Jessell, T. M., et al. (2000a). Principles of neural science. In The cells of the nervous system (Vol. 4). New York: McGraw-Hill.
46. Kandel, E. R., Schwartz, J. H., Jessell, T. M., et al. (2000b). Principles of neural science. In Overview of synaptic transmission (Vol. 4). New York: McGraw-Hill.
47. Kemp, C., Tenenbaum, J. B., Griffiths, T. L., Yamada, T., & Ueda, N.

(2006). Learning systems of concepts with an infinite relational model. In AAAI (Vol. 3) (p. 5.

48. Kim, M., & Leskovec, J. (2011). The network completion problem: Inferring missing nodes and edges in networks. In SDM (Vol. 11) (pp. 47–58). SIAM.
49. Kim, S., Putrino, D., Ghosh, S., & Brown, E. N. (2011). A granger causality measure for point process models of ensemble neural spiking activity. PLoS Computational Biology, 7(3), e1001110.
50. Knox, C. K. (1981). Detection of neuronal interactions using correlation analysis. Trends in Neurosciences, 4, 222–225.
51. Koyama, S., & Paninski, L.(2010). Efficient computation of the maximum a posteriori path and parameter estimation in integrate-and-fire and more general statespace models. Journal of Computational Neuroscience, 29(1–2), 89–105.
52. Kulkarni, J. E., & Paninski, L. (2007). Common-input models for multiple neural spike-train data. Network: Computation in Neural Systems, 18(4), 375–407.
53. Lang, E. W., Tomé, A., Keck, I. R., Górriz-Sáez, J., & Puntonet, C. (2012). Brain connectivity analysis: a short survey. Computational Intelligence and Neuroscience, 2012, 8.
54. LeCun, Y., Bottou, L., Bengio, Y., & Haffner, P.(1998). Gradient-based learning applied to document recognition. Proceedings of the IEEE, 86(11), 2278–2324.
55. Lee, Y., Lim, K. W., & Ong, C. S. (2016). Hawkes processes with stochastic excitations. In Proceedings of the 33rd international conference on machine learning (pp. 79– 88).
56. Linderman, S. W., & Adams, R. P. (2014). Discovering latent network structure in point process data. In ICML (pp. 1413–1421).
57. Linderman, S. W., & Adams, R. P. (2015). Scalable Bayesian inference for excitatory point process networks. ArXiv preprint arXiv:1507.03228.
58. Linderman, S. W., Johnson, M. J., Datta, S. R., & Adams, R. P. (2015). Discovering switching autoregressive dynamics in neural spike train recordings. In Computational and systems neuroscience (Cosyne).
59. Linderman, S., Stock, C. H., & Adams, R. P. (2014). A framework for studying synaptic plasticity with neural spike train data. In Advances in neural information processing systems (pp. 2330–2338).

60. Lindner, G., Staudt, C. L., Hamann, M., Meyerhenke, H., & Wagner, D. (2015). Structure-preserving sparsification of social networks. In Proceedings of the 2015 IEEE/ACM International conference on advances in social networks analysis and mining 2015 (pp. 448–454). ACM.
61. Liu, H., & Motoda, H.(2012). Feature selection for knowledge discovery and data mining (Vol. 454). Springer Science & Business Media.
62. Liu, J. S., & Chen, R. (1998). Sequential Monte Carlo methods for dynamic systems. Journal of the American Statistical Association, 93(443), 1032–1044.
63. Lizier, J. T. (2014). JIDT: an information-theoretic toolkit for studying the dynamics of complex systems. ArXiv preprint arXiv:1408.3270.
64. Lizier, J. T., Heinzle, J., Horstmann, A., Haynes, J.-D., & Prokopenko, M. (2011). Multivariate information-theoretic measures reveal directed information structure and task relevant changes in fMRI connectivity. Journal of Computational Neuroscience, 30(1), 85–107.
65. Looger, L. L., & Griesbeck, O. (2011). Genetically encoded neural activity indicators. Current Opinion in Neurobiology, 22, 18–23.
66. Lütcke, H., & Helmchen, F. (2011). Two-photon imaging and analysis of neural network dynamics. Reports on Progress in Physics, 74(8), 086602.
67. Magrans, I., & Nowe, A. (2014). Supervised neural network structure recovery. In ECML workshop -"neural connectomics: from imaging to connectivity". Nancy, France.
68. Mahmud, M., Pulizzi, R., Vasilaki, E., & Giugliano, M. (2015). Qspike tools: a generic framework for parallel batch preprocessing of extracellular neuronal signals recorded by substrate microelectrode arrays. Recent Advances and the Future Generation of Neuroinformatics Infrastructure, 248.
69. Markram, H. (2012). The human brain project. Scientific American, 306(6), 50–55.
70. Martin, C. L., & Chun, M. (2016). The {BRAIN} initiative: Building, strengthening, and sustaining. Neuron, 92(3), 570–573. URL http://www.sciencedirect.com/ science/article/pii/S0896627316307899.
71. McCoy, B. (2010). Ising model: exact results. Scholarpedia, 5(7), 10313. Revision number 123804.

72. Mishchenko, Y., Vogelstein, J. T., & Paninski, L. (2011). A Bayesian approach for inferring neuronal connectivity from calcium fluorescent imaging data. Annals of Applied Statistics, 5(2B), 1229–1261.
73. Mohammed, A. I., Gritton, H. J., Tseng, H.-a., Bucklin, M. E., Yao, Z., & Han, X. (2016). An integrative approach for analyzing hundreds of neurons in task performing mice using wide-field calcium imaging. Scientific Reports, 6.
74. Murphy, K. P. (2002). Dynamic bayesian networks. Probabilistic Graphical Models, M. Jordan, 7.
75. Murphy, K. P. (2012). Machine learning: a probabilistic perspective. MIT Press.
76. Nguyen, J. P., Shipley, F. B., Linder, A. N., Plummer, G. S., Liu, M., Setru, S. U., et al. (2016).Whole-brain calcium imaging with cellular resolution in freely behaving caenorhabditis elegans. Proceedings of the National Academy of Sciences, 113(8), E1074–E1081.
77. Oba, S., Nakae, K., Ikegaya, Y., Aki, S., Yoshimoto, J., & Ishii, S. (2016). Empirical Bayesian significance measure of neuronal spike response. BMC Neuroscience, 17(1), 1.
78. Obozinski, G., Taskar, B., & Jordan, M. (2006). Multi-task feature selection, Tech. rep. 2. Statistics Department, UC Berkeley.
79. Okano, H., & Mitra, P. (2015). Brain-mapping projects using the common marmoset. Neuroscience Research, 93, 3–7. URL http://www.ncbi.nlm.nih.gov/pubmed/ 25264372.
80. Okano, H., Sasaki, E., Yamamori, T., Iriki, A., Shimogori, T., Yamaguchi, Y., et al. (2016). Brain/minds: A japanese national brain project for marmoset neuroscience.
81. Neuron, 92(3), 582–590. URL http://www.sciencedirect.com/science/article/pii/ S089662731630719X.
82. Okatan, M., Wilson, M. A., & Brown, E. N. (2005). Analyzing functional connectivity using a network likelihood model of ensemble neural spiking activity. Neural Computation, 17(9), 1927–1961.
83. Orlandi, J. G., Ray, B., Battaglia, D., Guyon, I., Lemaire, V., & Saeed, M. (2014). First connectomics challenge: From imaging to connectivity. In ECML workshop "Neural connectomics: From imaging to connectivity". Nancy, France.
84. Orlandi, J. G., Stetter, O., Soriano, J., Geisel, T., & Battaglia, D. (2014). Transfer entropy reconstruction and labeling of neuronal connections

from simulated calcium imaging. PloS One, 9(6), e98842.

85. Paninski, L., Pillow, J., & Lewi, J. (2007). Statistical models for neural encoding, decoding, and optimal stimulus design. In T. D. Paul Cisek, & J. F. Kalaska (Eds.), Progress in brain research: Vol. 165. Computational neuroscience: Theoretical insights into brain function (pp. 493–507). Elsevier, URL http://www.sciencedirect. com/science/ article/pii/S0079612306650310.
86. Paninski, L., Pillow, J. W., & Simoncelli, E. P. (2004). Maximum likelihood estimation of a stochastic integrate-and-fire neural encoding model. Neural Computation, 16(12), 2533–2561.
87. Patnaik, D., Laxman, S., & Ramakrishnan, N. (2011). Discovering excitatory relationships using dynamic bayesian networks. Knowledge and Information Systems, 29(2), 273–303.
88. Pillow, J. W., Paninski, L., Uzzell, V. J., Simoncelli, E. P., & Chichilnisky, E. (2005). Prediction and decoding of retinal ganglion cell responses with a probabilistic spiking model. The Journal of Neuroscience, 25(47), 11003–11013.
89. Pillow, J. W., Shlens, J., Paninski, L., Sher, A., Litke, A. M., Chichilnisky, E. J., et al. (2008). Spatio-temporal correlations and visual signalling in a complete neuronal population. Nature, 454(7207), 995–999. URL http://www.ncbi.nlm. nih.gov/pubmed/18650810.
90. Pnevmatikakis, E. A., Soudry, D., Gao, Y., Machado, T. A., Merel, J., Pfau, D., et al. (2016). Simultaneous denoising, deconvolution, and demixing of calcium imaging data. Neuron, 89, 285–299.
91. Potjans, T. C., & Diesmann, M. (2014). The cell-type specific cortical microcircuit: relating structure and activity in a full-scale spiking network model. Cerebral Cortex, 24(3), 785–806.
92. Raftery, A. E. (1996). Approximate Bayes factors and accounting for model uncertainty in generalised linear models. Biometrika, 83(2), 251–266.
93. Rezende, D. J., Wierstra, D., & Gerstner, W. (2011). Variational learning for recurrent spiking networks. In Advances in neural information processing systems (pp. 136– 144).
94. Romaszko, L. (2014). Signal correlation prediction using convolutional neural networks. In ECML workshop - "Neural connectomics: From imaging to connectivity". Nancy, France.

95. Roudi, Y., Dunn, B., & Hertz, J. (2015). Multi-neuronal activity and functional connectivity in cell assemblies. Current Opinion in Neurobiology, 32, 38–44.
96. Schreiber, T. (2000). Measuring information transfer. Physical Review Letters, 85(2), 461.
97. Schrynemackers, M., Küffner, R., & Geurts, P. (2013). On protocols and measures for the validation of supervised methods for the inference of biological networks. Frontiers in Genetics, 4.
98. Shannon, I. C. (1948). A mathematical theori of cornm unifications. Bell System Technical Journal.
99. Shimono, M., & Beggs, J. M. (2014). Functional clusters, hubs, and communities in the cortical microconnectome. Cerebral Cortex, bhu252.
100. Smith, A. C., Fall, C. P., & Sornborger, A. T. (2011). Near-real-time connectivity estimation for multivariate neural data. In Engineering in medicine and biology society, EMBC, 2011 annual international conference of the IEEE (pp. 4721–4724).
101. Sohn, Y., Choi, M.-K., Ahn, Y.-Y., Lee, J., & Jeong, J. (2011). Topological cluster analysis reveals the systemic organization of the Caenorhabditis elegans connectome. PLoS Computational Biology, 7(5), e1001139.
102. Song, D., Wang, H., Tu, C., Marmarelis, V., Hampson, R., Deadwyler, S., et al. (2013). Identification of sparse neural functional connectivity using penalized likelihood estimation and basis functions. Journal of Computational Neuroscience, 35(3), 335–357.
103. Sporns, O. (2012). Discovering the human connectome. MIT press.
104. Sporns, O. (2013). The human connectome: origins and challenges. Neuroimage, 80, 53–61.
105. Sporns, O., Tononi, G., & Kötter, R. (2005). The human connectome: a structural description of the human brain. PLoS Computational Biology, 1(4), e42.
106. Stetter, O., Battaglia, D., Soriano, J., & Geisel, T. (2012). Model-free reconstruction of excitatory neuronal connectivity from calcium imaging signals. PLoS Computational Biology, 8(8), e1002653.
107. Stevenson, I., & Koerding, K. (2011). Inferring spike-timing-dependent plasticity from spike train data. In Advances in neural information processing systems (pp. 2582–2590).

108. Stevenson, I. H., Rebesco, J. M., Hatsopoulos, N. G., Haga, Z., Miller, L. E., & Kording, K. P. (2009). Bayesian inference of functional connectivity and network structure from spikes. IEEE Transactions on Neural Systems Rehabilitation Engineering, 17(3), 203–213.
109. Su, R. Q., Lai, Y. C., Wang, X., & Do, Y. (2014). Uncovering hidden nodes in complex networks in the presence of noise. Scientific Reports, 4, 3944. URL http://www. ncbi.nlm.nih.gov/pubmed/24487720.
110. Sutera, A., Joly, A., François-Lavet, V., Qiu, Z. A., Louppe, G., & Ernst, D. (2014). Simple connectome inference from partial correlation statistics in calcium imaging. In ECML workshop - "Neural connectomics: From imaging to connectivity". Nancy, France.
111. Tateno, T., & Jimbo, Y. (1999). Activity-dependent enhancement in the reliability of correlated spike timings in cultured cortical neurons. Biological Cybernetics, 80(1), 45–55.
112. Tkacik, G., Mora, T., Marre, O., Amodei, D., Berry, I., & Michael, J. (2014). Thermodynamics for a network of neurons: Signatures of criticality. ArXiv preprint arXiv:1407.5946.
113. Valdes-Sosa, P. A., Roebroeck, A., Daunizeau, J., & Friston, K.(2011). Effective connectivity: influence, causality and biophysical modeling. Neuroimage, 58(2), 339– 361.
114. Valdés-Sosa, P. A., Sánchez-Bornot, J. M., Lage-Castellanos, A., Vega-Hernández, M., Bosch-Bayard, J., Melie-García, L., et al. (2005). Estimating brain functional connectivity with sparse multivariate autoregression. Philosophical Transactions of the Royal Society Series B (Biological Sciences), 360(1457), 969–981.
115. Van Drongelen, W. (2006). Signal processing for neuroscientists: an introduction to the analysis of physiological signals. Academic Press.
116. Veeriah, V., Durvasula, R., & Qi, G.-J. (2015). Deep learning architecture with dynamically programmed layers for brain connectome prediction. In Proceedings of the 21th ACM SIGKDD international conference on knowledge discovery and data mining (pp. 1205–1214). ACM.
117. Vidne, M., Ahmadian, Y., Shlens, J., Pillow, J. W., Kulkarni, J., Litke, A. M., et al. (2012). Modeling the impact of common noise inputs on the network activity of retinal ganglion cells. Journal of Computational Neuroscience, 33(1), 97–121.
118. Vogelstein, J. T.(2009). OOPSI: A family of optimal optical spike inference algorithms for inferring neural connectivity from population

calcium imaging. The Johns Hopkins University.

119. Vogelstein, J. T., Watson, B. O., Packer, A. M., Yuste, R., Jedynak, B., & Paninski, L. (2009). Spike inference from calcium imaging using sequential Monte Carlo methods. Biophysical Journal, 97(2), 636–655.
120. Wainwright, M. J., & Jordan, M. I. (2007). Graphical models, exponential families, and variational inference. Foundations and Trends in Machine Learning, 1(1–2), 1–305.
121. Wang, S., Chang, X., Li, X., Sheng, Q. Z., & Chen, W. (2016). Multi-task support vector machines for feature selection with shared knowledge discovery. Signal Processing, 120, 746–753.
122. White, J., Southgate, E., Thomson, J., & Brenner, S. (1986). The structure of the nervous system of the nematode Caenorhabditis elegans: the mind of a worm. Philosophical Transactions of the Royal Society of London, 314, 1–340.
123. Wollstadt, P., Martínez-Zarzuela, M., Vicente, R., Díaz-Pernas, F. J., & Wibral, M. (2014). Efficient transfer entropy analysis of non-stationary neural time series. PloS One, 9(7), e102833.
124. Wu, M.-Y., Zhang, X.-F., Dai, D.-Q., Ou-Yang, L., Zhu, Y., & Yan, H. (2016). Regularized logistic regression with network-based pairwise interaction for biomarker identification in breast cancer. BMC Bioinformatics, 17(1), 1.
125. Yang, H. H., & St-Pierre, F. (2016). Genetically encoded voltage indicators: opportunities and challenges. Journal of Neuroscience, 36(39), 9977–9989. URL https: //www.ncbi.nlm.nih.gov/pubmed/27683896.
126. Yeh, F.-C., Tang, A., Hobbs, J. P., Hottowy, P., Dabrowski, W., Sher, A., et al. (2010). Maximum entropy approaches to living neural networks. Entropy, 12(1), 89– 106.
127. Yook, C., Druckmann, S., & Kim, J. (2013). Mapping mammalian synaptic connectivity. Cellular and Molecular Life Sciences, 70(24), 4747–4757.
128. Yu, S., Huang, D., Singer, W., & Nikolić, D. (2008). A small world of neuronal synchrony. Cerebral Cortex, 18(12), 2891–2901.
129. Zhou, Y., Jin, R., & Hoi, S. C. (2010). Exclusive lasso for multi-task feature selection. In AISTATS (Vol. 9) (pp. 988–995).

130. Ziv, Y., Burns, L. D., Cocker, E. D., Hamel, E. O., Ghosh, K. K., Kitch, L. J., et al. (2013). Long-term dynamics of ca1 hippocampal place codes. Nature Neuroscience, 16(3), 264–266. URL http://www.ncbi.nlm.nih.gov/pubmed/23396101.

131. Zou, H., & Zhang, H. H. (2009). On the adaptive elastic-net with a diverging number of parameters. Annals of Statistics, 37(4), 1733.

CHAPTER 7

GENERATIVE MODELS FOR NETWORK NEUROSCIENCE: PROSPECTS AND PROMISE

Richard F. Betzel[1] and Danielle S. Bassett[1,2]

[1] Department of Bioengineering

[2] Department of Electrical and Systems Engineering, University of Pennsylvania, Philadelphia, PA 19104, USA

ABSTRACT

Network neuroscience is the emerging discipline concerned with investigating the complex patterns of interconnections found in neural systems, and identifying principles with which to understand them. Within this discipline, one particularly powerful approach is network generative modelling, in which wiring rules are algorithmically implemented to produce synthetic network architectures with the same properties as observed in empirical network data. Successful models can highlight the principles by

CITATION: Betzel, R. F., & Bassett, D. S. (2017). Generative models for network neuroscience: Prospects and promise. Journal of the Royal Society Interface, 14(136), 20170623. (17 pages)

which a network is organized and potentially uncover the mechanisms by which it grows and develops. Here, we review the prospects and promise of generative models for network neuroscience. We begin with a primer on network generative models, with a discussion of compressibility and predictability, and utility in intuiting mechanisms, followed by a short history on their use in network science, broadly. We then discuss generative models in practice and application, paying particular attention to the critical need for cross-validation. Next, we review generative models of biological neural networks, both at the cellular and large-scale level, and across a variety of species including *Caenorhabditis elegans*, *Drosophila*, mouse, rat, cat, macaque and human. We offer a careful treatment of a few relevant distinctions, including differences between generative models and null models, sufficiency and redundancy, inferring and claiming mechanism, and functional and structural connectivity. We close with a discussion of future directions, outlining exciting frontiers both in empirical data collection efforts as well as in method and theory development that, together, further the utility of the generative network modelling approach for network neuroscience.

INTRODUCTION

Many complex systems are composed of elements that interact dyadically with one another and can therefore be represented as graphs (networks) composed of nodes interconnected by edges. The network framework can be applied to systems across a range of disciplines, from sociology and psychology to molecular biology and genomics, making it possible to leverage a common mathematical language and set of analytic tools to investigate the topological organization of systems that, outwardly, might appear dissimilar to one another [1].

In neuroscience, network-based analyses have become common. This is due in part to initiatives for sharing large, multimodal neuroimaging datasets [2,3], the availability of easy-to-use software packages for computing graph-theoretic metrics [4,5], and the fact that networks are natural vehicles for representing and analysing complex spatio-temporal interactions among neural elements, including neurons, populations and brain areas [6].

Though the scope of topics studied in network neuroscience is broad, the typical study involves characterizing the structure of a network with a series of summary statistics. Each statistic describes a particular feature of the network, ranging from simple to complex and operating over all topological scales. For example, *degree* is a local (node-level) property that simply counts a node›s total number of incoming and outgoing connections. On the other hand, *characteristic path length* is a global (whole-network) measure of the average length of all pairwise shortest paths. In general, summary statistics offer succinct descriptions of a network›s organizational features, especially those that are not immediately apparent given a network›s list of nodes and edges.

The application of summary statistics to better understand the structure and function of biological neural networks has been fruitful. Over a decade or so, evidence from networks across different organisms and spatial scales [7] has converged onto a small set of properties and summary statistics that, collectively, can be used to describe the organization of most biological neural networks. These include indices of small-worldness [8], heavy-tailed degree and edge-weight distributions [9,10], a diverse meso-scale structure that includes segregated modules but also core–periphery structure [11–13], hubs and rich clubs [14,15], and economic spatial layouts favouring the formation of short-range (low-cost) connections [16,17]. Further, such core organizational principles also include functional constraints, like the need to balance properties that support either segregated or integrated brain function [18], but also emphasize the trade-off between the cost of such properties and their functionality [19]. These properties, collectively, create a caricature of neural system organization and function.

While illuminating, the process of describing networks in terms of their topological properties amounts to an exercise in 'fact collecting'. Though summary statistics might be useful for comparing individuals [20] and as biomarkers of disease [21], they offer limited insight into the mechanisms by which a network functions, grows, and evolves. Arguably, one of the overarching goals of neuroscience (and biology, in general) is to manipulate or perturb networks in targeted and deliberate ways that result in repeatable and predictable outcomes [22]. For network neuroscience to take steps towards addressing this goal, it must shift its current emphasis beyond network taxonomy—i.e. studying subtle individual- or population-level differences in summary statistics—towards a science of mechanisms and processes [23,24].

While there exist many methodological approaches for seeking mechanisms in networks and a range of spatial, topological and temporal scales at which those methods can be deployed [25], the focus of this article is on network generative modelling. Network generative modelling is a flexible framework for generating synthetic networks from a set of parametrized wiring rules. Generative models figure prominently in the network science canon [26–29], and have recently been deployed in domain-specific scenarios to study the evolution of protein interaction networks [30–32], the worldwide web [33], and social systems [34]. Importantly, and provided that the wiring rule is sufficiently informed and biologically grounded, generative models can be used to test and identify potential mechanisms that underlie the growth and evolution of biological neural networks. With mechanisms in hand, it becomes possible to distinguish the topological features that drive a network's growth from those that emerge as mere byproducts [35], and to pursue deliberate and targeted interventions [36,37].

In the following sections, we present a primer on network generative models, highlighting their past use, their interpretation and several open methodological considerations. We review current applications of generative models to neural systems, emphasizing several outstanding questions and implementation details. Finally, we plot a course for future studies. While our review discusses generative models of biological neural networks, in general, we focus our discussions on structural networks that represent the physical pathways among neural elements—e.g. synapses, axonal projections and white-matter fibre bundles.

GENERATIVE MODELS: A PRIMER

This article deals with the topic of generative models. Broadly, a generative model is a statistical process that outputs a synthetic set of data or observations. Usually, these synthetic data and the generative process are designed to have some properties in common with empirical data and with the process believed to have generated those data. Generative models are often parametrized, and those parameters can be chosen so as to minimize the discrepancy between observed and synthetic data. The models, themselves, can be compared against one another using standard model comparison techniques, including goodness-of-fit criteria and cross-validation approaches.

In the context of network science, generative models represent algorithmically implemented wiring rules or causal processes that output synthetic networks with a particular set of topological properties or that perform a particular set of functions. While a network's nodes and edges encode all of its structural properties, studying generative models shifts focus away from those structural properties and instead onto wiring rules and the process of network formation. This shift in emphasis confers a number of distinct advantages:

(1) Generative models compress our descriptions of networks and highlight regularities in their organization.
(2) They make predictions about out-of-sample and unobserved network data.
(3) Under the best circumstances, generative models can uncover network mechanisms.

We discuss these topics in greater detail throughout the following subsections.

Compressibility of Networks

Generative models compress our descriptions of a network, encoding the network's topology in a set of wiring rules and parameters. Naively, we could describe a network exactly given a list of its nodes and edges: that is, by consulting the list, we could correctly connect nodes that are supposed to be connected and avoid connecting nodes that are not supposed to be connected. However, connections in many networks are not independent of one another, but instead exhibit statistical regularities such that, given the wiring rule that matches those regularities, we could predict the presence/absence of connections ahead of time. In this case, it becomes unnecessary to consult the list of nodes and edges to describe the network. More importantly, we can often interpret the wiring rule itself to uncover the network's organizing principles.

As an example, consider real-world spatial networks, where the probability of observing an edge between two nodes decays as a function of distance [38]. Often, these kinds of networks can be well approximated by a simple geometric model whose wiring rule mimics the network's distance-dependent connection formation [39]. To perfectly describe a spatial network, we could generate a long and possibly unwieldy list of its nodes and edges. However, if the geometric model is a good approximation, e.g. synthetic networks generated by the model recapitulate many observed edges, then the model can be used to replace those edges in the list, effectively shortening our description of the network. The geometric model naturally mimics the distance dependencies of the spatial network. For many networks, however, the statistical regularities among links may not be obvious, in which case selecting the appropriate model may not be straightforward. We discuss this issue of model selection later in this section.

Predictability

Besides compressing our descriptions of a network, generative models also have predictive capacity and can be used as forward models of unobserved and out-of-sample data. Returning to the example of spatial networks, we might hypothesize the relevance of a generative model in which the probability of connection formation is given by a decaying exponential. If we let $A_{ij} \in \{0, 1\}$ indicate the presence or absence of an edge between nodes i and j, we can write this connection probability as: $P(A_{ij} = 1) \propto \exp(-\beta \cdot D_{ij})$, where D_{ij} is the distance between nodes i and j and $\beta \geq 0$ is a parameter to be fit [40]. If we were given a network G, we could fit the parameter β so that the discrepancy between synthetic networks generated by the model and G is minimized. Having fit the model, we could use it to make predictions about a second network, G', whose connectivity pattern is unknown but whose nodes' spatial locations are given.

As another example, consider the stochastic blockmodel [27,41], in which nodes are assigned membership to one of K communities, $z_i \in \{1, \ldots, K\}$, and where the probability of two nodes, i and j, being connected to one another depends only on their community assignments: $P(A_{ij} = 1) = \omega z i_{,zj}$($\omega$ is a $K \times K$ matrix that encodes community-to-community connection probabilities). Fitting this model to a network G entails inferring nodes› communities and connection probabilities. If we encountered a second network, G', with an unknown connectivity pattern but whose nodes correspond to those in G, e.g. the same set of neurons or brain regions, then we could use the model to predict the configuration of nodes and edges in that network.

Mechanisms

Finally, provided that it incorporates sufficient system-specific details (in our case, neurobiological information), a generative model can be used to gain insight into the mechanisms that guide the formation and growth of a system. This last point is critical. A generative model, under ideal circumstances, is a recipe for building a network. Having such a recipe opens new avenues for interrogating a network. It allows us to identify structural features of a network that emerge as a direct result of the wiring rule, versus those that emerge spontaneously as a consequence of constraints imposed by a given wiring rule [35]. For example, a geometric model will generate networks with high levels of clustering even though the wiring rule never explicitly optimizes for this property. Importantly, a recipe for building a network also gives us the ability to explore alternative ingredients. What happens if we change a parameter slightly? Does the model generate networks of vastly different character? Can we control the trajectory of a network's growth and guide it into a desired target configuration [42]? The ability to selectively drive the growth of a network is a tantalizing prospect, and one with profound implications for the treatment of psychiatric disease and neurological disorders.

Canonical Generative Models for Networks

Before engaging neuroscience-specific questions, it is useful to discuss examples of generative models as they have been applied in network science and other fields. In the remainder of this section we review some canonical generative models, emphasizing the properties that they share with one another as well as those that make them distinct. While the models discussed in this section certainly fit the definition of generative models, we emphasize that the space of all possible models is broad and includes models that share few characteristics with those discussed here.

Generative models have a long history in network science and mathematics. One of the earliest examples is the so-called Erdo″s-Rényi (ER) model [26], in which connections are formed independently between pairs of N nodes with probability P (another version exists where, instead of P, a fixed number of edges, M, are added uniformly at random). While the ER model has interesting combinatoric and mathematical properties, e.g. binomially distributed node degree [43], it is a poor approximation of most real-world networks. That is, the random and independent process by which connections are formed in the ER model results in networks with no real structure (poor compressibility) and does not resemble any of the mechanisms by which many real-world networks grow. Accordingly, if we wish to model networks in the real world, we need a set of models that generate networks with realistic properties.

Initial explorations into generative models for real-world data resulted in two models that, collectively, helped spark broad interest in complex networks. The first, introduced by Duncan Watts and Steven Strogatz, sought the origin of empirically observed 'small world' topologies, in which a network simultaneously exhibits greater-than-expected clustering and shorter-than-expected path length [28]. Broadly speaking, the model supposed that small-world networks are an interpolation between two extreme configurations: a ring lattice network (nodes arranged on the circumference of a circle and linked to their k clockwise and anti-clockwise neighbours) and an ER network. To move from one extreme to the other, the authors introduced a tuning parameter, p, which governed the probability that an edge in the lattice network would be rewired randomly. When p is small, the model generates networks that have mostly lattice-like properties, but when $p$is large, the model generates networks whose properties are indistinguishable from those produced by the ER model. Between those extremes, however, is a 'sweet spot'—a region of parameter space yielding networks with properties of both extremes, namely high clustering and short path length. This model is referred to as the Watts–Strogatz (WS) model.

At around the same time, a second group sought an explanation for why many real-world networks exhibited heavy-tailed degree distributions. The proposed model, by Réka Albert and Albert-László Barabási, was based on a growth rule [29]. Starting with a small set of fully connected nodes, the model adds new nodes to the network by forming connections preferentially to already-existing nodes with higher degrees. This growth mechanism is a sort of 'rich get richer' process; nodes that have existed for a long time accumulate many connections, which further increases their likelihood of being connected to newly added nodes. The result of this process is a network with an approximately power-law degree distribution, mimicking those frequently observed in real networks [44]. This model is identical to that defined by Price in 1976 with a single value change to one parameter [45], and is generally referred to as the Barabási–Albert (BA) or the preferential attachment (PA) model.

Generative Models in Practice and Application

The WS and BA models generate synthetic networks with properties *qualitatively* similar to those observed in real-world networks (small-worldness and heavy-tailed degree distribution). If we wanted to make the similarity of empirical and synthetic networks quantitative and more precise, how would we do so? Supposing that a model yields networks that repeatably and exactly recapitulate *all* properties of an empirical network, can we equate the model with mechanism? Both of these questions are difficult to answer, and represent some of the technical challenges associated with generative modelling.

Choosing an Objective Function

We will first address the issue of how to perform quantitative comparisons between synthetic and empirical networks. Fortunately, there exists a plurality of approaches for quantitatively comparing networks. The challenge is selecting the approach that is best suited to a given research question.

Typically, we wish to answer the question of whether an empirically observed network could have been produced by some generative model. One strategy for addressing this question involves defining a likelihood function over the space of all possible networks, and evaluating that function for the observed network. Stochastic blockmodels are a good example of this strategy in action [27,41,46]. The probability of a connection forming between nodes i and j, P_{ij}, depends on their community assignments, z_i and z_j: $P(A_{ij} = 1) = \omega_{zi,zj}$. The probability that i and j are disconnected, is therefore $1 - P_{ij}$ and the likelihood that the observed network was generated by this model is given by

$$\mathcal{L} = \prod_{i,j>i} P_{ij}^{A_{ij}} (1 - P_{ij})^{1-A_{ij}}. \tag{2.1}$$

Blockmodels are convenient in that this likelihood function can be written in closed form. This approach can be generalized for other models—even when the precise likelihood function is unknown—by generating a sample of networks from a given step of parameters and estimating, from those samples, the probability of any connection existing.

This approach is similar to others in the literature [47], in that it links the model's fitness with its ability to correctly account for the empirical network's exact configuration of nodes and edges. While this approach seems useful, it is not difficult to envision scenarios where even near-perfect prediction of an empirical network's connections nonetheless fails to account for some of its critical topological properties. For example, consider the canonical small-world network—a ring lattice plus a few random (shortcut) connections that reduce the network's characteristic path length. The ring lattice and small-world network have nearly perfect edge overlap. If we were to regard edge overlap as the definitive measure of fitness, we might be inclined to treat the lattice network as a good approximation of the small-world network. In other words, from a strictly structural point of view, these two networks are almost perfect matches; from a functional perspective, however, the two networks are highly dissimilar; because of its longer characteristic path length, the ring lattice will lack efficient (short) routes that would be useful for communication or transportation.

Comparing synthetic and empirical networks on the basis of their edge configuration is useful, but has some shortcomings that motivate the exploration of alternative approaches. Another approach, and one that has been used in several recent studies [48,49], eschews the edgewise comparison of two networks, instead simultaneously comparing them along several topological dimensions (e.g. their efficiency, clustering, modularity, etc.), and calculating a statistic of average dissimilarity. For example, Betzel *et al.* [49] defined the energy function $E = \max(KS_K, KS_C, KS_B, KS_E)$, where each term is a Kolmogorov–Smirnov statistic comparing degree (K), clustering (C), betweenness centrality (B) and edge length (E) distributions of synthetic and observed networks [49]. Intuitively, smaller energy implies greater fitness.

This approach is flexible and can be adapted to include virtually any set of metrics. It is important to note, however, that many network measures are correlated with one another, so the choice of which to include should take this into account. Also, there might be synthetic networks that match an empirical network in terms of network statistics but not its precise set of connections. Irrespective of how the objective function is defined, having one makes it possible to perform different kinds of comparisons. For a given model, we can perform model fitting by selecting the parameter values that optimize the objective function. We can also leverage an objective function to compare different generative models to one another. For example, we may wish to discount a model that is incapable of generating networks that resemble our real-world network of interest.

Cross-validation

Suppose that we fit a generative model by optimizing some objective function so that the model generates synthetic networks that share some set of properties with an empirical network. As in any model-fitting exercise, we can continue adding layers of complexity and free parameters to the model so that it matches our real-world network to some arbitrary degree of precision. It is often the case, however, that we are less interested in predicting the organization of a single network, but of a class of networks. For example, we might wish to identify wiring rules that can recapitulate the organization of structural brain networks, on average, rather than the network of any one individual. Even if our aim was to predict subject-specific networks, it might be unsurprising (in a statistical sense) that our models reproduce many of the features of those networks; after all, the model's parameters were selected only after an optimization procedure.

In both cases (fitting models to empirical network data based on edge- or property-matching), it is essential that we perform a cross-validation procedure. This procedure might entail taking the best-fitting parameters from one model and using them to generate estimates of a second network not involved in the model-fitting process. We can compare the goodness-of-fit to that of a random (ER) model, to ensure that our model performs above chance. This type of cross-validation ensures that a generative model is identifying general wiring rules and not overfitting. A second type of cross-validation involves testing whether synthetic networks have properties in common with real-world networks that they were not explicitly optimized to possess. In other words, does a generative model give us certain properties 'for free?' This type of cross-validation ensures that our objective function is sufficiently general rather than emphasizing a specific subset of network properties. However, even in this case it is important to note that many network properties are correlated with one another, and so the emergence of one may necessarily imply the emergence of another. For example, if a wiring rule incidentally results in assortative modules, it is also likely that the network will exhibit a greater-than-expected clustering coefficient. In short, the network properties used for cross-validation must be carefully chosen and ideally would be orthogonal to one another. One possible solution is to cross-validate using the principal components of a list of network features rather than any individual feature [50].

The Space of Generative Models

What distinguishes one generative model from another? Is it possible to delineate classes of generative models based on their functions or characteristics? Arguably, one of the distinguishing features of any generative model is the timescale over which it operates (figure 1). On one extreme are models with no timescale at all, like stochastic blockmodels [27,41] or the family of exponential random graphs, which uses a regression-based framework to predict a network's link structure from node- and edge-level attributes, and which has recently been applied to brain network data [52–55]. These kinds of models are 'single-shot' generators of networks, and can therefore be quite poor representations of real-world networks that grow and evolve over time. On the other extreme are models whose internal timescale matches that of the real system. Nodes and edges are added or rewired on a realistic timescale to match known properties of the system. The growth model of *C. elegans* presented by Nicosia *et al.* [51] is a good example [51]. In this model, nodes and edges are added according to their empirically measured birth times (time of cell division); a feature that contributed to the success of that model in predicting different properties of the *C. elegans* connectome.

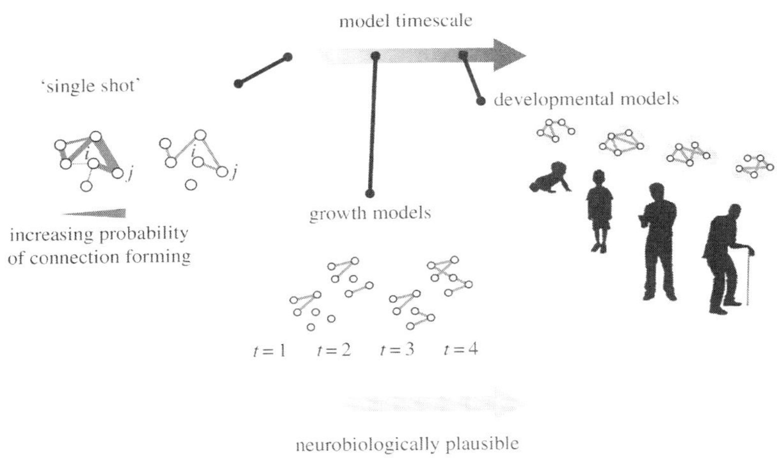

Figure 1: The space of generative models. Generative models can be differentiated from one another along several dimensions, one of which is the timescale over which they operate. A model's timescale is related to its neurobiological plausibility. Models whose timescale is nearer to that of development can incorporate more realistic and interpretable features and, in turn, have the chance of uncovering realistic growth mechanisms (e.g. the model of *C. elegans* proposed in [51]). At the opposite end of the spectrum are 'single shot' models, where connection probabilities are initialized early on and all connections and weights are generated in a single algorithmic step (e.g. stochastic blockmodels). Situated between these two extremes are growth models that exhibit intrinsic timescales over which connections and/or nodes are added to the network, but where the timescale has no clear biological interpretation. (Online version in colour.).

Between these two extremes—where models operate either without any timescale or with a biologically plausible timescale—is where most generative models are situated. In this middle ground, edges and nodes are added to or rewired in an existing network, but the timescale over which these processes occur is arbitrary. A good example is the BA model, in which new nodes are linked to an existing network over a series of steps. These steps are ordered, so the addition of one node precedes or follows that of another. However, time is measured in arbitrary units (steps) and direct comparison to biological timescales, e.g. human development, might be inappropriate. Ordering generative models based on their internal timescales is similar to ordering them according to their plausibility and mechanistic understanding. Blockmodels and models with arbitrary timescales can do a good job compressing our description of a network and might identify general organizational principles [35]. However, if our aim is to develop realistic mechanistic models of network growth and development, it is essential that we include the necessary components that ground the model in reality.

While a generative model's intrinsic timescale naturally results in a stratification of models according to their neurobiological plausibility, it is essential to note that increased plausibility does not necessarily imply improved fit or increased model performance. Entirely implausible models with many parameters that receive a variety of metadata as input can conceivably outperform more neurobiologically grounded mechanistic models, simply due to increased complexity [56].

GENERATIVE MODELS OF BIOLOGICAL NEURAL NETWORKS

Now that we have an intuition for what a generative model is, and what the goals are for building a generative model, we turn to a brief review of existing generative models for biological networks observed in neural systems. We note that this review is not comprehensive, but instead focuses on areas in which significant work has been accomplished, or areas that motivate important current and future frontiers. We also refer readers elsewhere for additional details on the mechanisms of connectome development [24], biophysical models of neural dynamics [57], and modelling mesoscale structure in dynamic networks [58] and multiscale networks [25].

Finally, we note that this review focuses mostly on generative models of structural and not functional networks (the distinction is in how edges are defined; in structural networks they represent physical connections, e.g. synapses, projections and fibre tracts, whereas in functional networks they represent statistical associations among neural elements' activity, e.g. correlation, coherence, etc.). Because of differences in how structural and functional networks are generated and evolve, certain classes of models that are appropriate for one type may be wholly inappropriate for the other. For example, functional networks are not generated through an edge addition process—they emerge from constrained dynamical processes. We discuss the implications of these differences in more detail later in this section.

The Requisite Ingredients

An open and important question that scientists face when embarking on a study to develop a generative model is: 'what features are required to build good network models?' Perhaps the simplest feature one requires is a target network topology, the organization of the network that one is trying to recapitulate and ultimately explain. Yet, a single network topology can be built in many different ways, with strikingly different underlying mechanisms [**59**]. Thus one might also wish to have a deep understanding of (i) the constraints on anatomy, from physical distance [**60**] to energy consumption [61], (ii) the rules of neurobiological growth, from chemical gradients [62] to genetic specification [63], and (iii) the pressures of normal or abnormal development, and their relevance for functionality. Moreover, each of these constraints, rules and pressures can change as the system grows, highlighting the importance of developmental timing [63]. Of course, one might also wish to choose which of these details to include in the model, with model parsimony being one of the key arguments in support of building models with fewer details.

Generative Models at the Cellular Level

Recent efforts to model cellular level network architecture have had the benefit of building on rich empirical observations made over the last several decades. At one of the smallest spatial scales of neuronal connectivity, evidence suggests that the arbors of single neurons can be characterized by both local [64] and global [65] optimization rules to more strongly minimize volume than length, signal propagation speed or surface area. Within the confines of relative volume cost minimization, there is also evidence for a maximization of the repertoire of possible connectivity patterns between dendrites and surrounding axons: in basal dendritic arbours of pyramidal neurons, arbour size scales with the total dendritic length, the spatial correlation of arbour branches appears to have a single functional form, and small sections of an arbour display self-similarity [66].

The morphology of dendritic arbours specifically and other parts of the cell more generally have a direct bearing on the degree of connectivity that can take place between neurons [67]. Like dendritic arbours, synaptic connectivity appears to be organized in a highly non-random manner [68], with unexpectedly high density in relation to its volume [67]. Interestingly, both synaptic connectivity and neuronal morphology appear to experience some similar constraints, including principles of wiring optimization [60,69]. Some suggest that constraints on synaptic wiring may be the more fundamental of the two, explaining the degree of separation between cortical neurons [60], as well as the placement of cell bodies [70]. Others suggest that it is in fact the combination of wiring economy and volume exclusion that can determine neuronal placement [71].

In either case, the highly non-random nature of synaptic connectivity has been the subject of several recent generative modelling efforts. Initial observations that this non-random organization could be parsimoniously described as small-world [8,72] have motivated the question of how this particular type of network complexity is combined with pressures for wiring minimization. Nicosia *et al.* [51] suggest that the growth rules shaping cellular nervous systems balance an economical trade-off between wiring cost and the functionality of network topology (figure 2). Using a dynamic economical model incorporating a continuously negotiated trade-off between wiring cost and network topology, they recapitulate an empirically observed phase transition in the proportion of nodes to links present over the developmental time period of *C. elegans* [51]. The authors speculate that such dynamically negotiated trade-offs may be characteristic of other complex systems, whether biological or man-made. It will be interesting in the future to consider scenarios in which such trade-offs may be negotiated over shorter time periods, such as in the alteration of the prevalence of autaptic connections posited to play a role in homeostatic network control of bursting [**73**].

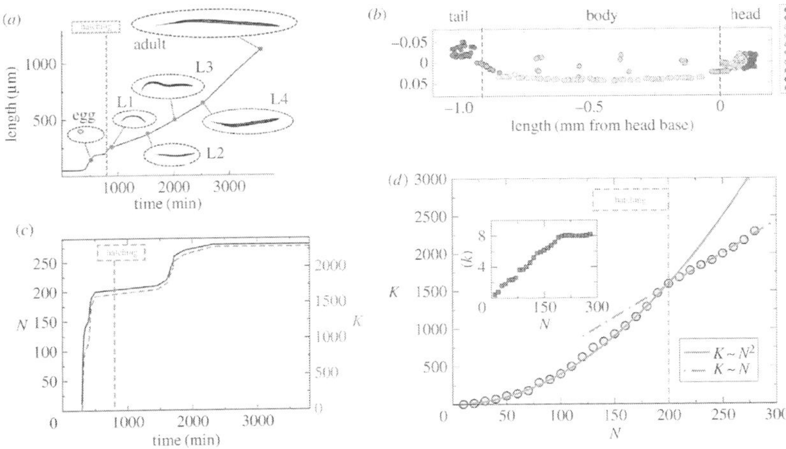

Figure 2: Development of the *C. elegans* nervous system. (*a*) *Caenorhabditis elegans* reaches adulthood approximately 63 h after fertilization, over which time its body increases appreciably in length. (*b*) In the adult hermaphrodite worm, neurons are distributed unevenly across the body, with more than 60% being located in the head and approximately 15% being located in the tail tip. Here, neurons are colour-coded according to their membership to the following ganglia: anterior [A], dorsal [B], lateral [C], ventral [D], retrovesicular [E], ventral cord [G], posterior lateral [F], preanal [H], dorsorectal [J] and lumbar [K]. (*c*) The total number of neurons (*N*, solid black), and connections (*K*, dashed blue) grows nonlinearly but monotonically with time. (*d*) A phase transition is evident in the number of synapses as a function of the number of neurons (yellow circles): before hatching, *K*grows as N^2 (solid blue line), whereas after hatching, *K* grows linearly with *N* (dashed green line). (Inset) Plot of the average nodal degree, *K*, versus number of nodes, *N*. Reproduced with permission from [51]. (Online version in colour.).

The incorporation of a dynamic economic trade-off is an example of the broader importance of incorporating biophysically accurate features in generative models of cellular neural systems. Another example of such a biophysical feature is axon and dendrite geography, which has been shown to predict the specificity of synaptic connections in a functioning spinal cord network of hatchling frog tadpoles [74]. Some generative models have also sought to determine the role of neuron type in observed network topology and function, for example by building models of sensory neurons, sensory pathway interneurons, central pattern generator (CPG) interneurons and motoneurons, and then linking them in a network with known inter-type connectivity [75]. By adding knowledge about development including chemical gradients and physical barriers [62], a cell-type specific model of 2000 neurons in the spine of a young *Xenopus* tadpole can produce swimming behaviour in response to sensory stimulation [76]. These and related efforts demonstrate the ability of generative network models built with neuron and synapse resolution, and incorporating biophysical phenomena, to reproduce behaviours observed in whole organisms. Such findings are reminiscent of other biophysical modelling efforts at the large scale of human areal networks [77,78], where the biophysics of regional rhythms and inter-regional synchronization inform our understanding of human cognition [79].

Increasing in Scale: Generative Models of Large-scale Connectomes in Non-human Animals

In the previous section, we reviewed some of the literature supporting the notion that cellular network organization in neural systems is characterized by pressures of wiring economy and topological complexity. Such pressures are similarly thought to play a role in the organization of networks at the meso- and large scale in both human and non-human mammalian brains [19]. Computational studies suggest that trade-offs between wiring economy and topological complexity [80] support the formation of network modules, offering relative segregation of function, and network hubs, offering relative integration of function [81]. The role of topological complexity and the presence of unusually high wiring costs in some parts of cortex suggests that simple notions of spatial embedding are not sufficient to explain the observed organization of the connectome. This limitation has motivated models deriving a latent (rather than physical) space from which to predict missing links [82], or incorporating information about cytoarchitecture [83] such that cytoarchitectonically similar cortical areas in the two hemispheres have an unexpectedly high probability of connecting with one another [84].

A particularly salient example of a generative model of areal connectivity in a mammalian brain that incorporates many of these considerations is the recent predictive model of Beul *et al.* [83] (figure 3). In this paper, the authors study mesoscale structural connectivity between 49 areas of the cat cerebral cortex as estimated by tract tracing techniques [83]. They test the predictive utility of three separate wiring rules: (i) a *structural* rule in which the laminar patterns of origins and terminations of inter-areal projections vary according to the relative cytoarchitectonic differentiation of the projection sources and targets, (ii) a *distance* rule in which connections are more frequent, and more dense, among neighbouring regions and sparser or absent between remote regions, and (iii) a *hierarchical* rule in which differences in the functional hierarchical levels of source and target areas are inversely related to the degree of connectivity between them. While the latter rule did not accurately fit the data, the first two rules (structure and distance) explained significant variance in the observed connectivity patterns, with a linear combination of the two predicting the existence of connections with more than 85% accuracy.

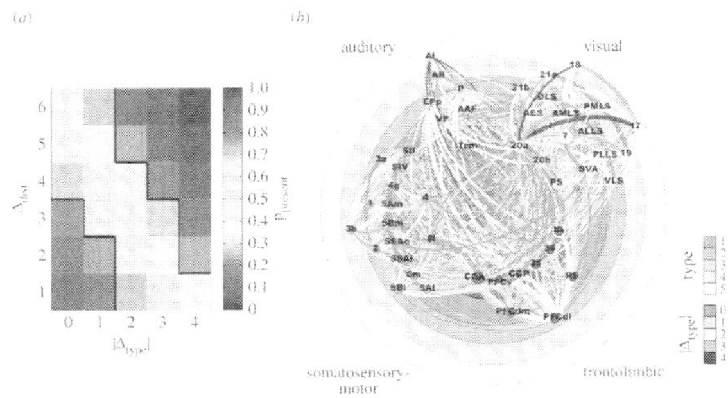

Figure 3: A predictive model of the cat cortical connectome based on cyto-architecture and distance. (*a*) Results of linear discriminant analysis showing posterior probabilities (*p*) for the presence of projections across the two-dimensional predictive variable space defined by the border distance between regions (Δ_{dist}) and the absolute type difference between regions ($|\Delta_{type}|$). Black borders enclose ranges of *p* present greater than 0.75 and *p* present less than 0.25. (*b*) Visualization of the corticocortical connections collated in Scannell *et al.* [85]. All present projections between cortical areas for which a structural type was defined (49 of 65 areas) are displayed. Circles correspond to structural types; cortical areas are placed accordingly. Structural type increases from the centre to the periphery. Projections are colour-coded according to the absolute structural type difference of the connected areas. Ordinal projection strength (sparse, intermediate or dense) is coded by increasing projection width. Nodes are grouped and colour-coded according to anatomical modules as indicated. Node sizes indicate the areas' (unweighted) degree. Hub-module areas, as classified by Zamora-López *et al.* [14], are marked by a white outline. Reproduced with permission from [83]. (Online version in colour.).

Work in non-human primates generally and the macaque cortex specifically recapitulates many of the same motifs from work in lesser mammals. Early work suggested that cortical components are optimally placed so as to minimize the costs of their interconnections [86], facilitating a global optimal cerebral cortex layout [87]. Later work suggested that component placement did not maximally minimize wiring, but also tended to favour short processing paths, due to long-distance projections [16]. Indeed, separate from where components are placed, it has been noted that there appear to be successfully arbitrated optimization problems in the organization of inter-areal connectivity, for example favouring near-minimization of distance [17,88] and increased support for connectivity between areas with similar topological properties [47]. In an extension of the model described above for the cat, Beul and colleagues similarly demonstrate the striking utility of the structural rule of architectonic similarity, where similarity in the laminar pattern of projection origins, and the absolute number of cortical connections of an area, demonstrated the strongest and most consistent influence on connection features [89]. In this case, the distance rule was surprisingly not predictive. Future extensions of this model may include explicit nonlinear growth rules, which have previously been linked to the emergence of network hubs [90].

Generative Models of Large-scale Connectomes in Humans

Efforts in humans support the notions of wiring economy [91,92] and topological complexity [49], and further add new considerations such as the geometric segregation of the brain into grey and white matter, enabling the relative minimization of conduction delays [93]. While one-shot models have been the most commonly exercised generative models for human structural networks, relatively new evaluation criteria for them include an assessment of their controllability profiles [94] and homological features [95]. Moreover, there has been a recent and growing interest in developing network growth models that incorporate biologically motivated rules for the probability of connections [96,97]. For example, spatially constrained adaptive rewiring creates small-world network architectures with spatially localized modules [97], while wiring rules based on topological affinities recapitulate known scaling laws of physical network topology [96]. It would be interesting in future work to determine how these rules could be adapted to explain the patterns of conserved and variable architecture of white matter networks across individual humans [98].

The recent paper by Betzel *et al.* [49] represents one of the first attempts at subject-level generative modelling [49]. In this study, the authors fit 13 generative models to white-matter networks acquired from three independent datasets, totalling 380 subjects (figure 4). The model generated synthetic networks using an edge-addition algorithm, in which connections were added probabilistically and one at a time according to a set of parametrized wiring rules. Each of the 13 models was fit in two stages: first by matching distributional statistics of the white-matter networks and later cross-validated on a separate set of network measures. The best fitting models across all three datasets featured wiring rules based on wiring cost reduction and homophilic attraction principles, the severity of each controlled by a separate parameter. Because the models were fit to individual subjects, it was possible to explore individual variability in model fit. When applied to lifespan data from the Nathan Kline Institute, the authors found that the parameter governing the severity of the wiring cost reduction weakened systematically with age, as did the model goodness of fit. These findings suggest that generative models are sensitive to changes in network organization with development and ageing, and may be useful tools in studying variation across individuals [99].

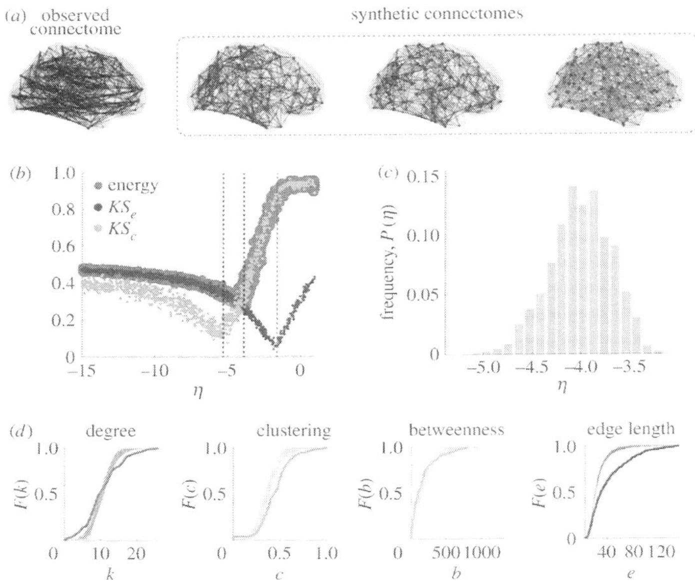

Figure 4: Summary of a geometric model for human white matter networks estimated from diffusion imaging. (*a*) Observed (black) and synthetic (colours) networks generated at different points in a predefined parameter space of interest. (*b*) Energy landscape where the energy function is defined as $E = \max(KS_K, KS_C, KS_B, KS_E)$, where each term is a Kolmogorov–Smirnov statistic comparing the degree (*K*), clustering (C), betweenness centrality (B) and edge length (E) distribution between the observed and synthetic networks

[49]. Specifically, here we show the behaviour of *KSe*, *KSc*, and energy as a function of η, a parameter that controls the characteristic connection length such that when $\eta < 0$, short-range connections are favoured, while when $\eta > 0$ long-range connections are favoured. The dashed vertical lines indicate the parameter values at which the example synthetic networks were generated. (*c*) Distribution of η parameter of top 1% lowest-energy synthetic networks aggregated across all human participants in the study. (*d*) Cumulative distributions of degree (orange), clustering coefficient (green), betweenness centrality (yellow), and edge length (purple) for observed connectome (darker line) and best fitting synthetic networks (lighter lines) for a representative participant. Reproduced with permission from [49]. (Online version in colour.).

Interestingly, Vértes *et al.* [48] found that a similar model also reproduced many of the topological features of brain networks reconstructed from function MRI data. In that study, the authors extended their analyses to show that, based on the optimal parameters fit to networks from individual subjects, the parameter space could be partitioned so as to distinguish brain networks of patients with schizophrenia from those of healthy controls.

The observation that basically the same model, when applied to structural or functional brain networks, outperformed all other models is intriguing. On one hand, it suggests that the brain's physical wiring and functional architecture might be organized according to similar principles. On the other hand, this similarity could simply be coincidental. While both Betzel *et al.* [49] and Vértes *et al.* [48] tested and compared a range of generative models, testing the space of all possible models and generative mechanisms is unfeasible. It is likely that other models not tested in either study would result in improved performance. Yet another explanation for the convergence of these two studies is that they model the brain at roughly the same organizational scale (inter-areal networks estimated from MRI data), and that deviations from this scale would uncover a different set of optimal models.

In a more recent study, Tang and colleagues study individual variation in youth by examining the white-matter networks of 882 individuals between the ages of 8 and 22 years [100]. Here, the authors posited that over this developmental time period, structural brain networks become optimized for a greater diversity of neural dynamics, as instantiated by recently defined metrics of network controllability [42]. They tested the hypothesis that an observed trajectory of network change over youth could be recapitulated by a generative model that increased average controllability (predicted ease of transitioning between nearby network states—the level of activity in each region, across the entire brain), increased modal controllability (predicted ease of transitioning between distant network states) and decreased synchronizability (predicted capacity for global synchronization). The model was initiated with a given brain network, and then evolved *in silico* according to a rewiring rule such that an existing edge was randomly chosen to take the place of an edge that did not exist, and this edge swap was retained only if the new network advanced the Pareto front, the set of all network configurations that were optimal in their trade-off between average and modal controllability (figure 5). As rewiring progressed forward in time, a course was charted in which networks increased in controllability and decreased in synchronizability; while as rewiring progressed backwards in time, networks decreased in controllability and increased in synchronizability. The simulated developmental trajectories displayed a striking similarity in functional form to the observed developmental trajectories, suggesting a possible mechanism of human brain development that preferentially optimizes dynamic network control over static network architecture.

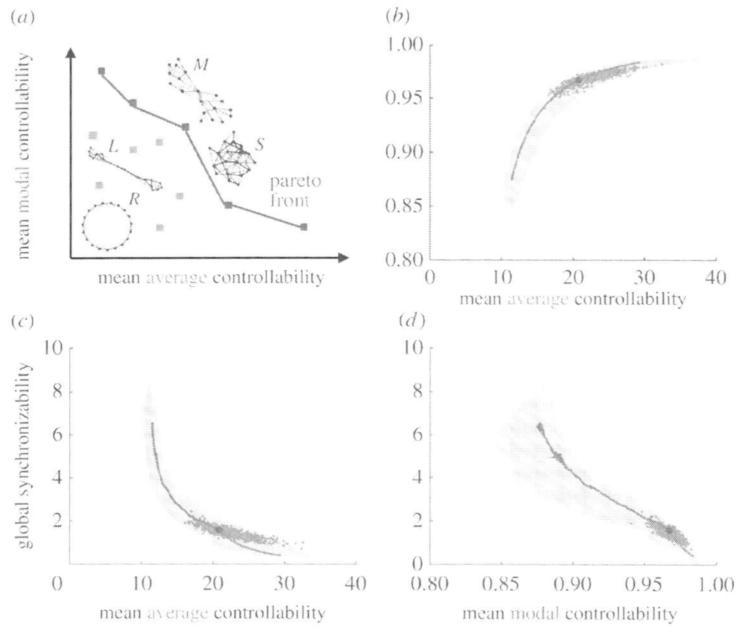

Figure 5: Brain networks are optimized for diverse dynamics over a fixed developmental time period between the age of 8 years and the age of 22 years. (*a*) Pareto optimization explores a family of networks with different topologies and

hence varying mean controllability and synchronizability (a few toy models illustrate this including a ring lattice *R*, regular lattice *L*, modular network *M* and small-world network *S*). Pareto optimal networks (purple dots) are the networks where these properties are most efficiently distributed, i.e. it is impossible to increase one property without decreasing another property—unlike in the non-optimal networks (green dots). The boundary connecting the Pareto-optimal networks forms the Pareto front (purple line). (*b,c,d*) Rewiring along Pareto fronts is a generative network process, here used to model developmental changes in network architecture according to a rewiring rule that maximizes two sorts of network controllability while minimizing synchronizability. Beginning from empirically measured brain networks of youth between the ages of 8 years and 22 years (purple dots), the authors swapped edges to modify the topology and test if the modified network advances the Pareto front. This procedure charts a course of network evolution characterized by increasingly optimal features: here the authors increased the mean average controllability and the mean modal controllability, and decrease global synchronizability, in 1500 edge swaps (yellow curves). For comparison, the authors also evolved the network in the opposite direction (to decrease controllability and increase synchronzability, pink curves). The trajectory for one subject (blue dot) is highlighted (orange and red). The simulated trajectories defined by the rewiring rules of increasing network controllability and decreasing synchronizability provided a suprisingly good fit to the observed data, suggesting that they are sufficient mechanisms for the empirical observations. Reproduced with permission from [100]. (Online version in colour.).

A FEW RELEVANT DISTINCTIONS

In this section, we describe a few important distinctions that are particularly relevant to the understanding and further development of generative network models for neural systems. First, we will explore the relations between generative models that seek mechanisms and explanations, and null models for statistical testing of hypotheses. Second, we will discuss the important trade-off in sufficiency of a generative model versus redundancy. Third, we will seek to disambiguate between inferring a possible mechanism versus claiming proof of a mechanism. And finally, we will describe some relevant considerations when building or evaluating generative models of structural versus functional connectivity.

Generative Models and Null Models

The stated goals of the generative modelling approach, as described in the early sections of this review, include the identification of putative mechanisms of observed network architecture, and intuitive explanations for some of the features that characterize that architecture. Yet, depending on their degree of biological realism, such models can also be used as statistical null models, potentially enabling the dismissal of a null hypothesis. In general, topological and spatially informed null models play a critical role in network science broadly [101–103], and network neuroscience specifically [92,104,105]. One could consider using a generative model to test the hypothesis that the topology of an empirically measured neural network was consistent with a topology of an artificial network built on a fixed set of rules or principles. In this case, one would need to be careful in the exposition of the study to distinguish between when the model was being used to propose a generative mechanism, and when the model was being used in a statistical sense to dismiss a null hypothesis.

Sufficiency and Redundancy

When building generative network models of neural systems, a common observation is that the models often fit topological signatures that they were designed to fit, but do not fit topological signatures that they were not designed to fit [96] (although see also [90,97,106]). It is important to ask whether such a model is sufficient, or whether one should seek a model that also predicts a topological signature that has not been hard coded into the objective function and/or generative algorithm. In addition to sufficiency, one might also wish to consider model redundancy: does the model combine two or more wiring rules that both induce the same topoogical signature? Such a scenario can be quite common, as there exist whole families of graphs that display similar graph metric values [107], community structure [108], controllability profiles [94] and homological features [95]. Broadly, one might wish to build generative models that balance a trade-off between (i) sufficiency, potentially enabled by a greater number of wiring rules, and (ii) redundancy, whose relative minimization is supported by notions of biological efficiency and parsimony.

Inferring and Claiming Mechanism

Suppose that one is thoroughly successful, and creates a generative model that beautifully reproduces an empirically observed network structure. Do the rules that compose the generative model provide a mechanism explaining the empirical network's architecture [109]? Even more brazenly, can such a generative model help us to develop a theory of brain network organization and resultant behaviour [110]? In seeking answers to these questions, it is important to disambiguate between inferring a possible mechanism and claiming proof of a mechanism. If a generative network model built upon rule *a* recapitulates the network structure of interest, one can say that rule *a* is a possible mechanism, but one cannot claim that it is *the* mechanism. To provide a more concrete example embedded in network neuroscience, let us consider the topological feature of Rentian scaling, an isometric scaling relationship between the number of processing elements and the number of connections, which is often found in systems that are built upon the principle of wiring reduction, and is observed in brain networks [111] as well as other transmission systems such as computer circuits [112], transportation systems [113] and vasculature [114]. Given the scaling relationship, one might infer that the network's structure is given by a mechanism that operates uniformly across all scales such as wiring minimization. However, such an inference would neglect the fact that many scale-heterogeneous mechanisms also produce topological scaling relationships [59]. In future work, it will be important to concretely discuss support for possible mechanisms separately from exact claims that such mechanisms have been proven.

Functional Connectivity and Structural Connectivity

This review has focused on mostly generative models for structural networks, where links represent physical pathways among neural elements. Generative network models can also be built for functional connectivity data, with some caveats and limitations [48,55,115]. Posited drivers of functional network organization across species include similar notions of cost-efficiency [116–118], small-world architecture [119] and spatial clustering [120]. However, the appropriate growth mechanisms that such generative models employ face different constraints in the functional domain from those in the structural domain [121]. Functional connectivity is not generated piece by piece, as instantiated by a discrete placement of edges in a network [122]. Instead, functional connectivity is a consequence of dynamical processes constrained by many factors [123], including but not limited to anatomical structure [124–126], the activity elicited by a particular task [127], the distance between brain areas [123], genetics [128–130] and any stimulation or other input to the system [131,132]. Many good models of brain dynamics exist, ranging from the biologically realistic to the heavily idealized [57].

However, growth models built from the placement of independent edges are conceptually more appropriate for structural networks than for functional networks.

A particularly powerful generative model for producing a functional network topology from a structural network topology (or for inferring a structural network topology from a functional network topology) is the pairwise maximum entropy model [133–135]. The technique was initially applied to neural spiking data to demonstrate that pairwise interactions give an excellent approximation of the full correlation network [136]. More recently, the technique has been used to accurately fit fMRI BOLD collected in humans [137,138], dynamic functional connectivity patterns [139] and patterns of transitions between brain states [140].

While the pairwise maximum entropy model has proven useful in inferring structural network organization from functional network organization, and vice versa, it is certainly true that non-pairwise interactions may nevertheless play a non-trivial role in neural population function. Intuitively, beyond-pairwise interactions can occur via common input [141], producing multiway synchrony [142] with varying prevalence across different length scales in the system [143]. Generative models of such high-order relations include beyond-pairwise maximum entropy models [144] and dichotomous Gaussian models [145]. Another way in which beyond-pairwise functional interactions can occur is if neurons themselves do not only display pairwise connections, but also higher-order connections. This possibility highlights a complementary challenge in describing the presence of such higher-order relations in structural networks from a topological point of view, with the goal of building generative models that account for them. We will discuss initial efforts to address these challenges using notions from algebraic topology in the next section.

FUTURE DIRECTIONS

In this section, we discuss future directions in efforts to develop, extend and apply network generative modelling approaches to questions of import to neuroscience. We begin by offering a description of what generative models could accomplish, with a particular focus on clinical applications.

We next consider dream datasets and experiments whose acquisition and open sharing would inherently change the sorts of questions that network generative modelling approaches could tackle. Finally, we discuss a few natural directions in which to increase the sophistication of network generative models, including the consideration of multilayer networks and simplicial complexes. This section is purposefully more forward-looking and speculative than the previous sections, but we nevertheless offer a generous helping of appropriate citations to the relevant domain-specific efforts.

What would a generative model accomplish?

In practice, many of the current approaches for studying biological neural networks involve computing and comparing summary statistics between groups or continuously across individuals. While this approach is useful in identifying 'what' is different, it fails to explain 'how' those differences come to be, in the first place. In this review, we echo other recent reviews [23,24] and call for a shift in emphasis away from 'fact collecting' studies and towards uncovering the mechanisms that explain the organization of neural systems. We argue that network generative modelling represents a framework that can help us move towards addressing these lofty goals.

Suppose that—with the right dataset and the right modelling approach—we can devise a model that, to a reasonable approximation, can successfully mimic the growth or evolution of a real-world neural system. In other words, the model results in a network that changes over time (where time has a clear developmental or biological interpretation) and whose topology evolves in a way that is consistent with known facts about the real-world growth of that network. What does having such a model buy us? On the one hand, we could simply maintain the *status quo*, fit the model's parameters to individual subjects and compute statistical relationships between parameters and behavioural measures (figure 6*a*) using machine learning techniques to partition the model's parameter space into regions associated with clinical and control populations (figure 6*b*). While useful, these approaches are quite similar to the current state of the field.

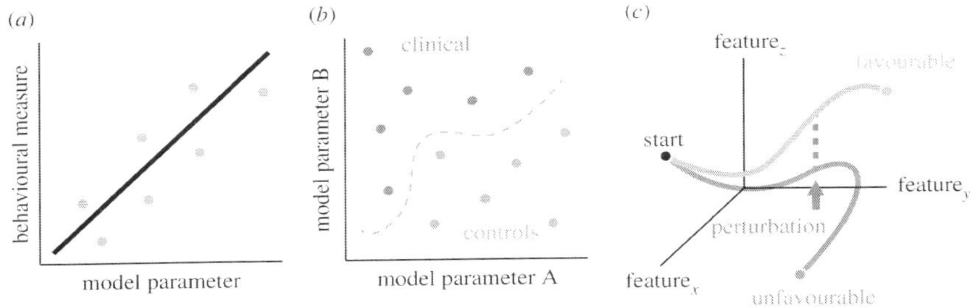

Figure 6: Applications using generative models. Model parameters can be fit to individual subjects and those parameters compared to some behavioural measures (*a*) or used to classify different populations from one another (*b*). Generative models can also be used to simulate the development of a biological neural network (*c*). These simulations can be used as forecasting devices to identify individuals at risk of developing maladaptive network topologies. They can also be used to explore possible interventions, e.g. perturbations to parameters or wiring rules, that drive an individual away from an unfavourable, maladaptive network topology towards a more favourable state. (Online version in colour.).

On the other hand, another more novel possibility is to use the model for disease simulation. Many psychiatric [146] and neurodegenerative diseases [147] are manifest at the network level in the form of miswired or dysconnected systems [148], but it is unclear what predisposes an individual to evolve into a disease state. The generative model can be used to propagate individuals from one time point to another and identify those that are likely to evolve into a state similar to that of the disease phenotype and perhaps likely to develop that disease. In this way, the model has a clear role as a forecaster (figure 6*c*).

Similarly, the generative model can be used to explore *in silico* the effect of potential intervention strategies. We can think of biological neural networks as living in a high-dimensional space based on their topological characteristics, where some regions (of this space; not of the brain) are associated with neurological disease and considered maladaptive (and perhaps even deadly) [149,150]. In this context, the generative model represents an evolution operator that propagates a network from one point to another, tracing out a trajectory through this space. If we can identify individuals who are predisposed to travel near those maladative regions, we can begin to identify perturbations—changes to model parameters or wiring rules—that steer those trajectories towards regions not associated with disease (figure 6*c*). These goals are in line with current theoretical work, applying tools from network control theory to neuroimaging data [42,151,152].

Dream Datasets and Experiments

Generative models have clear utility in furthering our capacity to predict disease and identify the mechanisms that shape the development, growth and evolution of biological neural networks. A major hindrance in realizing these goals, however, is the absence of data tailored for generative models. The ideal data would (i) be longitudinal, enabling one to track and incorporate individual-level changes over time in the model, and (ii) include multiple data modalities, such as functional and structural connectivity, and genetics, along with other select factors that could influence network-level organization. In short, any metadata that could theoretically be incorporated into a model would be valuable and possibly worth collecting. Ideally, these data would be acquired at the earliest possible time point *in utero* [153] and proceed through maturity.

Clearly, collecting and curating such a dataset represents a massive undertaking. Though recent large-scale studies have made it possible to image thousands of individuals over a short period of time [2,3,154] and a small number of individuals over a long period of time [155–157], the duration and scale of a longitudinal study of the nature proposed here seems, at present, out of reach. Furthermore, the studies that have come closest to acquiring these kinds of data have relied on MRI due to its non-invasive nature. However, this same advantage also limits the fidelity and kinds of data that can be acquired from an individual (e.g. region-specific gene transcription levels can only be acquired post-mortem [158]).

An attractive alternative, then, is to consider building generative models of data from non-human model organisms. Not only are the life cycles of several model organisms much shorter than that of humans (making it possible to track an individual over the course of its entire life), but new advances in network reconstruction techniques [159–161] and the ability to make recordings of activity in unprecedented detail [162,163] ensure that any generative model will be endowed with sufficiently rich data to probe for novel wiring rules. Moreover, working with model organisms also makes it possible to collect data modalities that, otherwise, would be inaccessible, including details about gene expression [164].

Increasing Sophistication of Network Generative Models

Finally, given ideal data, there are also exciting and important future directions in increasing the mathematical sophistication of network generative models. One particularly accessible extension of current methods lies in multilayer network generative models. A multilayer network consists of multiple single-layer networks, e.g. representing a neural system's structural connectivity, functional connectivity and gene co-expression [165,166], that are linked across layers to one another. A generative model for these types of data is one that, instead of single-layer networks, generates multilayer networks [167], and the rules of generation can apply to a single layer, to multiple layers or to the interconnectivity between layers [168]. One potentially useful place to start would be to construct multilayer generative models where the neural connectivity evolves with a specific set of dynamics (or network growth rules) that are explicitly coupled to the underlying tissue growth or to the inervating vasculature growth [169]. At the larger scale, one could also consider developing multilayer generative models that couple brain network growth with social network growth, a coupling that has recently been postulated to occur through processes of development and learning [170].

Indeed, it is likely that there are other ways in which our brain network topology and changes in that topology are coupled to our experiences. Such experiences could be defined by our environment, for example as partially stipulated by our socio-economic status [171], or by our practices, for example as instantiated in our practice of curiosity [172]. Indeed, it is interesting to speculate that network generative models may be useful in understanding the relations between brain network architecture and the architecture of knowledge networks, which are physically instantiated in the brain [173], as well as semantic networks [174], which can be tuned by our attention [175]. Semantic networks, social networks, brain networks, vasculature networks and tissue networks may all evolve with one another in intertwined multilayer network systems, an understanding of any pair of which will require concerted efforts in extending the sophistication of current network generative modelling techniques.

In addition to multilayer approaches, network generative modelling could also benefit from incorporating methods to address the existence and growth of non-pairwise relations between nodes. A useful language with which to meet this challenge is the language of algebraic topology and specifically simplicial complexes [176] whose fundamental units are simplices: a

0-simplex is a node, a 1-simplex is a dyad, a 2-simplex is a face, a 3-simplex is a tetrahedron, a 4-simplex is a 5-cell, etc. A collection of simplices—called a simplicial complex—can include many interesting features including cliques (i.e. fully connected subgraphs) and cavities (collections of n-simplices arranged so that they have an empty geometric boundary). In patterns of correlations among the activity of pyramidal neurons in rat hippocampus, the topology of cliques and cavities demonstrates geometric organization consistent with a generative model of simplicial complexes related to random geometric graphs [177]. This higher-order structure has also enabled the identification of unexpectedly long structural loops linking regions of early and late evolutionary origin, underscoring their unique role in controlling brain function [178]. Indeed, the topology of cliques and cavities has specific implications for local processing (cliques) versus processing in which information may flow in either diverging or converging patterns (cavities) [178], and can support efficient coding by enabling inference of neural codes even in highly undersampled set of patterns [179]. While generative models of simplicial complexes based on random geometric graphs have shown some utility in explaining these structures, further work is needed to understand the extent of their applicability, and to consider models for growing simplicial complexes [180].

Another way in which the network generative modelling framework can be extended is to consider the effect of generative mechanisms on not only the structural properties of networks, but also the dynamics that they support [181]. Different growth mechanisms result in particular network motifs and topological features. In doing so, they shape flows over the network, organize dynamics and contribute to determining a network's functional fingerprint [182,183]. Incorporating information about dynamics into objective functions or the generative process itself facilitates the exploration of a richer tapestry of models. This theoretical work could eventually be used to model the functional effect of perturbations to the network's structure, retuning of a model's parameters or alterations to the wiring rule itself [184].

CONCLUSION

As the field of network neuroscience matures, efforts in data description and statistical characterization are being complemented by efforts to infer principles, to predict unobserved data and to perturb the system with theoretically grounded expectations about the results of those perturbations. Generative modelling is a particularly powerful approach for moving beyond description towards prediction, mechanism and eventually theory. In this article, we have offered a simple primer on generative models, a review of recent efforts in generative models of biological neural networks, and a discussion of current frontiers in empirical data collection and mathematical sophistication. We look forward with anticipation to efforts in the coming years that use generative models to understand human development, and to potentially inform interventions in psychiatric disease or neurological disorders in which wiring patterns have gone awry.

ACKNOWLEDGEMENTS

The authors thank Lia Papadopoulos and Evelyn Tang for helpful comments on earlier versions of this manuscript.

REFERENCES

1. Newman M. 2010 Networks: an introduction. Oxford, UK: Oxford university press.
2. Nooner KB et al. 2012 The NKI-Rockland sample: a model for accelerating the pace of discovery science in psychiatry. Front. Neurosci. 6, 152. (doi:10.3389/ fnins.2012.00152)
3. VanEssen DC, Smith SM, Barch DM, Behrens TE, Yacoub E, Ugurbil K, WU-Minn HCP Consortium. 2013 The WU-Minn human connectome project: an overview. Neuroimage 80, 62 – 79. (doi:10.1016/j. neuroimage.2013.05.041)
4. Rubinov M, Sporns O. 2010 Complex network measures of brain connectivity: uses and interpretations. Neuroimage 52, 1059– 1069. (doi:10.1016/j.neuroimage.2009.10.003)
5. Sizemore AE, Bassett DS. 2017 Dynamic graph metrics: tutorial, toolbox, and tale. Neuroimage S1053– 8119, 30 564 – 300 565.
6. Bassett DS, Sporns O. 2017 Network neuroscience. Nat. Neurosci. 20, 353– 364. (doi:10.1038/nn.4502)
7. van den Heuvel MP, Bullmore ET, Sporns O. 2016 Comparative connectomics. Trends Cogn. Sci. 20, 345– 361. (doi:10.1016/j. tics.2016.03.001)
8. Bassett DS, Bullmore E. 2006 Small-world brain networks. Neuroscientist 12, 512– 523. (doi:10. 1177/1073858406293182)
9. Hagmann P, Cammoun L, Gigandet X, Meuli R, Honey CJ, Wedeen VJ, Sporns O. 2008 Mapping the structural core of human cerebral cortex. PLoS Biol. 6, e159. (doi:10.1371/journal.pbio.0060159)
10. Markov NT et al. 2012 A weighted and directed interareal connectivity matrix for macaque cerebral cortex. Cereb. Cortex 24, 17 – 36. (doi:10.1093/ cercor/bhs270)
11. Sporns O, Betzel RF. 2016 Modular brain networks. Annu. Rev. Psychol. 67, 613 – 640. (doi:10.1146/ annurev-psych-122414-033634)
12. Markov NT, Ercsey-Ravasz M, Van Essen DC, Knoblauch K, Toroczkai Z, Kennedy H. 2013 Cortical high-density counterstream architectures. Science 342, 1238406. (doi:10.1126/science.1238406)
13. Bassett DS, Wymbs NF, Rombach MP, Porter MA, Mucha PJ, Grafton ST. 2013 Task-based coreperiphery organization of human brain

dynamics. PLoS Comput. Biol. 9, e1003171. (doi:10.1371/ journal.pcbi.1003171)

14. Zamora-Lo´pez G, Zhou C, Kurths J. 2010 Cortical hubs form a module for multisensory integration on top of the hierarchy of cortical networks. Front. Neuroinform. 4, 1. (doi:10.3389/neuro.11.001.2010)
15. Van Den Heuvel MP, Sporns O. 2011 Rich-club organization of the human connectome. J. Neurosci. 31, 15 775– 15 786. (doi:10.1523/JNEUROSCI.3539- 11.2011)
16. Kaiser M, Hilgetag CC. 2006 Nonoptimal component placement, but short processing paths, due to longdistance projections in neural systems. PLoS Comput. Biol. 2, e95. (doi:10.1371/journal.pcbi.0020095)
17. Horva´t S et al. 2016 Spatial embedding and wiring cost constrain the functional layout of the cortical network of rodents and primates. PLoS Biol. 14, e1002512. (doi:10.1371/journal.pbio.1002512)
18. Sporns O. 2013 Network attributes for segregation and integration in the human brain. Curr. Opin. Neurobiol. 23, 162– 171. (doi:10.1016/j.conb.2012. 11.015)
19. Bullmore E, Sporns O. 2012 The economy of brain network organization. Nat. Rev. Neurosci. 13, 336– 349. (doi:10.1038/nrn3214)
20. Pen˜a-Go´mez C, Avena-Koenigsberger A, Sepulcre J, Sporns O. 2017 Spatiotemporal network markers of individual variability in the human functional connectome. Cereb. Cortex 11, 1– 13. (doi:10.1093/ cercor/bhx170)
21. Fornito A, Zalesky A, Breakspear M. 2015 The connectomics of brain disorders. Nat. Rev. Neurosci. 16, 159– 172. (doi:10.1038/nrn3901)
22. Schiff SJ. 2012 Neural control engineering: the emerging intersection between control theory and neuroscience. Cambridge, MA: MIT Press.
23. Ve´rtes PE, Bullmore ET. 2015 Annual research review: growth connectomics—the organization and reorganization of brain networks during normal and abnormal development. J. Child Psychol. Psychiatry 56, 299– 320. (doi:10.1111/jcpp.12365)
24. Kaiser M. 2017 Mechanisms of connectome development. Trends. Cogn. Sci. 21, 703– 717. (doi:10.1016/j.tics.2017.05.010)
25. Betzel RF, Bassett DS. 2016 Multi-scale brain networks. Neuroimage 160, 73 – 83. (doi:10.1016/j. neuroimage.2016.11.006)
26. Erdos P, Re´nyi A. 1960 On the evolution of random graphs. Publ. Math. Inst. Hung. Acad. Sci. 5, 17 – 60.

27. Wasserman S, Anderson C. 1987 Stochastic a posteriori blockmodels: construction and assessment. Social Netw. 9, 1– 36. (doi:10.1016/0378-8733(87)90015-3)
28. Watts DJ, Strogatz SH. 1998 Collective dynamics of small-world networks. Nature 393, 440– 442. (doi:10.1038/30918)
29. Baraba´si AL, Albert R. 1999 Emergence of scaling in random networks. Science 286, 509– 512. (doi:10. 1126/science.286.5439.509)
30. Sole´ RV, Pastor-Satorras R, Smith E, Kepler TB. 2002 A model of large-scale proteome evolution. Adv. Complex Syst. 5, 43 – 54. (doi:10.1142/ S021952590200047X)
31. Va´zquez A, Flammini A, Maritan A, Vespignani A. 2003 Modeling of protein interaction networks. Complexus 1, 38 – 44. (doi:10.1159/000067642)
32. MiddendorfM, Ziv E, Wiggins CH. 2005 Inferring network mechanisms: the Drosophila melanogaster protein interaction network. Proc. Natl Acad. Sci. USA 102, 3192–3197. (doi:10.1073/pnas.0409515102)
33. Kumar R, Raghavan P, Rajagopalan S, Sivakumar D, Tomkins A, Upfal E. 2000 Stochastic models for the web graph. In Proc. 41st IEEE Symposium on Foundations of Computer Science (FOCS) 2000, pp. 57– 65. Piscataway, NJ: IEEE.
34. Kumar R, Novak J, Tomkins A. 2010 Structure and evolution of online social networks. In Link mining: models, algorithms, and applications (eds PS Yu, J Han, C Faloutsos), pp. 337– 357. Berlin, Germany: Springer.
35. Rubinov M. 2016 Constraints and spandrels of interareal connectomes. Nat. Commun. 7, 13812. (doi:10.1038/ncomms13812)
36. Bassett DS, Khambhati AN. 2017 A network engineering perspective on probing and perturbing cognition with neurofeedback. Ann NY Acad. Sci. 1396, 126 – 143. (doi:10.1111/nyas.13338)
37. Murphy AC, Bassett DS. 2017 A network neuroscience of neurofeedback for clinical translation. Curr. Opin. Biomed. Eng. 1, 63 – 70. (doi:10.1016/j.cobme.2017.03.003)
38. Barthe´lemy M. 2011 Spatial networks. Phys. Rep. 499, 1– 101. (doi:10.1016/j.physrep.2010.11.002)
39. Dall J, Christensen M. 2002 Random geometric graphs. Phys. Rev. E 66, 016121. (doi:10.1103/ PhysRevE.66.016121)

40. Kaiser M, Hilgetag CC. 2004 Spatial growth of realworld networks. Phys. Rev. E 69, 036103. (doi:10. 1103/PhysRevE.69.036103)
41. Snijders TA, Nowicki K. 1997 Estimation and prediction for stochastic blockmodels for graphs with latent block structure. J. Classif. 14, 75 – 100. (doi:10.1007/s003579900004)
42. Gu S et al. 2015 Controllability of structural brain networks. Nat. Commun. 6, 8414. (doi:10.1038/ ncomms9414)
43. Bolloba´s B. 1998 Random graphs. In Modern graph theory, pp. 215–252. Berlin, Germany: Springer.
44. Clauset A, Shalizi CR, Newman ME. 2009 Power-law distributions in empirical data. SIAM Rev. 51, 661– 703. (doi:10.1137/070710111)
45. Price DJdS. 1976 A general theory of bibliometric and other cumulative advantage processes. J. Am. Soc. Inform. Sci. 27, 292–306. (doi:10.1002/asi. 4630270505)
46. Holland PW, Laskey KB, Leinhardt S. 1983 Stochastic blockmodels: first steps. Social Netw. 5, 109– 137. (doi:10.1016/0378-8733(83)90021-7)
47. Costa LF, Kaiser M, Hilgetag CC. 2007 Predicting the connectivity of primate cortical networks from topological and spatial node properties. BMC. Syst. Biol. 1, 16. (doi:10.1186/1752-0509-1-16)
48. Ve´rtes PE, Alexander-Bloch AF, Gogtay N, Giedd JN, Rapoport JL, Bullmore ET. 2012 Simple models of human brain functional networks. Proc. Natl Acad. Sci. USA 109, 5868–5873. (doi:10.1073/ pnas.1111738109)
49. Betzel RF et al. 2016 Generative models of the human connectome. Neuroimage 124, 1054–1064. (doi:10.1016/j.neuroimage.2015.09.041)
50. Costa LDF, Zawadzki K, Miazaki M, Viana MP, Taraskin SN. 2010 Unveiling the neuromorphological space. Front. Comput. Neurosci. 4, 150. (doi:10. 3389/fncom.2010.00150)
51. Nicosia V, Vertes PE, Schafer WR, Latora V, Bullmore ET. 2013 Phase transition in the economically modeled growth of a cellular nervous system. Proc. Natl Acad. Sci. USA 110, 7880 – 7885. (doi:10.1073/ pnas.1300753110)
52. Simpson SL, Hayasaka S, Laurienti PJ. 2011 Exponential random graph modeling for complex brain networks. PLoS ONE 6, e20039. (doi:10.1371/ journal.pone.0020039)
53. Simpson SL, Moussa MN, Laurienti PJ. 2012 An exponential random graph modeling approach to creating group-based representative

whole-brain connectivity networks. Neuroimage 60, 1117 – 1126. (doi:10.1016/j.neuroimage.2012.01.071)

54. Stillman PE, Wilson JD, Denny MJ, Desmarais BA, Bhamidi S, Cranmer SJ, Lu ZL. 2017 Statistical modeling of the default mode brain network reveals a segregated highway structure. Sci. Rep. 7, 11694. (doi:10.1038/s41598-017-09896-6)
55. Obando C, Fallani FDV. 2017 A statistical model for brain networks inferred from large-scale electrophysiological signals. J. R. Soc. Interface 14, 20160940. (doi:10.1098/rsif.2016.0940)
56. Goldenberg A, Zheng AX, Fienberg SE, Airoldi EM. 2010 A survey of statistical network models. Found. Trendsw Mach. Learn. 2, 129 – 233. (doi:10.1561/ 2200000005)
57. Breakspear M. 2017 Dynamic models of large-scale brain activity. Nat. Neurosci. 20, 340– 352. (doi:10. 1038/nn.4497)
58. Khambhati AN, Sizemore AE, Betzel RF, Bassett DS. 2017 Modeling and interpreting mesoscale network dynamics. Neuroimage S1053–8119, 30 500 – 30 501. (doi:10.1016/j.neuroimage.2017.06.029)
59. Stumpf MP, Porter MA. 2012 Mathematics. Critical truths about power laws. Science 335, 665– 666. (doi:10.1126/science.1216142)
60. Karbowski J. 2001 Optimal wiring principle and plateaus in the degree of separation for cortical neurons. Phys. Rev. Lett. 86, 3674– 3677. (doi:10. 1103/PhysRevLett.86.3674)
61. Laughlin SB, de Ruyter van Steveninck RR, Anderson JC. 1998 The metabolic cost of neural information. Nat. Neurosci. 1, 36–41. (doi:10.1038/236)
62. Borisyuk R, Kalam al Azad A, Conte D, Roberts A, Soffe S. 2014 A developmental approach to predicting neuronal connectivity from small biological datasets: a gradient based neuron growth model. PLoS ONE 9, e89461. (doi:10.1371/journal. pone.0089461)
63. Kennedy H, Dehay C. 1993 The importance of developmental timing in cortical specification. Perspect. Dev. Neurobiol. 1, 93 – 99.
64. Cherniak C. 1992 Local optimization of neuron arbors. Biol. Cybern. 66, 503 – 510. (doi:10.1007/ BF00204115)
65. Cherniak C, Changizi M, Kang D. 1999 Large-scale optimization of neuron arbors. Phys. Rev. E 59, 6001 – 6009. (doi:10.1103/ PhysRevE.59.6001)

66. Wen Q, Stepanyants A, Elston GN, Grosberg AY, Chklovskii DB. 2009 Maximization of the connectivity repertoire as a statistical principle governing the shapes of dendritic arbors. Proc. Natl Acad. Sci. USA 106, 12 536 – 12 541. (doi:10.1073/ pnas.0901530106)
67. Chklovskii DB. 2004 Synaptic connectivity and neuronal morphology: two sides of the same coin. Neuron 43, 609– 617. (doi:10.1016/j. neuron.2004. 08.012)
68. Song S, Sjostrom PJ, Reigl M, Nelson S, Chklovskii DB. 2005 Highly nonrandom features of synaptic connectivity in local cortical circuits. PLoS Biol. 3, e68. (doi:10.1371/journal.pbio.0030068)
69. Cherniak C. 2012 Neural wiring optimization. Prog. Brain Res. 195, 361– 371. (doi:10.1016/B978-0- 444-53860-4.00017-9)
70. Rivera-Alba M, Peng H, de Polavieja GG, Chklovskii DB. 2014 Wiring economy can account for cell body placement across species and brain areas. Curr. Biol. 24, R109 –R110. (doi:10.1016/j.cub.2013.12.012)
71. Rivera-Alba M, Vitaladevuni SN, Mishchenko Y, Lu Z, Takemura SY, Scheffer L, Meinertzhagen IA, Chklovskii DB, de Polavieja GG. 2011 Wiring economy and volume exclusion determine neuronal placement in the Drosophila brain. Curr. Biol. 21, 2000– 2005. (doi:10.1016/j. cub.2011.10.022)
72. Bassett DS, Bullmore ET. 2016 Small-world brain networks revisited. Neuroscientist. 23, 499– 516.
73. Wiles L, Gu S, Pasqualetti F, Parvesse B, Gabrieli D, Bassett DS, Meaney DF. 2017 Autaptic connections shift network excitability and bursting. Sci. Rep. 7, 44006. (doi:10.1038/srep44006)
74. Li WC, Cooke T, Sautois B, Soffe SR, Borisyuk R, Roberts A. 2007 Axon and dendrite geography predict the specificity of synaptic connections in a functioning spinal cord network. Neural. Dev. 2, 17. (doi:10.1186/1749-8104-2-17)
75. Sautois B, Soffe S, Li WC, Roberts A. 2007 Role of type-specific neuron properties in a spinal cord motor network. J. Comput. Neurosci. 23, 59– 77. (doi:10.1007/s10827-006-0019-1)
76. Roberts A, Conte D, Hull M, Merrison-Hort R, Kalam al Azad A, Bhul E, Borisyuk R, Soffe SR. 2014 Can simple rules control development of a pioneer vertebrate neuronal network generating behaviour? J. Neurosci. 34, 608 – 621. (doi:10.1523/JNEUROSCI. 3248-13.2014)

77. Jones SR, Pritchett DL, Sikora MA, Stufflebeam SM, Hamalainen M, Moorem CI. 2009 Quantitative analysis and biophysically realistic neural modeling of the MEG mu rhythm: rhythmogenesis and modulation of sensory-evoked responses. J. Neurophysiol. 102, 3554 – 3572. (doi:10.1152/jn. 00535.2009)
78. Jones SR. 2016 When brain rhythms aren't 'rhythmic': implication for their mechanisms and meaning. Curr. Opin. Neurobiol. 40, 72 – 80. (doi:10. 1016/j.conb.2016.06.010)
79. Kopell NJ, Gritton HJ, Whittington MA, Kramer MA. 2014 Beyond the connectome: the dynome. Neuron 83, 1319–1328. (doi:10.1016/j. neuron.2014.08.016)
80. Tononi G, Sporns O, Edelman GM. 1994 A measure for brain complexity: relating functional segregation and integration in the nervous system. Proc. Natl Acad. Sci. USA 91, 5033– 5037. (doi:10.1073/pnas. 91.11.5033)
81. Chen Y, Wang S, Hilgetag CC, Zhou C. 2013 Tradeoff between multiple constraints enables simultaneous formation of modules and hubs in neural systems. PLoS Comput. Biol. 9, e1002937. (doi:10.1371/ journal.pcbi.1002937)
82. Hinne M, Meijers A, Bakker R, Tiesinga PH, Morup M, van Gerven MA. 2017 The missing link: predicting connectomes from noisy and partially observed tract tracing data. PLOS Comput. Biol. 13, e1005374. (doi:10.1371/journal.pcbi. 1005374)
83. Beul SF, Grant S, Hilgetag CC. 2015 A predictive model of the cat cortical connectome based on cytoarchitecture and distance. Brain Struct. Funct. 220, 3167– 3184. (doi:10.1007/s00429-014-0849-y)
84. Goulas A, Uylings HB, Hilgetag CC. 2017 Principles of ipsilateral and contralateral cortico-cortical connectivity in the mouse. Brain Struct. Funct. 222, 1281– 1295. (doi:10.1007/s00429-016-1277-y)
85. Scannell JW, Blakemore C, Young MP. 1995 Analysis of connectivity in the cat cerebral cortex. J. Neurosci. 15, 1463 – 1483.
86. Cherniak C. 1994 Component placement optimization in the brain. J. Neurosci. 14, 2418– 2427.
87. Cherniak C, Mokhtarzada Z, Rodriguez-Esteban R, Changizi K. 2004 Global optimization of cerebral cortex layout. Proc. Natl Acad. Sci. USA 101, 1081– 1086. (doi:10.1073/pnas.0305212101)

88. Ercsey-Ravasz M, Markov NT, Lamy C, Van Essen DC, Knoblauch K, Toroczkai Z, Kennedy H. 2013 A predictive network model of cerebral cortical connectivity based on a distance rule. Neuron 80, 184– 197. (doi:10.1016/j.neuron.2013.07.036)
89. Beul SF, Barbas H, Hilgetag CC. 2017 A predictive structural model of the primate connectome. Sci. Rep. 7, 43176. (doi:10.1038/srep43176)
90. Bauer R, Kaiser M. 2017 Nonlinear growth: an origin of hub organization in complex networks. R. Soc. open sci. 4, 160691. (doi:10.1098/rsos.160691)
91. Raj A, Chen YH. 2011 The wiring economy principle: connectivity determines anatomy in the human brain. PLoS ONE 6, e14832. (doi:10.1371/journal. pone.0014832)
92. Samu D, Seth AK, Nowotny T. 2014 Influence of wiring cost on the large-scale architecture of human cortical connectivity. PLoS Comput. Biol. 10, e1003557. (doi:10.1371/journal.pcbi.1003557)
93. Wen Q, Chklovskii DB. 2005 Segregation of the brain into gray and white matter: a design minimizing conduction delays. PLoS Comput. Biol. 1, e78. (doi:10.1371/journal.pcbi.0010078)
94. Wu-Yan E, Betzel RF, Tang E, Gu S, Pasqualetti F, Bassett DS. 2017 Benchmarking measures of network controllability on canonical graph models. (https://arxiv.org/abs/1706.05117)
95. Sizemore A, Giusti C, Bassett DS. 2017 Classification of weighted networks through mesoscale homological features. J. Complex Netw. 5, 245 – 273. (doi:10.1093/comnet/cnw013)
96. Klimm F, Bassett DS, Carlson JM, Mucha PJ. 2014 Resolving structural variability in network models and the brain. PLoS Comput. Biol. 10, e1003491. (doi:10.1371/journal.pcbi.1003491)
97. Jarman N, Trengove C, Steur E, Tyukin I, van Leeuwen C. 2014 Spatially constrained adaptive rewiring in cortical networks creates spatially modular small world architectures. Cogn. Neurodyn. 8, 479 – 497. (doi:10.1007/s11571-014-9288-y)
98. Bassett DS, Brown JA, Deshpande V, Carlson JM, Grafton ST. 2011 Conserved and variable architecture of human white matter connectivity. Neuroimage 54, 1262– 1279. (doi:10.1016/j.neuroimage.2010. 09.006)
99. Zuo XN, He Y, Betzel RF, Colcombe S, Sporns O, Milham MP. 2016 Human connectomics across the life span. Trends Cogn. Sci. 21, 32 – 45. (doi:10. 1016/j.tics.2016.10.005)

100. Tang E et al. 2017 Developmental increases in white matter network controllability support a growing diversity of brain dynamics. Nat. Commun. 8, 1252. (doi:10.1038/s41467-017-01254-4)

101. Bazzi M, Jeub LGS, Arenas A, Howison SD, Porter MA. 2016 Generative benchmark models for mesoscale structure in multilayer networks. (http:// arxiv.org/abs/1608.06196)

102. Fosdick BK, Larremore DB, Nishimura J, Ugander J. 2016 Configuring random graph models with fixed degree sequences. arXiv (http://arxiv.org/ abs/1608.00607)

103. Wiedermann M, Donges JF, Kurths J, Donner RK. 2016 Spatial network surrogates for disentangling complex system structure from spatial embedding of nodes. Phys. Rev. E 93, 042308. (doi:10.1103/PhysRevE.93.042308)

104. Bassett DS, Porter MA, Wymbs NF, Grafton ST, Carlson JM, Mucha PJ. 2013 Robust detection of dynamic community structure in networks. Chaos 23, 013142. (doi:10.1063/1.4790830)

105. Rubinov M, Sporns O. 2011 Weight-conserving characterization of complex functional brain networks. Neuroimage 56, 2068– 2079. (doi:10. 1016/j.neuroimage.2011.03.069)

106. Rubinov M, Sporns O, Thivierge JP, Breakspear M. 2011 Neurobiologically realistic determinants of self-organized criticality in networks of spiking neurons. PLoS Comput. Biol. 7, e1002038. (doi:10. 1371/journal.pcbi.1002038)

107. da F Costa L, Rodrigues FA, Travieso G, Villas Boas PR. 2007 Characterization of complex networks: a survey of measurements. Adv. Phys. 56, 167– 242. (doi:10.1080/00018730601170527)

108. Onnela JP, Fenn DJ, Reid S, Porter MA, Mucha PJ, Fricker MD, Jones NS. 2012 Taxonomies of networks from community structure. Phys. Rev. E 86, 036104. (doi:10.1103/PhysRevE.86.036104)

109. Craver CF. 2005 Beyond reduction: mechanisms, multifield integration and the unity of neuroscience. Stud. Hist. Philos. Biol. Biomed. Sci. 36, 373– 395. (doi:10.1016/j.shpsc.2005.03.008)

110. Bassett DS, Mattar MG. 2017 A network neuroscience of human learning: potential to inform quantitative theories of brain and behavior. Trends Cogn. Sci. 21, 250– 264. (doi:10.1016/j.tics.2017. 01.010)

111. Bassett DS, Greenfield DL, Meyer-Lindenberg A, Weinberger DR, Moore SW, Bullmore ET. 2010 Efficient physical embedding of

topologically complex information processing networks in brains and computer circuits. PLoS Comput. Biol. 6, e1000748. (doi:10.1371/journal.pcbi.1000748)

112. Landman BS, Russo RL. 1971 On a pin versus block relationship for partitions of logic graphs. IEEE Trans. Comput. C–20, 1469 – 1479. (doi:10.1109/TC.1971.223159)
113. Sperry MM, Telesford QK, Klimm F, Bassett DS. 2017 Rentian scaling for the measurement of optimal embedding of complex networks into physical space. J. Complex Netw. 5, 199– 218. (doi:10.1093/ comnet/cnw010)
114. Papadopoulos L, Blinder P, Ronellenfitsch H, Klimm F, Katifori E, Kleinfeld D, Bassett DS. 2016 Embedding of biological distribution networks with differing environmental constraints. arXiv (http:// arxiv.org/abs/1612.08058)
115. Vertes PE, Alexander-Bloch A, Bullmore ET. 2014 Generative models of rich clubs in Hebbian neuronal networks and large-scale human brain networks. Phil. Trans. R. Soc. B 369, 20130531. (doi:10.1098/rstb.2013.0531)
116. Liang X, Hsu LM, Lu H, Sumiyoshi A, He Y, Yang Y. 2017 The rich –club organization in rat functional brain network to balance between communication cost and efficiency. Cereb. Cortex. (doi:10.1093/cercor/bhw416)
117. Bassett DS, Bullmore ET, Meyer-Lindenberg A, Apud JA, Weinberger DR, Coppola R. 2009 Cognitive fitness of cost-efficient brain functional networks. Proc. Natl Acad. Sci. USA 106, 11 747– 11 752. (doi:10.1073/pnas.0903641106)
118. Fornito A et al. 2011 Genetic influences on costefficient organization of human cortical functional networks. J. Neurosci. 31, 3261– 3270. (doi:10. 1523/JNEUROSCI.4858-10.2011)
119. Bettencourt LM, Stephens GJ, Ham MI, Gross GW. 2007 Functional structure of cortical neuronal networks grown in vitro. Phys. Rev. E. 75, 021915. (doi:10.1103/PhysRevE.75.021915)
120. Feldt Muldoon S, Soltesz I, Cossart R. 2013 Spatially clustered neuronal assemblies comprise the microstructure of synchrony in chronically epileptic networks. Proc. Natl Acad. Sci. USA 110, 3567–3572. (doi:10.1073/pnas.1216958110)

121. Feldt S, Bonifazi P, Cossart R. 2011 Dissecting functional connectivity of neuronal microcircuits: experimental and theoretical insights. Trends Neurosci. 34, 225–236. (doi:10.1016/j.tins.2011.02.007)

122. Bazzi M, Porter MA, Williams S, McDonald M, Fenn DJ, Howison SD. 2015 Community detection in temporal multilayer networks, and its application to correlation networks. arXiv (http://arxiv.org/abs/1501.00040)

123. Deco G, Jirsa V, McIntosh AR, Sporns O, Kotter R. 2009 Key role of coupling, delay, and noise in resting brain fluctuations. Proc. Natl Acad. Sci. USA 106, 10 302– 10 307. (doi:10.1073/pnas. 0901831106)

124. Honey CJ, Sporns O, Cammoun L, Gigandet X, Thiran JP, Meuli R, Hagmann P. 2009 Predicting human resting-state functional connectivity from structural connectivity. Proc. Natl Acad. Sci. USA 106, 2035– 2040. (doi:10.1073/pnas.0811168106)

125. Goni J et al. 2014 Resting-brain functional connectivity predicted by analytic measures of network communication. Proc. Natl Acad. Sci. USA 111, 833 – 838. (doi:10.1073/pnas. 1315529111)

126. Hermundstad AM et al. 2014 Structurallyconstrained relationships between cognitive states in the human brain. PLoS Comput. Biol. 10, e1003591. (doi:10.1371/journal.pcbi.1003591)

127. Hermundstad AM et al. 2013 Structural foundations of resting-state and task-based functional connectivity in the human brain. Proc. Natl Acad. Sci. USA 110, 6169– 6174. (doi:10.1073/pnas. 1219562110)

128. Richiardi J et al. 2015 Correlated gene expression supports synchronous activity in brain networks. Science 348, 1241– 1244. (doi:10.1126/science. 1255905)

129. Mills BD, Grayson D, Shunmugavel A, MirandaDominguez O, Feczko E, Earl E, Neve K, Fair D. 2017 Correlated gene expression and anatomical communication support synchronized brain activity in the mouse functional connectome. bioRxiv. (doi:10.1101/167304).

130. Betzel RF, Medaglia JD, Kahn AE, Soffer J, Schonhaut DR, Bassett DS. 2017 Inter-regional ECoG correlations predicted by communication dynamics, geometry, and correlated gene expression. (http:// arxiv.org/abs/1706.06088)

131. Gratton C, Lee TG, Nomura EM, D'Esposito M. 2013 The effect of theta-burst TMS on cognitive control networks measured with resting

state fMRI. Front. Syst. Neurosci. 7, 124. (doi:10.3389/fnsys.2013.00124)

132. Muldoon SF, Pasqualetti F, Gu S, Cieslak M, Grafton ST, Vettel JM, Bassett DS. 2016 Stimulation-based control of dynamic brain networks. PLoS Comput. Biol. 12, e1005076. (doi:10.1371/journal.pcbi. 1005076)

133. Nguyen HC, Zecchina R, Berg J. 2017 Inverse statistical problems: from the inverse Ising problem to data science. arXiv (http://arxiv.org/abs/1702.01522)

134. Tang A et al. 2008 A maximum entropy model applied to spatial and temporal correlations from cortical networks in vitro. J. Neurosci. 28, 505 – 518. (doi:10.1523/JNEUROSCI.3359-07.2008)

135. Shlens J, Field GD, Gauthier JL, Grivich MI, Petrusca D, Sher A, Litke AM, Chichilnisky EJ. 2006 The structure of multi-neuron firing patterns in primate retina. J. Neurosci. 26, 8254 – 8266. (doi:10.1523/JNEUROSCI.1282-06.2006)

136. Schneidman E, Berry MJn, Segev R, Bialek W. 2006 Weak pairwise correlations imply strongly correlated network states in a neural population. Nature 440, 1007– 1012. (doi:10.1038/nature04701)

137. Watanabe T, Hirose S, Wada H, Imai Y, Machida T, Shirouzu I, Konishi S, Miyashita Y, Masuda N. 2013 A pairwise maximum entropy model accurately describes resting-state human brain networks. Nat. Commun. 4, 1370. (doi:10.1038/ncomms2388)

138. Watanabe T, Masuda N, Megumi F, Kanai R, Rees G. 2014 Energy landscape and dynamics of brain activity during human bistable perception. Nat. Commun. 5, 4765. (doi:10.1038/ncomms5765)

139. Ashourvan A, Gu S, Mattar MG, Vettel JM, Bassett DS. 2017 The energy landscape underpinning module dynamics in the human brain connectome. Neuroimage 157, 364– 380. (doi:10.1016/j.neuroimage.2017.05.067)

140. Ezaki T, Watanabe T, Ohzeki M, Masuda N. 2017 Energy landscape analysis of neuroimaging data. Phil. Trans. R. Soc. A 375, 20160287. (doi:10.1098/ rsta.2016.0287)

141. Barreiro AK, Gjorgjieva J, Rieke F, Shea-Brown E. 2014 When do microcircuits produce beyondpairwise correlations? Front. Comput. Neurosci. 8, 10. (doi:10.3389/fncom.2014.00010)

142. Kelly RC, Kass RE. 2012 A framework for evaluating pairwise and multiway synchrony among stimulusdriven neurons. Neural Comput. 24, 2007 – 2032. (doi:10.1162/NECO_a_00307)
143. Ohiorhenuan IE, Mechler F, Purpura KP, Schmid AM, Hu Q, Victor JD. 2010 Sparse coding and high-order correlations in fine-scale cortical networks. Nature 466, 617– 621. (doi:10.1038/nature09178)
144. Ganmor E, Segev R, Schneidman E. 2011 Sparse low-order interaction network underlies a highly correlated and learnable neural population code. Proc. Natl Acad. Sci. USA 108, 9679– 9684. (doi:10. 1073/ pnas.1019641108)
145. Leen DA, Shea-Brown E. 2015 A simple mechanism for beyond-pairwise correlations in integrate-andfire neurons. J. Math. Neurosci. 5, 30. (doi:10.1186/ s13408-015-0030-9)
146. Stephan KE, Friston KJ, Frith CD. 2009 Dysconnection in schizophrenia: from abnormal synaptic plasticity to failures of self-monitoring. Schizophrenia Bull. 35, 509 – 527. (doi:10.1093/ schbul/ sbn176)
147. Seeley WW, Crawford RK, Zhou J, Miller BL, Greicius MD. 2009 Neurodegenerative diseases target largescale human brain networks. Neuron 62, 42 – 52. (doi:10.1016/j.neuron.2009.03.024)
148. Rubinov M, Bassett DS. 2011 Emerging evidence of connectomic abnormalities in schizophrenia. J. Neurosci. 31, 6263 – 6265.
149. Magnin B, Mesrob L, Kinkingne´hun S, Pe´le´griniIssac M, Colliot O, Sarazin M, Dubois B, Lehe´ricy S, Benali H. 2009 Support vector machine-based classification of Alzheimer's disease from wholebrain anatomical MRI. Neuroradiology 51, 73 – 83. (doi:10.1007/s00234-008-0463-x)
150. Shen H, Wang L, Liu Y, Hu D. 2010 Discriminative analysis of resting-state functional connectivity patterns of schizophrenia using low dimensional embedding of fMRI. Neuroimage 49, 3110– 3121. (doi:10.1016/j.neuroimage.2009.11.011)
151. Betzel RF, Gu S, Medaglia JD, Pasqualetti F, Bassett DS. 2016 Optimally controlling the human connectome: the role of network topology. Sci. Rep. 6, 30770. (doi:10.1038/srep30770)
152. Iturria-Medina Y, Carbonell FM, Sotero RC, Chouinard-Decorte F, Evans AC, Alzheimer's Disease Neuroimaging Initiative. 2017 Multifactorial causal model of brain (dis) organization and therapeutic

intervention: application to Alzheimer's disease. NeuroImage 152, 60 – 77. (doi:10.1016/j. neuroimage.2017.02.058)

153. van den Heuvel MI, Thomason ME. 2016 Functional connectivity of the human brain in utero. Trends Cogn. Sci. 20, 931–939. (doi:10.1016/j. tics.2016.10.001)
154. Sudlow C et al. 2015 UK biobank: an open access resource for identifying the causes of a wide range of complex diseases of middle and old age. PLoS Med. 12, e1001779. (doi:10.1371/journal.pmed. 1001779)
155. Laumann TO et al. 2015 Functional system and areal organization of a highly sampled individual human brain. Neuron 87, 657– 670. (doi:10.1016/j.neuron. 2015.06.037)
156. Braga RM, Buckner RL. 2017 Parallel interdigitated distributed networks within the individual estimated by intrinsic functional connectivity. Neuron 95, 457 – 471. (doi:10.1016/j.neuron.2017. 06.038)
157. Gordon EM et al. 2017 Precision functional mapping of individual human brains. Neuron 95, 791– 807. (doi:10.1016/j.neuron.2017.07.011)
158. Hawrylycz MJ et al. 2012 An anatomically comprehensive atlas of the adult human brain transcriptome. Nature 489, 391– 399. (doi:10.1038/ nature11405)
159. Zador AM, Dubnau J, Oyibo HK, Zhan H, Cao G, Peikon ID. 2012 Sequencing the connectome. PLoS Biol. 10, e1001411. (doi:10.1371/ journal.pbio. 1001411)
160. Chung K et al. 2013 Structural and molecular interrogation of intact biological systems. Nature 497, 332 – 337. (doi:10.1038/nature12107)
161. Helmstaedter M, Briggman KL, Turaga SC, Jain V, Seung HS, Denk W. 2013 Connectomic reconstruction of the inner plexiform layer in the mouse retina. Nature 500, 168– 174. (doi:10.1038/ nature12346)
162. Kim TH, Zhang Y, Lecoq J, Jung JC, Li J, Zeng H, Niell CM, Schnitzer MJ. 2016 Long-term optical access to an estimated one million neurons in the live mouse cortex. Cell Rep. 17, 3385 – 3394. (doi:10.1016/j. celrep.2016.12.004)
163. Eichler K et al. 2017 The complete connectome of a learning and memory centre in an insect brain. Nature 548, 175– 182. (doi:10.1038/ nature23455)

164. Lein ES et al. 2007 Genome-wide atlas of gene expression in the adult mouse brain. Nature 445, 168 – 176. (doi:10.1038/nature05453)

165. Battiston F, Nicosia V, Chavez M, Latora V. 2017 Multilayer motif analysis of brain networks. Chaos: Interdiscip. J. Nonlinear Sci. 27, 047404. (doi:10. 1063/1.4979282)

166. Bentley B, Branicky R, Barnes CL, Chew YL, Yemini E, Bullmore ET, Ve´rtes PE, Schafer WR. 2016 The multilayer connectome of Caenorhabditis elegans. PLoS Comput. Biol. 12, e1005283. (doi:10.1371/ journal.pcbi.1005283)

167. Kivela M, Arenas A, Barthelemy M, Gleeson JP, Moreno Y, Porter MA. 2014 Multilayer networks. J. Complex Netw. 2, 203 – 271. (doi:10.1093/ comnet/cnu016)

168. Nicosia V, Skardal PS, Arenas A, Latora V. 2017 Collective phenomena emerging from the interactions between dynamical processes in multiplex networks. Phys. Rev. Lett. 118, 138302. (doi:10.1103/ PhysRevLett.118.138302)

169. Ronellenfitsch H, Katifori E. 2016 Global optimization, local adaptation, and the role of growth in distribution networks. Phys. Rev. Lett. 117, 138301. (doi:10.1103/PhysRevLett.117.138301)

170. Falk EB, Bassett DS. 2017 Brain and social networks: fundamental building blocks of human experience. Trends Cogn. Sci. 21, 674– 690. doi:10.1016/j.tics. 2017.06.009)

171. Ursache A, Noble KG, Pediatric Imaging, Neurocognition and Genetics Study. 2016 Socioeconomic status, white matter, and executive function in children. Brain Behav. 6, e00531. (doi:10.1002/brb3.531)

172. Bassett DS. In press. A network science of the practice of curiosity. In Curiosity studies: toward a new ecology of knowledge (eds P Zurn, A Shankar). Minneapolis, MN: Minnesota Press.

173. Constantinescu AO, O'Reilly JX, Behrens TEJ. 2016 Organizing conceptual knowledge in humans with a gridlike code. Science 352, 1464– 1468. (doi:10. 1126/science.aaf0941)

174. Huth AG, de Heer WA, Griffiths TL, Theunissen FE, Gallant JL. 2016 Natural speech reveals the semantic maps that tile human cerebral cortex. Nature 532, 453– 458. (doi:10.1038/nature17637)

175. Cukur T, Nishimoto S, Huth AG, Gallant JL. 2013 Attention during natural vision warps semantic representation across the human brain. Nat. Neurosci. 16, 763– 770. (doi:10.1038/nn.3381)

176. Giusti C, Ghrist R, Bassett DS. 2016 Two's company, three (or more) is a simplex: algebraic-topological tools for understanding higher-order structure in neural data. J. Comput. Neurosci. 41, 1– 14. (doi:10. 1007/ s10827-016-0608-6)

177. Giusti C, Pastalkova E, Curto C, Itskov V. 2015 Clique topology reveals intrinsic geometric structure in neural correlations. Proc. Natl Acad. Sci. USA 112, 13 455– 13 460. (doi:10.1073/pnas.1506407112)

178. Sizemore A, Giusti C, Kahn A, Betzel RF, Bassett DS. In press. Cliques and cavities in the human connectome. J. Comput. Neurosci. (http:// arxiv.org/ abs/1608.03520)

179. Curto C, Itskov V, Veliz-Cuba A, Youngs N. 2013 The neural ring: an algebraic tool for analyzing the intrinsic structure of neural codes. Bull. Math. Biol. 75, 1571– 1611. (doi:10.1007/s11538-013-9860-3)

180. Courtney OT, Bianconi G. 2017 Weighted growing simplicial complexes. Phys. Rev. E 95, 062301. (doi:10.1103/PhysRevE.95.062301)

181. Schaub MT, Delvenne JC, Rosvall M, Lambiotte R. 2017 The many facets of community detection in complex networks. Appl. Netw. Sci. 2, 4. (doi:10. 1007/s41109-017-0023-6)

182. Lambiotte R, Delvenne JC, Barahona M. 2015 Laplacian dynamics and multiscale modular structure in networks. IEEE Trans. Netw. Sci. Eng. 1, 76 – 90. (http://arxiv.org/abs/0812.1770)

183. Delvenne JC, Yaliraki SN, Barahona M. 2010 Stability of graph communities across time scales. Proc. Natl Acad. Sci. USA 107, 12 755– 12 760. (doi:10.1073/ pnas.0903215107)

184. Avena-Koenigsberger A, Gon˜i J, Betzel RF, van den Heuvel MP, Griffa A, Hagmann P, Thiran JP, Sporns O. 2014 Using Pareto optimality to explore the topology and dynamics of the human connectome. Phil. Trans. R. Soc. B 369, 20130530. (doi:10.1098/ rstb.2013.0530)

SECTION III
GENE REGULATORY NETWORKS AND BIOLOGICAL SYSTEMS

CHAPTER 8

INFERRING REGULATORY NETWORKS FROM EXPRESSION DATA USING TREE-BASED METHODS

Va^n Anh Huynh-Thu[1,2], Alexandre Irrthum[1,2], Louis Wehenkel[1,2], Pierre Geurts[1,2]

[1] Department of Electrical Engineering and Computer Science, Systems and Modeling, University of Lie`ge, Lie`ge, Belgium

[2] GIGA-Research, Bioinformatics and Modeling, University of Lie`ge, Lie`ge, Belgium

ABSTRACT

One of the pressing open problems of computational systems biology is the elucidation of the topology of genetic regulatory networks (GRNs) using high throughput genomic data, in particular microarray gene expression data. The Dialogue for Reverse Engineering Assessments and Methods (DREAM) challenge aims to evaluate the success of GRN inference

CITATION (APA): Huynh-Thu VA, Irrthum A, Wehenkel L, Geurts P (2010) Inferring Regulatory Networks from Expression Data Using Tree-Based Methods. PLoS ONE 5(9): e12776. (10 pages)

algorithms on benchmarks of simulated data. In this article, we present GENIE3, a new algorithm for the inference of GRNs that was best performer in the DREAM4 *In Silico Multifactorial* challenge. GENIE3 decomposes the prediction of a regulatory network between p genes into p different regression problems. In each of the regression problems, the expression pattern of one of the genes (target gene) is predicted from the expression patterns of all the other genes (input genes), using tree-based ensemble methods Random Forests or Extra-Trees. The importance of an input gene in the prediction of the target gene expression pattern is taken as an indication of a putative regulatory link. Putative regulatory links are then aggregated over all genes to provide a ranking of interactions from which the whole network is reconstructed. In addition to performing well on the DREAM4 *In Silico Multifactorial* challenge simulated data, we show that GENIE3 compares favorably with existing algorithms to decipher the genetic regulatory network of *Escherichia coli*. It doesn't make any assumption about the nature of gene regulation, can deal with combinatorial and non-linear interactions, produces directed GRNs, and is fast and scalable. In conclusion, we propose a new algorithm for GRN inference that performs well on both synthetic and real gene expression data. The algorithm, based on feature selection with tree-based ensemble methods, is simple and generic, making it adaptable to other types of genomic data and interactions.

INTRODUCTION

Genetic regulatory networks (GRNs) [1] are central to all biological organisms, and their deciphering is crucial to understand the development, functioning and pathology of these organisms. Once a remote theoretical possibility, this deciphering is now made possible by advances in genomics, most notably high-throughput profiling of gene expression patterns with DNA microarrays. These advances have prompted the development of a plethora of models of GRNs and algorithms to reverse-engineer them from expression data [2]–[5].

The simplest models of genetic regulatory networks are based on Boolean logic. Because of their simplicity, these Boolean network models have provided high-level insights into the design principles and emerging properties of GRNs [6]. At the other end of the complexity spectrum are physical models mimicking the biological mechanisms at play, including promoter recognition, mRNA transcription and protein translation. These models, typically based on systems of ordinary or stochastic differential equations, can generate realistic behavior [7]. One of their main drawbacks is that they have high-dimensional parameter spaces, and thus a large number of experimental data are needed for their identification. Nevertheless, hybrid methods involving ordinary differential equations have shown good performances on real-life genome-wide GRN inference [8].

Models based on the statistical analysis of dependencies between expression patterns have an intermediate complexity, and have already been successfully applied to the inference of large GRNs. Early models used correlation coefficients between expression patterns of all pairs of genes to infer "coexpression networks" [9]. However, correlation coefficients fail to capture more complex statistical dependencies (e.g. non-linear ones) between expression patterns, and thus more general measures of dependency based on mutual information (MI), have been proposed. The simplest model based on this measure, the "relevance network", computes MI between all pairs of genes and infers the presence of a regulatory interaction when MI is larger than a given threshold [10]. Various refinements have been proposed to try to discriminate between direct and indirect interactions in relevance networks. The CLR algorithm [11] modifies the MI score based on the empirical distribution of all MI scores. The ARACNE algorithm [12]filters out indirect interactions from triplets of genes with the Data Processing Inequality [13]. Finally, MRNET [14] uses an iterative feature selection method based on a maximum relevance/minimum redundancy criterion.

Probabilistic graphical models have been widely used to model GRNs [15]. With respect to correlation or mutual information based approaches, these methods are potentially able to model higher-order dependencies between the expression patterns of genes. Among these methods, Bayesian networks have been used since the advent of microarray technologies for GRN modeling and inference [16]. A Bayesian network represents conditional dependencies between random variables with a directed acyclic graph. Learning the structure of a Bayesian network is a non trivial problem, both from a theoretical and computational point of view, and several sophisticated heuristics have been proposed in the context of GRN inference [17], [18]. One limitation of Bayesian networks for GRN inference is that these models do not allow the presence of cycles (feedback loops). While this limitation is partially circumvented by dynamic Bayesian networks [17], [19], these models can only be learned from time-series expression data. Another family of probabilistic models that gained interest recently for GRN inference are Gaussian graphical models. These methods assume that gene expression values are jointly Gaussian distributed and represent conditional dependencies between genes by an undirected graph. The estimation of this graph for high-dimensional data is difficult but several robust solutions have been proposed in the literature [20]–[23]. Although often very effective, the main limitations of these methods is of course the Gaussianity assumption, which also implies linear dependencies between variables, and the undirected nature of the inferred regulatory links (although some heuristics have been proposed to direct them [24]).

Within this context, this article presents GENIE3 (for "GEne Network Inference with Ensemble of trees"), a new GRN inference method based on variable selection with ensembles of regression trees. This method was best performer in the DREAM4 *In Silico Multifactorial*challenge [25]. Its main features with respect to existing techniques is that it makes very few assumptions about the nature of the relationships between the variables (which can thus be non-linear) and can potentially capture high-order conditional dependencies between expression patterns. It also produces a *directed* graph of regulatory interactions and naturally allows for the presence of feedback loops in the network. At the same time, it remains intuitive, computationally tractable, and easy to implement. In addition to its good performance on the synthetic data of the DREAM4 challenge, we show that GENIE3 compares favorably with existing algorithms to decipher the genetic regulatory network of *Escherichia coli*.

METHODS

Problem Definition

We address the problem of recovering regulatory networks from gene expression data. The targeted networks are directed graphs with p nodes, where each node represents a gene, and an edge directed from one gene i to another gene j indicates that gene i (directly) regulates the expression of gene j. We only consider unsigned edges; when gene i is connected to gene j, the former can be either an activator or a repressor of the latter.

The goal of (unsupervised) gene regulatory network inference is to recover the network solely from measurements of the expression of the genes in various conditions. Given the dynamic and combinatorial nature of genetic regulation, measurements of different kinds can be obtained, including steady-state expression profiles resulting from the systematic knockout or knockdown of genes or time series measurements resulting from random perturbations. In this paper, we focus on multifactorial perturbation data as generated for the DREAM4 *In Silico Size 100 Multifactorial* subchallenge. Multifactorial expression data are static steady-state measurements obtained by (slightly) perturbing all genes simultaneously. Multifactorial data might correspond for example to expression profiles obtained from different patients or biological replicates. Such data are easier and less expensive to obtain than knockout/knockdown or time series data and are thus more common in practice. They are however also less informative for the prediction of edge directionality [3], [26], [27] and therefore make the regulatory network inference task more challenging.

In what follows, we define a (multifactorial) learning sample from which to infer the network as a sample of N measurements:

$$LS = \{\mathbf{x}_1, \mathbf{x}_2, \ldots, \mathbf{x}_N\},$$

where $\mathbf{x}_k \in \mathbb{R}^p, k = 1, \ldots, N$ is a vector of expression values of all p genes in the kth experiment:

$$\mathbf{x}_k = (x_k^1, x_k^2, \ldots, x_k^p)^{\mathrm{T}}.$$

From this learning sample, the goal of network inference algorithms is to make a prediction of the underlying regulatory links between genes. Most network inference algorithms work first by providing a ranking of the potential regulatory links from the most to the less significant. A practical network prediction is then obtained by setting a threshold on this ranking. In this paper, we focus only on the first task, which is also targeted by the evaluation procedure of the DREAM4 challenge. The question of the choice of an optimal confidence threshold, although important, will be left open.

A network inference algorithm is thus defined in this paper as a procedure that exploits a LS to assign weights $w_{i,j} \geq 0, (i, j = 1, \ldots, p)$ to putative regulatory links from any gene i to any gene j, with the aim of yielding large values for weights which correspond to actual regulatory interactions.

Network Inference with Tree-based Methods

The basic idea of our procedure is to decompose the problem of recovering a network involving p genes into p different subproblems, where each of these subproblems consists in identifying the regulators of one of the genes of the network. Exploiting expression data, the identification of the regulatory genes for a given target gene is defined as determining the subset of genes whose expression directly influences or is predictive of the expression of the target gene. Within the framework of supervised learning, this problem is equivalent to a feature selection problem. In this context, our solution will exploit the embedded feature ranking mechanism of tree-based ensemble methods.

We first describe our procedure to solve the network inference problem using feature selection techniques and then specialize it to the case of tree-based ensemble methods.

Network Inference as a Feature Selection Problem

Our method makes the assumption that the expression of each gene in a given condition is a function of the expression of the other genes in the network (plus some random noise). Denoting by $\mathbf{x}_k^{-j}$ the vector containing the expression values in the *k*th experiment of all genes except gene *j*:

$$\mathbf{x}_k^{-j} = (x_k^1, \ldots, x_k^{j-1}, x_k^{j+1}, \ldots, x_k^p)^{\mathrm{T}},$$

we assume that we can write:

$$x_k^j = f_j(\mathbf{x}_k^{-j}) + \varepsilon_k, \forall k \tag{1}$$

where e_k is a random noise with zero mean (conditionally to $\mathbf{x}_k^{-j}$). We further make the assumption that the function f_j only exploits the expression in x^{-j} of the genes that are direct regulators of gene j, i.e. genes that are directly connected to gene j in the targeted network. Recovering the regulatory links pointing to gene j thus amounts at finding those genes whose expression is predictive of the expression of the target gene. In machine learning terminology, this can be considered as a feature selection problem (in regression) for which many solutions exist [28]. We assume here the use of a feature ranking technique that, instead of directly returning a feature subset, yields a ranking of the features from the most relevant to the less relevant for predicting the output.

The proposed network inference procedure is illustrated in Figure 1 and works as follows:

- For $j=1$ to p:
    - Generate the learning sample of input-output pairs for gene *j*: $LS^j = \{(\mathbf{x}_k^{-j}, x_k^j), k = 1, \ldots, N\}$.

- Use a feature selection technique on LS^j to compute confidence levels $w_{i,j}, \forall i \neq j$, for all genes except gene j itself.
- Aggregate the p individual gene rankings to get a global ranking of all regulatory links.

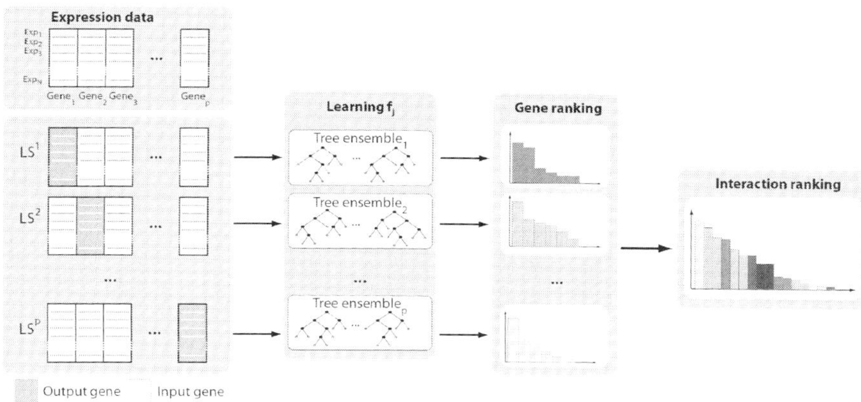

Figure 1: GENIE3 procedure. For each gene j=1, ... ,p, a learning sample LS^j is generated with expression levels of j as output values and expression levels of all other genes as input values. A function f_j is learned from LS^j and a local ranking of all genes except j is computed. The p local rankings are then aggregated to get a global ranking of all regulatory links.

https://doi.org/10.1371/journal.pone.0012776.g001

Note that depending of the interpretation of the weights $w_{i,j}$, their aggregation to a get a global ranking of regulatory links is not trivial. We will see in the context of tree-based methods that it requires to normalize each expression vector appropriately.

Gene Ranking with Tree-based Methods

The nature of the problem and the proposed solution put some constraints on candidate feature selection techniques. The nature of the functions f_i is unknown but they are expected to involve the expression of several genes (combinatorial regulation) and to be non-linear. The number of input features in each of these problems is typically much greater than the number of observations. Computationally, since the identification of a network involving p genes requires to rerun the algorithm p times, it is also of interest for this algorithm to be fast and to require as few manual tuning as possible. Tree-based ensemble methods are good candidates for that purpose. These methods do not make any assumption about the nature of the target function, can potentially deal with interacting features and non-linearity. They work well in the presence of a large number of features, are fast to compute, scalable, and essentially parameter-free (see [29] for a review).

We first briefly describe these methods and their built-in feature ranking mechanism and then discuss their use in the context of the network inference procedure described in the previous section.

Tree-based Ensemble Methods

Each subproblem, defined by a learning sample LS^j , is a supervised (non-parametric) regression problem. Using square error loss, each problem amounts at finding a function f_j that minimizes the following error:

$$\sum_{k=1}^{N} (x_k^j - f_j(\mathbf{x}_k^{-j}))^2. \tag{2}$$

Regression trees [30] solve this problem by developing tree structured models. The basic idea of this method is to recursively split the learning sample with binary tests based each on one input variable (selected in x^{-j}), trying to reduce as much as possible the variance of the output variable (x_j) in the resulting subsets of samples. Candidate splits for numerical variables typically compare the input variable values with a threshold which is determined during the tree growing.

Single trees are usually very much improved by ensemble methods, which average the predictions of several trees. In our network inference procedure, we compare two tree-based ensemble methods based on randomization, namely Random Forests [31] and Extra-Trees [32]. In a Random Forests ensemble, each tree is built on a bootstrap sample from the original learning sample and, at each test node, K attributes are selected at random among all candidate attributes before determining the best split. In the Extra-Trees method on the other hand, each tree is built from the original learning sample and at each test node, the best split is determined among K random splits, each determined by randomly selecting one input (without replacement) and a threshold. For these two methods, we will grow ensembles of 1000 trees and we will consider two values of the main parameter of these methods: $K=\sqrt{p-1}$ and $K=p-1$, where p-1 is the number of inputs, equal to the number of potential regulators of each gene. Empirical validations in [32] have shown that these two values of K were near-optimal in terms of predictive accuracy on several problems. Note however that we do not exclude that better results could be obtained with other settings of K.

Variable Importance Measure

One of the most interesting characteristics of tree-based methods is that it is possible to compute from a tree a variable importance measure that allows to rank the input features according to their relevance for predicting the output. Several variable importance measures have been proposed in the literature for tree-based methods. In our experiment, we consider a measure which at each test node $\mathcal{N}$ computes the total reduction of the variance of the output variable due to the split, defined by [30]:

$$I(\mathcal{N}) = \#S\text{Var}(S) - \#S_t\text{Var}(S_t) - \#S_f\text{Var}(S_f), \tag{3}$$

where S denotes the set of samples that reach node $\mathcal{N}$, S_t (resp. S_f) denotes its subset for which the test is true (resp. false), Var(.) is the variance of the output variable in a subset, and # denotes the cardinality of a set of samples. For a single tree, the overall importance of one variable is then computed by summing the I values of all tree nodes where this variable is used to split. Those attributes that are not selected at all obtain a zero value of their importance, and those that are selected close to the root node of the tree typically obtain high scores. Attribute importance measures can be easily extended to ensembles, simply by averaging importance scores over all trees in the ensemble. The resulting importance measure is then even more reliable because of the variance reduction effect resulting from this averaging.

Breiman [31] proposed an alternative measure that computes the average reduction of the tree accuracy on out-of-bag samples (i.e. training objects that are not present in the bootstrap sample used to build each tree) when the values of the corresponding variable are randomly permuted. While this procedure has some advantages with respect to the variance reduction based measure of (3) [33], it gives in most practical applications very similar results but is much more computationally demanding. Furthermore, it does not extend to methods like the Extra-Trees method which do not consider bootstrap sampling.

Regulatory Link Ranking

Each tree-based model yields a separate ranking of the genes as potential regulators of a target gene in the form of weights $w_{i,j}$ computed as sums of total variance reductions in the form (3). The sum of the importances of all variables for a tree is equal to the total variance of the output variable explained by the tree, which in the case of unpruned trees (as they are in the case of Random Forests and Extra-Trees ensembles) is usually very close to the initial total variance of the output:

$$\sum_{i \neq j} w_{i,j} \approx N \operatorname{Var}(S), \quad (4)$$

where S is the learning sample from which the tree was built (i.e. LS^j for the Extra-Trees method and a bootstrap sample for the Random Forests method) and where Var(S) is the variance of the target gene estimated in the corresponding learning sample. As a consequence, if we trivially order the regulatory links according to the weights $w_{i,j}$, this is likely to introduce a positive bias for regulatory links towards the more highly variable genes. To avoid this bias, we have first normalized the gene expressions so that they all have a unit variance in the training set, before applying the tree-based ensemble methods. This normalization indeed implies that the different weights inferred from different models predicting the different gene expressions are comparable.

Computational Complexity

The computational complexity of the Random Forests and Extra-Trees algorithms is on the order of O(TKN log N), where T is the number of trees, N is the learning sample size and K is the main parameter of the two tree-based methods. GENIE3's complexity is thus on the order of O(pTKN log N) since it requires to build an ensemble of trees for each of the p genes. The complexity of the whole procedure is thus log linear with respect to the number of measurements and, at worst, quadratic with respect to the number of genes (when K=p-1).

To fix ideas, with our MatLab implementation of GENIE3, it takes 6.5 minutes to infer the five networks of the DREAM4 challenge and 7 hours to infer the E. coli network (with known transcription factors), in both cases with Random Forests and $K = \sqrt{n_{TF}}$, where n_{TF} is the number of potential regulators (see later for the details of these experiments). These computing times where measured on a 16GB RAM, Intel L5420 2.50 GHz computer.

Note that, if needed, the algorithm can be easily parallelized as the p feature selection problems, as well as the different trees in an ensemble, are independent of each other.

RESULTS

Datasets

We report below two series of experiments: first on the DREAM4 *In Silico Multifactorial*challenge and then on the *Escherichia coli* regulatory network.

DREAM4 Datasets

The DREAM (for "Dialogue for Reverse Engineering Assessments and Methods") initiative organizes an annual reverse engineering competition called the DREAM challenge [34]–[37]. We report here our results on the DREAM4 edition of this competition, where one challenge concerned *in silico* regulatory network inference [25]. This challenge was divided into three subchallenges, called *In Silico Size 10*, *In Silico Size 100*, and *In Silico Size 100 Multifactorial*. We only report here our result of this last subchallenge.

The goal of the *In Silico Size 100 Multifactorial* subchallenge was to infer five networks of $p=100$ genes each from multifactorial perturbation data. Multifactorial data are defined as static steady-state expression profiles resulting from slight perturbations of all genes simultaneously.

All networks and data were generated with Gene Net Weaver (GNW) version 2.0 [38]. Network topologies were obtained by extracting subnetworks from transcriptional regulatory networks of *E. coli* and *S. cerevisiae*. The subnetwork extraction method was adapted to preferentially include parts of the network with cycles but direct self-interactions were removed. The dynamics of the networks were simulated using a detailed kinetic model of gene regulation. Noise was added both in the dynamics of the networks and on the measurement of expression data. Multifactorial perturbations were simulated by slightly increasing or decreasing the basal activation of all genes of the network simultaneously by different random amounts. In total, the number of expression conditions N for each network was set to 100.

Escherichia coli Dataset

In addition, we carried out experiments with our method on the inference of the regulatory network of *Escherichia coli*, which has been used by several authors as a benchmark.

The dataset of expression profiles we used was retrieved from the Many Microbe Microarrays (M^{3D}) database [39] (version 4 build 6). It contains 907 *E. coli* microarray expression profiles of 4297 genes collected from different experiments at steady-state level. To validate the network predictions we used 3433 experimentally confirmed regulatory interactions among 1471 genes that have been curated in RegulonDB version 6.4 [40].

Performance Metrics

Our algorithm provides a ranking of the regulatory links from the most confident to the less confident. To evaluate such a ranking independently of the choice of a specific threshold, we used both precision-recall (PR) curve and receiver operating characteristic (ROC) curve. The former plots for varying thresholds on the importance scores the proportion of true positives among all predictions (precision) versus the percentage of true positives that are retrieved (recall), whereas a ROC curve plots the true positive rate versus the false positive rate.

To summarize these curves, the DREAM organizers proposed different statistics:

- AUPR: The area under the PR curve.
- AUROC: The area under the ROC curve.
- AUPR p-value: The probability that a given or larger AUPR is obtained by random ordering of the potential network edges.
- AUROC p-value: The probability that a given or larger AUROC is obtained by random ordering of the potential network edges.

An overall score was used to evaluate the predictions for the five networks of each subchallenge:

$$\text{overall score} = -0.5 \log_{10}(p_1 p_2),$$

where p_1 and p_2 are respectively the geometric means of AUPR pvalues and AUROC p-values taken over the five networks.

Results on the DREAM4 Multifactorial Data

Challenge

We took part in the DREAM4 *In Silico Multifactorial* challenge, where the goal was to provide the ranking of the potential (directed) regulatory interactions for five simulated networks. At the time of submission, the gold standard networks were unknown and it was thus impossible to choose the best one among several tree-based methods at our disposal. We thus submitted the rankings obtained by our GENIE3 procedure using the Random Forests algorithm with $K=\sqrt{p-1}$.

Among twelve challengers, GENIE3 got the best performance with an overall score of 37.428. As a comparison, the score of the first runner-up was 28.165.

Table 1 shows the AUPR and AUROC values of our predictions and those of the first runner-up, and Table 2 shows their associated p-values, indicating that our predictions were significantly better than random guessing. On all networks, these scores were the highest among the twelve challengers. Individual PR and ROC curves on each network are collected in Figure S1.

Table 1: AUPR and AUROC scores for DREAM4 Multifactorial challenge

https://doi.org/10.1371/journal.pone.0012776.t001

	Method	NET1	NET2	NET3	NET4	NET5
AUPR	GENIE3-RF-sqrt	0.154	0.155	0.231	0.208	0.197
	2nd best	0.108	0.147	0.185	0.161	0.111
AUROC	GENIE3-RF-sqrt	0.745	0.733	0.775	0.791	0.798
	2nd best	0.739	0.694	0.748	0.736	0.745

GENIE3-RF-sqrt: GENIE3 using Random Forests with $K=\sqrt{p-1}$. 2nd best: Second best performer in the DREAM4 Multifactorial challenge.
doi:10.1371/journal.pone.0012776.t001

Table 2: AUPR and AUROC p-values for DREAM4 Multifactorial challenge

https://doi.org/10.1371/journal.pone.0012776.t002

	Method	NET1	NET2	NET3	NET4	NET5	Overall p-value
AUPR p-value	GENIE3-RF-sqrt	3.3e-34	7.9e-54	1.8e-54	5.5e-47	4.6e-44	1.0e-46
	2nd best	5.6e-23	9.7e-50	6.6e-43	1.5e-35	4.4e-23	7.4e-35
AUROC p-value	GENIE3-RF-sqrt	3.3e-18	1.1e-28	9.7e-34	6.7e-33	1.9e-34	1.4e-29
	2nd best	1.7e-17	5.4e-21	4.9e-28	1.9e-23	1.1e-24	6.3e-23

GENIE3-RF-sqrt: GENIE3 using Random Forests with $K=\sqrt{p-1}$. 2nd best: Second best performer in the DREAM4 Multifactorial challenge.
doi:10.1371/journal.pone.0012776.t002

Comparison of Tree-based Methods

We have subsequently applied GENIE3 on these same datasets, using the Extra-Trees algorithm, and also setting K to its maximum value (K=p-1). Table 3 shows the overall scores obtained with the four different combinations. The Random Forests and the Extra-Trees algorithms gave comparable results, and the predictions were improved when the parameter K was increased, i.e. when the randomization was reduced. The overall best result was achieved when we used Random Forests with K=p-1, giving an overall score equal to 40.471. This result is slightly better than our initial submission to the challenge. Unless otherwise stated, all subsequent experiments in the paper will be carried out with this particular setting. Note that in this case, the algorithm simply corresponds to the Bagging method applied on standard regression trees [41].

Table 3: Overall scores of GENIE3 for DREAM4 networks

https://doi.org/10.1371/journal.pone.0012776.t003

	GENIE3-RF-all	CLR	ARACNE	MRNET	GGM
Overall score	36.736	35.838	32.632	34.124	26.846

Links (i,j) and (j,i) were both assigned the same weights by CLR, ARACNE, MRNET, and GGM, while the predictions of GENIE3 were symmetrized by assigning to each pair (i,j) the maximum between $w_{i,j}$ and $w_{j,i}$.
doi:10.1371/journal.pone.0012776.t006

Detailed Analysis of the Predictions

To have a more precise picture of the quality of the predictions obtained with GENIE3, Figure 2depicts the ranking of regulators for all genes, grouped according to their number of regulators, for the third network which was predicted with the highest AUPR score by our method. Similar plots for the other networks can be found in Figure S2.

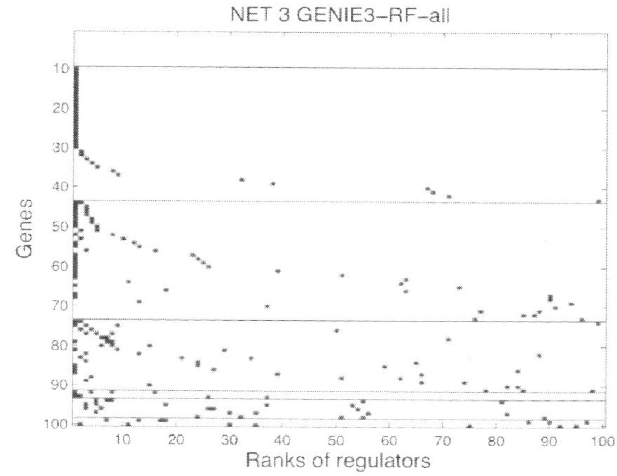

Figure 2: Detailed results on DREAM4 NET3. Ranking of the regulators for all genes. Each row corresponds to a gene. Dots in each row represent the positions in the Random Forests ranking of the regulators of this gene. Genes are ordered on the y-axis according to their number of regulators in the gold standard network; those having the same number of regulators are grouped inside an horizontal block (from no regulator at the top to 6 regulators at the bottom). Inside each block, genes are ordered according to the median rank of their regulators. The ranking of interactions was obtained with Random Forests and K=p-1.

https://doi.org/10.1371/journal.pone.0012776.g002

As can be seen from this figure, GENIE3 is able to retrieve the best regulator for about two thirds of the genes that have only one regulator. For genes with two regulators, the method retrieves one of the two regulators for about the same proportion of genes but is less good at retrieving the second regulator (only for one gene, the two regulators are at the top of the ranking). For genes with three or more regulators, even one regulator seems to be difficult to retrieve.

This suggests that the performance of GENIE3 at retrieving a regulator of one gene degrades as the number of regulators of this gene increases, as also observed from the analysis of the results of the DREAM3 challenge in [42]. To further check this hypothesis, we plotted in Figure 3 the median rank of the regulators of gene *j*, such that gene *j* is regulated by an increasing number of genes. The rank is presented here as a percentage, such that the first and last regulators of the ranking have a rank equal to 100% and 0% respectively. This plot clearly shows that the quality of the ranking monotonically decreases with the in-degree of the genes.

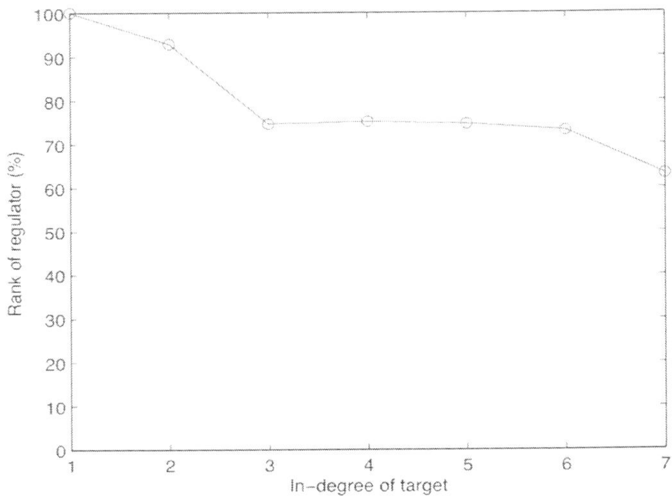

Figure 3: Rank of regulators as a function of the in-degree of the target.

The in-degree of a target is its number of regulators. The dot corresponding to in-degree *n* is the median rank of regulators that regulate a gene with in-degree *n*, over the five networks. The rank is presented here as a percentage, such that the first and last regulators of each ranking have a rank equal to 100% and 0% respectively. The ranking of interactions was obtained with Random Forests and K=p-1.

https://doi.org/10.1371/journal.pone.0012776.g003

Undirected Versus Directed Predictions

One interesting feature of GENIE3 is its potential ability to predict directed networks, while methods based on mutual information or correlation are only able to predict undirected networks.

To see to what extent the networks predicted by our method are asymmetric, we show in Table 4 for each network the proportion of predicted regulatory links for which the opposite link is not predicted. Notice that these predictions were obtained from the Random Forests ranking, by fixing a weight threshold such that the predicted network contains the same total number of edges as the gold standard. This percentage is compared with the same percentage computed for the gold standard. Our predicted networks are clearly more symmetric than the corresponding gold standards but they nevertheless contain a significant number of asymmetric predictions (52% of the links on the average).

Table 4: Asymmetry of predicted and gold standard networks

https://doi.org/10.1371/journal.pone.0012776.t004

	NET1	NET2	NET3	NET4	NET5
GENIE3-RF-all	50%	58%	48%	48%	58%
Gold standard	92%	94%	97%	96%	98%

The asymmetry of a network is measured by the proportion of regulatory links for which the opposite link is not predicted.
doi:10.1371/journal.pone.0012776.t004

Of course, the fact that GENIE3 predicts asymmetric networks does not ensure that the prediction of these asymmetric links is really informative; asymmetric predictions might precisely correspond to spurious predictions. To check this, we swapped the weights w_{ij} and w_{ji} for each pair of genes (i,j) and assessed the new resulting rankings. The overall score dropped from 40.471 to 14.674, suggesting that GENIE3 tends to correctly assign the highest weight to the true direction, given an undirected regulatory link.

To further assess the ability of our method to predict link directions, we computed the proportion of edges $i \rightarrow j$ in the gold standard network such that there is no edge $j \rightarrow i$ and for which our method wrongly predicts $w_{i,j} < w_{j,i}$. This can be considered as an error rate when our method is used for directing the edges of a known undirected network. Table 5 shows the average value of this error rate over the five networks, for increasing recall values. Given that there are only two choices for a given link, a random ranking of the directed interactions would yield an error rate close to 50%. For all recall values, the error rate is significantly lower than 50% suggesting that our method is a plausible approach for directing an undirected network. The error rate is smaller (20%) for the top ranked interactions but it remains quite good (27%) even when considering less confident predictions.

Table 5: Error rates on edge directionality on DREAM4 networks

https://doi.org/10.1371/journal.pone.0012776.t005

Recall	5%	25%	50%	75%	100%
Error rate	20%	28%	27%	27%	26%

The error rate is the proportion of edges $i \rightarrow j$ in the gold standard network such that there is no edge $j \rightarrow i$ and for which our method wrongly predicts $w_{i,j} < w_{j,i}$. Each column corresponds to one value of the number of considered directed links of the gold standard. These error rates were obtained with Random Forests and $K = p - 1$, and averaged over the five networks.
doi:10.1371/journal.pone.0012776.t005

Finally, we compared GENIE3 to three existing approaches based on the computation of mutual information (MI), namely CLR [11], ARACNE [12] and MRNET [14], and to one approach based on graphical Gaussian models (GGMs) [20]. All these four methods can only predict undirected networks. For these experiments, we used the original MatLab implementation of CLR [43] and the implementations of ARACNE and MRNET in the minet R package [44]. To compute mutual information, we used a B-spline smoothing and discretization, as implemented in the CLR package, with the parameter setting used in [11] (10 bins and third order B-splines). For ARACNE, the tolerance parameter was optimized between 0 and 15%, as advised in [12]. For GGMs, we used the GeneNet R package [45].

We carried two evaluations, the first one against the undirected gold standard (Table 6) and the second one against the directed gold standard (Table 7). In the first case, the predictions of GENIE3 were symmetrized by assigning to each pair (i,j) the maximum between $w_{i,j}$ and $w_{j,i}$. In the second case, links (i,j) and (j,i) were both assigned the same weights by the four undirected methods, while GENIE3 was used unmodified. In the undirected case, GGMs give the lowest score while all MI-based methods are equally good with only a slight advantage to our method. In the directed case, GENIE3 is significantly better than the four other methods that are constrained to predict undirected links.

Table 6: Overall scores for the undirected networks of DREAM4

https://doi.org/10.1371/journal.pone.0012776.t006

	GENIE3-RF-all	CLR	ARACNE	MRNET	GGM
Overall score	36.736	35.838	32.632	34.124	26.846

Links (i,j) and (j,i) were both assigned the same weights by CLR, ARACNE, MRNET, and GGM, while the predictions of GENIE3 were symmetrized by assigning to each pair (i,j) the maximum between $w_{i,j}$ and $w_{j,i}$.
doi:10.1371/journal.pone.0012776.t006

Table 7: Overall scores for the directed networks of DREAM4

https://doi.org/10.1371/journal.pone.0012776.t007

	GENIE3-RF-all	CLR	ARACNE	MRNET	GGM
Overall score	40.471	31.57	28.488	30.435	23.705

Links (i,j) and (j,i) were both assigned the same weights by CLR, ARACNE, MRNET, and GGM, while GENIE3 was used unmodified.
doi:10.1371/journal.pone.0012776.t007

Performance on Escherichia coli Dataset

As a first experiment on the real *E. coli* dataset, we adopted the same evaluation protocol as in [11] that assumes that we have prior knowledge about which genes of the gold standard (i.e. the experimentally confirmed interactions curated in RegulonDB) are transcription factors. In the context of our method, this makes each feature selection problem much easier as the regulators have to be identified among a much smaller set of genes. This also makes undirected and directed methods equally applicable (since all links are automatically directed from transcription factors to genes). Figure 4A shows the precision-recall curves for the four different settings of the tree-based procedure. Contrary to the DREAM4 networks, setting $K=\sqrt{n_{TF}}$, where n_{TF} is the number of potential regulators, improves the performance compared to $K=n_{TF}$, RF-sqrt leading to the best precision-recall curve. Figure 4B compares this method with the four undirected methods, CLR, ARACNE, MRNET, and GGMs, using exactly the same protocol. The predictions obtained using GENIE3 with Random Forests and

$K = \sqrt{n_{TF}}$ outperform those obtained from ARACNE and MRNET, and give a precision-recall curve comparable with CLR and GGMs (although less good for small recall values). Figure S3shows the ranking of the regulators with GENIE3-RF-sqrt for all genes grouped according to their number of regulators.

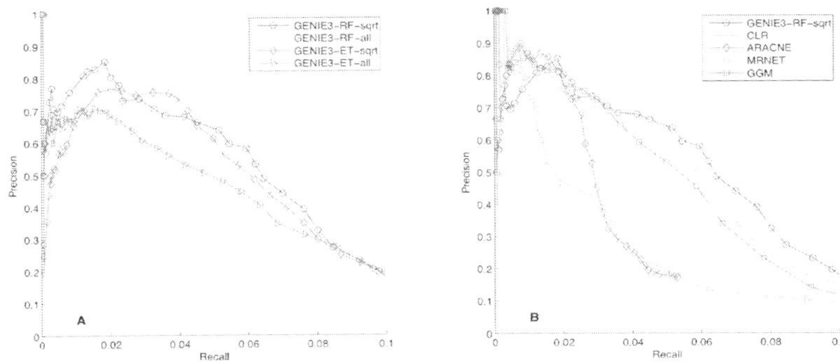

Figure 4: Precision-Recall curves for the *E. coli* **network.**

Only known transcription factors were used as input genes. A. Comparison between the four different settings of the tree procedure. B. Comparison to other approaches.

https://doi.org/10.1371/journal.pone.0012776.g004

As a second experiment, we simulated conditions similar to the DREAM4 challenge, where transcription factors were unknown and tried to infer the network using as input features in each step of our procedure all 1471 genes except the target gene itself. For this experiment, precision never exceeded 6%, even for the smallest values of recall. This indicates that the predictions are extremely poor and only slightly better than random guessing.

With respect to the results obtained in the DREAM4 challenge, these results are disappointing. The larger number of genes in this case does not explain everything since it also comes with an increase of the number of observations. Actually in both cases, the number of observations is comparable to the number of genes. However, since the *E. coli* dataset is a collection of expression data compiled from experiments carried out in different laboratories, there may be some redundancy among these experiments or some bias in their selection. They are thus probably not as statistically useful as the really randomized and i.i.d. perturbation data generated for the

DREAM4 multifactorial challenge. Other potential reasons for these poor results are the fact that the gold standard network is not complete and also the discrepancy that exists between the simulation model used to generate the DREAM4 data and the real regulation mechanism of *E. coli*.

DISCUSSION

We developed GENIE3, a procedure that aims at recovering a gene regulatory network from multifactorial expression data. This procedure decomposes the problem of inferring a network of size p into p different feature selection problems, where the goal is to identify the regulators of one of the genes of the network. Among different feature selection methods, we chose to use tree-based ensemble methods. These methods do not make any assumption about the nature of gene regulation, can potentially deal with combinatorial regulations and non-linearity. They work well in the presence of a large number of genes, are fast to compute and scalable.

GENIE3 got the best performance on the DREAM4 *In Silico Multifactorial* challenge and is competitive with existing algorithms to decipher the genetic regulatory network of *Escherichia coli* assuming that transcription factors are known. When no prior knowledge is available about transcription factors, our results on the *E. coli* network were however not better than random guessing. The reason of this discrepancy with respect to the results on the DREAM4 challenge deserves to be further analysed.

Our algorithm can be improved along several directions. As tree-based ensemble methods, we used the Random Forests and the Extra-Trees algorithms, that both gave comparable results. However, the performances of these methods depend to some extent on their main parameter, the number K of randomly selected attributes at each node of one tree. On the DREAM4 *Multifactorial* datasets, improved predictions were obtained by increasing K to its maximum value ($K = p - 1$), while on the *E. coli* dataset, the best ranking of interactions was obtained by using $K = \sqrt{n_{TF}}$. It would thus be of interest to find a way to automatically tune this parameter. A first solution could be to select the value of the parameter that leads to the best performance for the prediction of the expression values, i.e. that minimizes mean square error in (2) estimated by cross-validation. Unfortunately, this solution did not work on the *E. coli*dataset, where using $K = n_{TF}$ led to lower mean square error but a less good precision-recall curve.

There is also a potential room for improvement on the way variable importance scores are normalized. One apparent drawback of the measure we proposed is that it does not take into account the quality of the trees in generalization. Indeed since our trees are fully grown, importance weights satisfy equation (4) which, given our normalization, attributes equal weights to all tree models irrespective of their quality when used to predict the expression values of the target gene. We tried to correct for this bias by normalizing the variable importance scores by the effective variance reduction brought by the model as estimated by cross-validation but it actually deteriorated the performances. The question of the optimal normalization remains thus open at this stage.

In this paper, we focused on providing a ranking of the regulatory interactions. In some practical applications however, one would like to determine a threshold on this ranking to obtain a practical predicted network. To address this question, we have tried to exploit cross-validation estimates of the mean-square error as a criterion to determine such a threshold but we have not been successfull so far. As future work, we therefore would like to extend the technique developed in [46] to better assess the significance of the predicted regulatory links and thus help determining a threshold.

Our experiments on the DREAM4 dataset show that GENIE3 is able to predict the direction of the edges to some extent, even though it only exploits steady-state measurements. This is an interesting result as this is commonly admitted to be a difficult problem. Bayesian networks also potentially allow to predict edge directionality. A comparison with this family of methods would be an interesting future work direction. Note that with respect to our approach, Bayesian networks do not allow for the presence of cycles in the predicted network, which could be a limiting factor for networks such as those in DREAM4 that contain cycles by construction.

Several procedures using regression trees have already been proposed to solve the regulatory network inference problem. Most of these procedures exploits other kinds of data in addition to expression data, e.g. counts of regulatory motifs that serve as binding sites for transcription factors [47], [48], or ChIP-based binding data [49]. The closest work to ours is the procedure developed by Segal et al. [50], that recovers module networks from expression data, so that the genes in each module share the same regulators in the network and the same conditional probability distribution, represented by a (single) regression tree.

Finally, although we exploited tree-based ensemble methods, our framework is general and other feature selection techniques could have been used as well. Actually, several existing methods for network inference can be interpreted as special instances of this framework. In particular, mutual information as used in Relevance Networks [10] or CLR [11] is a common dependency measure exploited in filter-kind approaches for feature selection [28]. MRNET [14]also considers each gene in turn as the target output and exploits the maximum relevance/minimum redundancy feature selection method to rank its candidate regulators. Like Relevance Networks and CLR, this method reduces all the information contained in the expression data to mutual information between all pairs of genes, while our approach is by nature multivariate. Meinshausen and Bühlmann [21] show that finding the zero entries in the inverse covariance matrix of a multivariate Gaussian distribution can be solved by applying the LASSO embedded feature selection mechanism using each gene in turn as the target output, which links Gaussian graphical models with our approach. While the latter assumes that the functions f_i in (1) are linear, our approach can be seen as a relaxation of this assumption by exploiting a non-parametric supervised learning method. Whether or not this is an advantage in practice for inferring regulatory networks is still an open question that deserves to be studied.

AUTHOR CONTRIBUTIONS

Conceived and designed the experiments: VAHT PG. Performed the experiments: VAHT. Analyzed the data: VAHT AI LW PG. Wrote the paper: VAHT AI LW PG.

REFERENCES

1. Bolouri H (2008) Computational Modeling of Gene Regulatory Networks - a Primer. Imperial College Press.
2. Gardner TS, Faith JJ (2005) Reverse-engineering transcription control networks. Physics of Life Reviews 2: 65–88.
3. Bansal M, Belcastro V, Ambesi-Impiombato A, di Bernardo D (2007) How to infer gene networks from expression profiles. Mol Syst Biol 3: 78.
4. Markowetz F, Spang R (2007) Inferring cellular networks–a review. BMC Bioinformatics 8: Suppl 6S5.
5. Lee WP, Tzou WS (2009) Computational methods for discovering gene networks from expression data. Brief Bioinform 10: 408–423.
6. Kauffman SA (1993) The Origins of Order: Self-Organization and Selection in Evolution. Oxford University Press.
7. Gardner TS, di Bernardo D, Lorenz D, Collins JJ (2003) Inferring genetic networks and identifying compound mode of action via expression profiling. Science 301: 102–105.
8. Bonneau R, Reiss DJ, Shannon P, Facciotti M, Hood L, et al. (2006) The inferelator: an algorithm for learning parsimonious regulatory networks from systems-biology data sets de novo. Genome Biol 7: R36.
9. Eisen MB, Spellman PT, Brown PO, Botstein D (1998) Cluster analysis and display of genome-wide expression patterns. Proc Natl Acad Sci U S A 95: 14863–14868.
10. Butte AJ, Kohane IS (2000) Mutual information relevance networks: functional genomic clustering using pairwise entropy measurements. Pac Symp Biocomput 418–429b.
11. Faith JJ, Hayete B, Thaden JT, Mogno I, Wierzbowski J, et al. (2007) Large-scale mapping and validation of *Escherichia coli* transcriptional regulation from a compendium of expression profiles. PLoS Biology 5: e8.
12. Margolin AA, Wang K, Lim WK, Kustagi M, Nemenman I, et al. (2006) Reverse engineering cellular networks. Nature Protocols 1: 663–672.
13. Cover TM, Thomas JA (2006) Elements of Information Theory 2nd Edition. Wiley-Interscience.
14. Meyer PE, Kontos K, Lafitte F, Bontempi G (2007) Information-

theoretic inference of large transcriptional regulatory networks. EURASIP J Bioinform Syst Biol 2007: 79879.

15. Friedman N (2004) Inferring cellular networks using probabilistic graphical models. Science 303: 799–805.
16. Friedman N, Linial M, Nachman I, Pe'er D (2000) Using bayesian networks to analyze expression data. Journal of computational biology 7: 601–620.
17. Yu J, Smith V, Wang P, Hartemink A, Jarvis E (2004) Advances to bayesian network inference for generating causal networks from observational biological data. Bioinformatics 20: 3594–603.
18. Auliac C, Frouin V, Gidrol X, D'Alché-Buc F (2008) Evolutionary approaches for the reverse-engineering of gene regulatory networks: A study on a biologically realistic dataset. BMC Bioinformatics 9: 91.
19. Perrin B, Ralaivola L, Mazurie A, Bottani S, Mallet J, et al. (2003) Gene networks inference using dynamic bayesian networks. Bioinformatics 19: ii138–48.
20. Schafer J, Strimmer K (2005) A shrinkage approach to large-scale covariance matrix estimation and implications for functional genomics. Statistical applications in genetics and molecular biology 4: 1175.
21. Meinshausen N, Bühlmann P (2006) High-dimensional graphs and variable selection with the lasso. Ann Statist 34: 1436–1462.
22. Castelo R, Roverato A (2009) Reverse engineering molecular regulatory networks from microarray data with qp-graphs. Journal of Computational Biology 16: 213–227.
23. Ambroise C, Chiquet J, Matias C (2009) Inferring sparse gaussian graphical models with latent structure. Electronic Journal of Statistics 3: 205–238.
24. Opgen-Rhein R, Strimmer K (2007) From correlation to causation networks: a simple approximate learning algorithm and its application to high-dimensional plant gene expression data. BMC Systems Biology 1: 37.
25. The DREAM4 *In Silico* network challenge. http://wiki. c2b2. columbia. edu/dream09/index. php/D4c2.
26. Werhli AV, Grzegorczyk M, Husmeier D (2006) Comparative evaluation of reverse engineering gene regulatory networks with relevance networks, graphical gaussian models and bayesian networks. Bioinformatics 22: 2523–2531.

27. Pournara I, Wernisch L (2004) Reconstruction of gene networks using bayesian learning and manipulation experiments. Bioinformatics 20: 2934–2942.
28. Saeys Y, Inza I, Larranaga P (2007) A review of feature selection techniques in bioinformatics. Bioinformatics 23: 2507–2517.
29. Geurts P, Irrthum A, Wehenkel L (2009) Supervised learning with decision tree-based methods in computational and systems biology. Mol Biosyst 5: 1593–605.
30. Breiman L, Friedman JH, Olsen RA, Stone CJ (1984) Classification and Regression Trees. Wadsworth International (California).
31. Breiman L (2001) Random forests. Machine Learning 45: 5–32.
32. Geurts P, Ernst D, Wehenkel L (2006) Extremely randomized trees. Machine Learning 36: 3–42.
33. Strobl C, Boulesteix AL, Zeileis A, Horthorn T (2007) Bias in random forest variable importance measures: Illustrations, sources and a solution. BMC Bioinformatics 8: 5.
34. The DREAM project. http://wiki. c2b2. columbia. edu/dream/index. php/The_DREAM_Project.
35. Prill RJ, Marbach D, Saez-Rodriguez J, Sorger PK, Alexopoulos LG, et al. (2010) Towards a rigorous assessment of systems biology models: The DREAM3 challenges. PLoS ONE 5: e9202.
36. Stolovitzky G, Prill RJ, Califano A (2009) Lessons from the DREAM2 challenges. Annals of the New York Academy of Sciences 1158: 159–95.
37. Stolovitzky G, Monroe D, Califano A (2007) Dialogue on reverse-engineering assessment and methods: The DREAM of high-throughput pathway inference. Annals of the New York Academy of Sciences 1115: 11–22.
38. Marbach D, Schaffter T, Mattiussi C, Floreano D (2009) Generating realistic *in silico*gene networks for performance assessment of reverse engineering methods. Journal of Computational Biology 16: 229–239.
39. Faith JJ, Driscoll ME, Fusaro VA, Cosgrove EJ, Hayete B, et al. (2008) Many microbe microarrays database: uniformly normalized affymetrix compendia with structured experimental metadata. Nucleic Acids Research 36 (Database issue): D866–70.
40. Gama-Castro S, JimŽnez-Jacinto V, Peralta-Gil M, Santos-Zavaleta A, Pealoza-Spinola MI, et al. (2008) RegulonDB (version 6. 0): gene

regulation model of Escherichia coli K-12 beyond transcription, active (experimental) annotated promoters and Textpresso navigation. Nucleic Acids Research 36 (Database issue): D120–4.

41. Breiman L (1996) Bagging predictors. Machine Learning 24: 123–140.
42. Marbach D, Prill RJ, Schaffter T, Mattiussi C, Floreano D, et al. (2010) Revealing strengths and weaknesses of methods for gene network inference. Proceedings of the National Academy of Sciences 107: 6286–6291.
43. Faith JJ, Hayete B, Thaden JT, Mogno I, Wierzbowski J, et al. Supplemental website for: Large-scale mapping and validation of *Escherichia coli* transcriptional regulation from a compendium of expression profiles. http://gardnerlab. bu. edu/data/PLoS_2007/.
44. Meyer PE, Lafitte F, Bontempi G (2008) minet: A r/bioconductor package for inferring large transcriptional networks using mutual information. BMC Bioinformatics 9: 461.
45. Schäfer J, Opgen-Rhein R, Strimmer K (2006) Reverse engineering genetic networks using the GeneNet package. R News 6/ 5: 50–53.
46. Huynh-Thu VA, Wehenkel L, Geurts P (2008) Exploiting tree-based variable importances to selectively identify relevant variables. JMLR: Workshop and Conference proceedings 4: 60–73.
47. Phuong TM, Lee D, Lee KH (2004) Regression trees for regulatory element identification. Bioinformatics 20: 750–757.
48. Ruan J, Zhang W (2006) A bi-dimensional regression tree approach to the modeling of gene expression regulation. Bioinformatics 22: 332–340.
49. Xiao Y, Segal MR (2009) Identification of yeast transcriptional regulation networks using multivariate random forests. PLoS Computational Biology 5: e1000414.
50. Segal E, Pe'er D, Regev A, Koller D, Friedman N (2005) Learning module networks. Journal of Machine Learning Research 6: 557–588.

CHAPTER 9

HYBRID-CONTROLLED NEUROFUZZY NETWORKS ANALYSIS RESULTING IN GENETIC REGULATORY NETWORKS RECONSTRUCTION

Roozbeh Manshaei[1], Pooya Sobhe Bidari[1], Mahdi Aliyari Shoorehdeli[2], Amir Feizi[3], Tahmineh Lohrasebi[4], Mohammad Ali Malboobi[4], Matthew Kyan[1], and Javad Alirezaie[1]

[1]Electrical and Computer Engineering Department, Ryerson University, Toronto, ON, Canada M5B 2K3

[2]Electrical and Computer Engineering Department, K.N. Toosi University of Technology, Tehran 16315-1355, Iran

[3]Department of Chemical and Biological Engineering, Systems and Synthetic Biology Group, Chalmers University, 41296 Gutenberg, Sweden

[4]National Institute of Genetic Engineering and Biotechnology (NIGEB), Tehran 14965/161, Iran

CITATION (APA): Manshaei, R., Sobhe Bidari, P., Aliyari Shoorehdeli, M., Feizi, A., Lohrasebi, T., Malboobi, M. A., ... & Alirezaie, J. (2012). Hybrid-controlled neurofuzzy networks analysis resulting in genetic regulatory networks reconstruction. ISRN bioinformatics, 2012. (16 pages)

ABSTRACT

Reverse engineering of gene regulatory networks (GRNs) is the process of estimating genetic interactions of a cellular system from gene expression data. In this paper, we propose a novel hybrid systematic algorithm based on neurofuzzy network for reconstructing GRNs from observational gene expression data when only a medium-small number of measurements are available. The approach uses fuzzy logic to transform gene expression values into qualitative descriptors that can be evaluated by using a set of defined rules. The algorithm uses neurofuzzy network to model genes effects on other genes followed by four stages of decision making to extract gene interactions. One of the main features of the proposed algorithm is that an optimal number of fuzzy rules can be easily and rapidly extracted without overparameterizing. Data analysis and simulation are conducted on microarray expression profiles of S. cerevisiae cell cycle and demonstrate that the proposed algorithm not only selects the patterns of the time series gene expression data accurately, but also provides models with better reconstruction accuracy when compared with four published algorithms: DBNs, VBEM, time delay ARACNE, and PF subjected to LASSO. The accuracy of the proposed approach is evaluated in terms of recall and F-score for the network reconstruction task.

INTRODUCTION

Biological systems are inherently stochastic, uncertain, and fuzzy [1]. Therefore, research in bioinformatics and computational biology, where computer technologies are applied to manage and analyze biological data and make computational models, is faced with a great deal of uncertainty. For instance, growth and development as well as environmental stresses can all contribute to change in gene expression levels. In addition, under such conditions, some genes influence the expression of other genes and their functionalities.

With the advent of high-throughput technologies in transcriptomics, proteomics, and metabolomics, now, biologists have the ability to investigate the expression of genes and consequences on a genome-wide scale. Gene expression data in the form of high-throughput microarray experiments measure the amounts of RNA associated with each of thousands of genes in parallel. Time-series microarrays have attracted biologists' interests for deciphering the dynamic and complex nature of biological networks. Time-series microarrays record multiple expression profiles at discrete time points (i.e., hours or days) of a continuous cellular process. Thus, analytical methods are needed to handle many genes with uncertain functions based on discrete datasets of continuous biological processes. The methodological areas range from experimental design [2] to data normalization [3, 4], missing value imputation [5], cluster analysis [6, 7], classification [8], identification of differentially expressed genes [9], and network modelling [10, 11].

As a challenging concept, reverse engineering can be employed to estimate gene regulatory networks from high-throughput expression data. Genetic regulatory network reconstruction provides a concise representation of the interactions between multiple genes at the system level. In addition, it confers a broader insight for biologists about the manner in which genes interact with one another, and about the roles that they play in various biological functions.

Viability of an organism down to the cells is essentially controlled by gene expression regulation at the transcript level. This concept leads one to ponder how the changes in the expression patterns of genes during the ordinary and the stressful conditions of cells may infer the way genes are affected by environmental conditions (e.g., lack of nutrients) [12]. Such studies of gene expression patterns can improve our understanding of biological systems, and may enhance our ability to combat undesired situations (e.g., diseases such as cancer) with the hope of improving human life quality.

There are several limitations in the study of time-series gene expression data such as small sample size (due to the time-consuming nature in which samples are produced and the high costs associated with microarray experiments, especially in clinical studies), genes with low level expression and noisy data structure [13, 14]. Problems related to high dimensionality accompanied by a small sample size, such as matrix singularity, model over-fitting and model over-parameterization become more pronounced in the case of the most available data [15]. Also, the unavoidable presence of noise has more influence on the analysis of short term rather than long term data series. This enhances the difficulty in distinguishing actual patterns from random data, thereby raising the potential of misleading analyses [16].

Reconstruction of gene regulatory networks based on expression data should therefore: be able to handle constrained data; should be robust to, and compensate for noise and incompleteness of data; and should be capable of providing interpretable results.

In view of recent advances, a wide spectrum of reverse engineering approaches have been proposed to infer gene regulatory networks (GRNs), including: Boolean networks [17–19]; Bayesian network models [20–23]; Hidden Markov Models (HMM) [24]; Graphical Gaussian models [25]. In addition, state space models [26, 27], Kalman filter (KF) [28], extended Kalman filter (EKF) [29], and Particle Filter (PF) subjected to LASSO [30] have also been employed to model gene regulatory networks. The goal of these methods is to explore a high-fidelity representation to determine possible cause-effect gene regulatory interactions, which are ultimately represented as a graph [31].

Modelling based on Boolean Networks is one of the common methods employed in GRNs inference [32]. The goal of these models is basically to infer rules based on their computational simplicity and ability to handle noisy experimental data [17]. Even though these models can be easily applied, much information is lost in binary encoding and, in practice, the derived models have insufficient dynamic resolution because they depend on arbitrary discretizations of the gene expression values [33–37]. In general, Boolean networks are limited by their definition.

Bayesian network modelling is based on probabilistic transitions between network states and assumes that there is no feedback in a network; in spite of the fact that cycles of events are the major mechanism to ensure robustness of the biological systems [21].

HMM also has been applied for analyzing time-series gene expression data [24]. However, there are several problems with HMMs. The number of parameters that need to be set in an HMM is quite high. As a result, the amount of data required to train an HMM is very large. Also, concepts learnt by an HMM are framed in terms of emission and transition probabilities. If one is trying to understand the concept learnt by the HMM, then this concept representation is difficult to understand.

Dynamic Bayesian Network (DBN) as another method combines the features of hidden Markov models to incorporate feedback [38]. Models based on Bayesian networks, despite attractiveness due to their ability to deal with stochastic aspects of gene expression and noisy measurements, have the disadvantage of minimizing the dynamical aspects of gene regulation [20]. A graphical Gaussian model (GGM) is an undirected probabilistic graphical model [25]. This model allows the identification of conditional independence relations among genes, under the assumption of a multivariate Gaussian distribution of gene expression data. The GGM does not identify the direction of gene relationships, but rather only calculates the correlations between their gene expression data. The Kalman filter (KF) [28] is only applicable to linear models and the Gaussian posterior density probability. Since position information is linear, standard Kalman filtering can be easily applied to the tracking problem without much difficulty. However, gene regulatory networks pose nonlinear information, requiring a modification to the KF. To overcome this problem, much research has been reported on nonlinear filtering methods such as extended Kalman filter (EKF) [29], unscented Kalman filter (UKF), and Particle filter (PF) [30]. Currently, considerable research is being devoted to introduce improvements in the working of these algorithms and enhance our understanding about gene interactions.

In this paper, we describe a novel algorithm that benefits from using rule-based neurofuzzy networks (RBNFNs) to extract information from time-series gene expression data. The suggested algorithm combines neural networks with fuzzy systems and allows for mapping the dynamics of gene expression data. Fuzzy logic [39–41] and artificial neural networks [42, 43] are complementary technologies in the design of an intelligent system, and their combination appears to be a promising path, since neural networks are essentially low-level, computational algorithms that sometimes offer a good performance in pattern-recognition tasks; whilst fuzzy logic provides a structural framework that uses and exploits those low-level capabilities of neural networks. Thus, the combination seems to offer potential for capturing subtle effects of genes. Neural networks can learn from data sets while fuzzy logic solutions are easy to verify and optimize. Fuzzy logic and neural networks generally approach the design of intelligent systems from quite different angles, and the combined system can have advantages from both sides: neural networks are implicit; although the system is not easily interpreted or modified, it trains itself by data sets. Fuzzy logic is explicit; thus the system verification and optimization is more efficient.

Here, we present a two-level algorithm based on RBNFNs for extracting gene interaction networks. This algorithm is advantageous as the network is never overparameterized, avoids redundant fuzzy sets, decreases the redundancy of the model, thus it is simpler and drastically reduces the computational cost. Also, in this network, the training process does not depend on the number of inputs and the sample size. The proposed networks are trained with a subset of experimental samples and tested with the remaining samples. Then, the constructed fuzzy functions in the gene interaction reconstruction (referred as edges) depend on the dynamics of input-output patterns of the networks. We applied the network to yeast cell-cycle regulation data and the resulted interactions from our model were compatible with previous experimentally verified interactions.

ALGORITHM

The method proposed in this paper is a quantitative computational approach consisting of five main stages shown in Figure 1. Our methodology includes two levels. In the first level (stage 1), RBNFNs are used for time series prediction of gene expression; in the second level (stage 2 to 5), the rules created by the RBNFNs for reconstructing gene regulatory networks are employed. The detailed descriptions of these stages are outlined as follows:(1) Training RBNFNs by gene expression data and creating IF-THEN rules (Section 2.1),(2)Extracting fuzzy IF-THEN rules of each RBNFN and the related weights (Section 2.2),(3)Sorting the obtained fuzzy rules of each RBNFN to have further analyses (Section 2.3),(4)Extracting similarity rules as a connectivity matrix, from sorted fuzzy rules of all RBNFNs, and generating the interactions from each rule (Section 2.4),(5)Analyzing the extracted interactions of stage 4 to reconstruct the final model (Section 2.5).

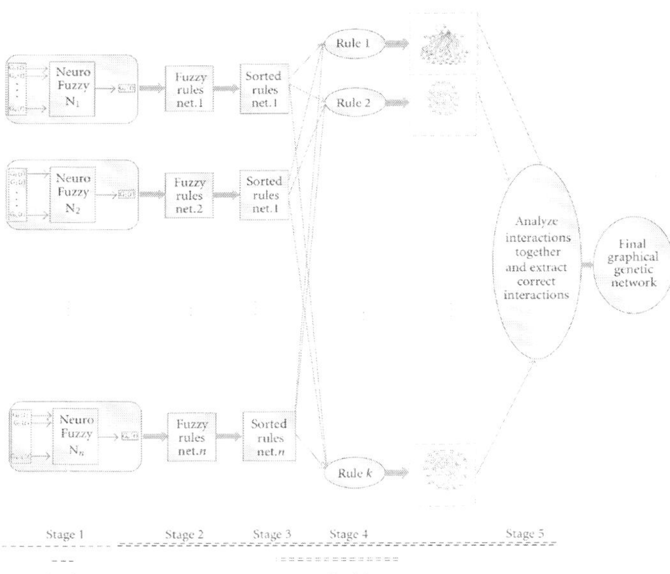

Figure 1: Proposed Hybrid Rule-Based Neurofuzzy (HRBNF) algorithm for extracting the gene interactions based on rules governing the gene expressions.

The algorithm includes training with expression data (Stage 1) to extract the rules (Stage 2) which are sorted (Stage 3) to compare the rules and prior to be used for gene interaction analysis (Stage 4) and modelling the final gene network (Stage 5).

Rule-Based Neurofuzzy Networks (RBNFNs)

Let us first assume that there are N genes in the expression dataset we are studying. In the first stage, Nindependent RBNFNs are generated with the following structure: in each RBNFN, the expression data for all genes are loaded as inputs except for one gene, which is considered as an output. Training each RBNFN with expression data gives rise to a network capable of predicting the expression pattern of the output as the result of input genes.

RBNFNs are employed in this stage because of their ability to overcome the drawbacks of pure neural networks. By incorporating elements of fuzzy reasoning processes, the RBNFN gives meaning and function as part of a fuzzy rule to each node via an associated weight. We have utilized the least square (LS) technique to learn the parameters of these RBNFNs. The architecture of a typical RBNFN (illustrated in Figure 2) includes the following steps:

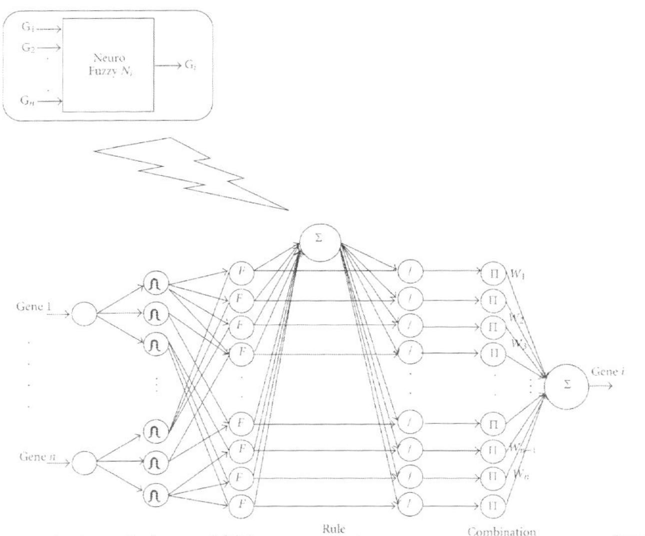

Figure 2: The architecture of a typical rule-based neurofuzzy network (RB-NFN). Expression data for each gene is normalized (input layer) before being

fuzzified using a three-state Gaussian matrix (fuzzier layer), then the preceding part of fuzzy rules are combined using a product operator (rule base layer), before the rules are normalized (rule normalization layer) and the fuzzy target value is defuzzified (combination & defuzzifier layer).

Input Layer (Step 1)

In this layer, the input expression dataset is normalized using (1). Each node, corresponding to each input variable, normalizes each input value to the scale of [0, 1] for the next layer, to facilitate in fuzzification:

$$O_i^{(1)} = \frac{G_i - \min(G_i)}{\max(G_i) - \min(G_i)}, \tag{1}$$

where G_i is the expression values of the ith gene, and $O_i^{(1)}$ is the ith output of input layer.

Fuzzier Layer (Step 2)

In layer 2, the normalized expression values are fuzzified using a three-state Gaussian matrix, with linguistic values of "Low," "Medium," and "High." These linguistic labels represent the data in fuzzy logic terms. Figure 3depicts the Gaussian membership functions (MFs) for the three-state model employed to find the maximum degree that each sample of normalized input belongs to the respective label. In (2), the membership functions are represented in Gaussian form:

$$O_{i,k}^{(2)} = \exp\left(-\frac{\left(O_{i,k}^{(1)} - m_{i,k}^{j}\right)^2}{\sigma_{i,k}^{j\,2}}\right), \tag{2}$$

where i, k, and j indicate the label of the input variable, the time point and the MF, respectively, m and σ are the mean and variance of the membership functions, while $O_{i,k}^{(1)}$ is the value of the ith input variable in the kth time point.

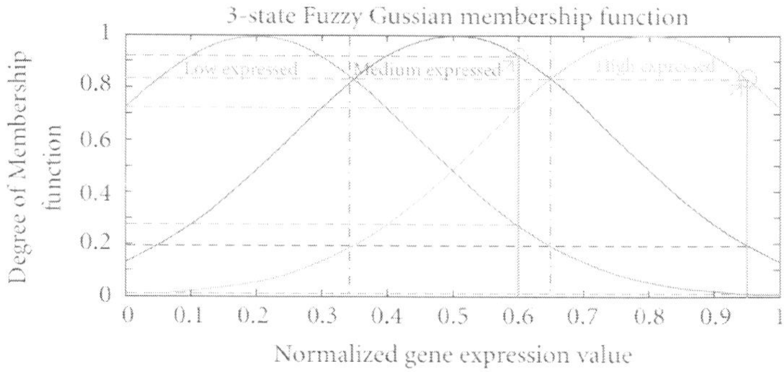

Figure 3: Three state fuzzy Gaussian membership functions. Two examples of normalized expression levels and their fuzzy representation for two genes are depicted. O_i(max), shown by green circle, is the maximum membership value of each input.

We have considered constant values for the means and variances of MFs in a manner that they cover the scale of [0, 1] in equal partitions. The constant means and variances are assumed in order to maintain an ability to compare and analyze the extracted rules in the next layers. In other words, if these values are different for each input, the concept of low, medium, and high expressed for each gene will not be the same as for the others. In addition, employing constant means and variances allows for fewer parameters to be trained in the network; an advantage of which is that, when encountering small sample size temporal genetic data, the network will not be overparameterized during training.

As described earlier, the algorithm in step 2 partitions the input/output space. This space is an n-dimensional unit hypercube [0, 1]n similar to Table 1. This hypercube results when membership functions map the points in the input/ output space to a degree of membership between 0 and 1. Table 1 presents the membership value of the normalized expression data of each gene assigned to low, medium, and high labels by (2). In this table, $O^j_{i,k}$ represents the membership value of kth data point for ith gene assigned to jth label.

Table 1: Membership matrix of a three-state fuzzy logic model

Labels	Genes				
	$G_{1,k}$	$G_{2,k}$	$G_{3,k}$	...	$G_{n,k}$
Low Expressed	$O^L_{1,k}$	$O^L_{2,k}$	$O^L_{3,k}$	...	$O^L_{n,k}$
Medium Expressed	$O^M_{1,k}$	$O^M_{2,k}$	$O^M_{3,k}$	...	$O^M_{n,k}$
High Expressed	$O^H_{1,k}$	$O^H_{2,k}$	$O^H_{3,k}$	...	$O^H_{n,k}$

Entries within the columns are the normalized expression value of genes in the kth time point applied to the membership functions of low, medium, and high expressed.

Rule Base Layer (Step 3)

Upon the fuzzified output of layer 2, a rule-based profile is set up in layer 3. This profile consists of fuzzy IF-THEN rules for determining the expression value of a target gene according to the expression levels of input genes. For instance, as illustrated in Figure 4 for a sample three-state model, a fuzzy IF-THEN rule may be presented as follows: IF "the expression level of G_1, G_2 and G_3 are respectively Low, High and Medium as input genes," THEN "the corresponding expression level for the target gene, GT, will be Low." The method used in our fuzzy system is based on a singleton output function, which assigns a single value to each of the N fuzzy states in the model.

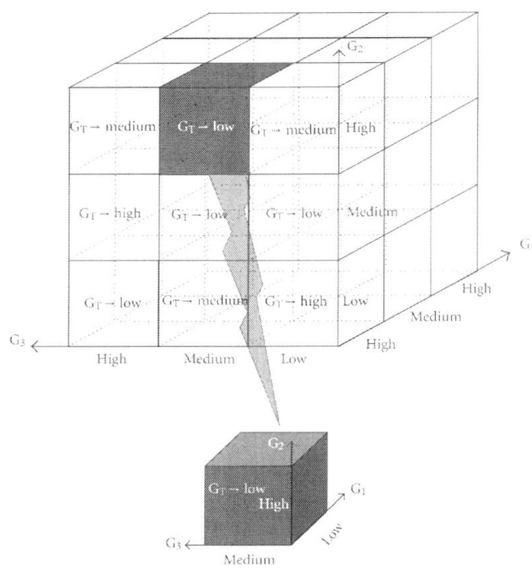

Figure 4: Decision matrix for the three-state rule based neuro fuzzy model. Entries within the decision matrix are the inferred levels of the target gene.

In order to reduce the search space of possible fuzzy rule combinations from the rule-based layer, we only create new rules associated with output states observed in the training sample set. As such, each new rule corresponds to interaction clusters discovered in the input/output space. This indicates that the number of created rules in the RBNFN depends on the number of sample sets of input/output genes. Thus, the more samples of genes presented, the greater number of rules created. The expected benefit of this approach is that there would be less contamination in knowledge extracted by not considering irrelevant states for gene interaction.

When a new set of input/output data is applied to the RBNFN, the network determines whether to generate a new rule for describing the incoming pattern (x,y) or not. This occurs after checking the similarities between the new rule and the previous ones. This process leads to a reduction in the number of fuzzy sets and avoids the existence of redundant rules.

Each node in layer 3 combines the antecedent part of a fuzzy rule using a T-norm operator. In this study, the T-norm operator is considered as a product operation. The output of each node represents the firing strength of the corresponding fuzzy rule, which is calculated by

$$O_k^{(3)} = \text{Prod}_r\left(\exp\left(-\left\{\delta_r^{-1}\left(O_r^{(1)} - C_r\right)\right\}^T \left\{\delta_r^{-1}\left(O_r^{(1)} - C_r\right)\right\}\right)\right),$$

$$\delta_r^{-1} = \text{diag}\left(\frac{1}{\sigma_{r1}}, \frac{1}{\sigma_{r2}}, \ldots, \frac{1}{\sigma_{rn}}\right),$$

$$C_r = (c_{r1}, c_{r2}, \ldots, c_{rn})^T, \qquad (3)$$

where r runs through all the selected nodes of step 2 corresponding to the kth rule.

Rule Normalization Layer (Step 4)

The number of nodes in this step is the same as step 3. The nodes in layer 4 calculate the ratio of the ith rule's firing power to the sum of all rules' firing strengths, which can be formulated by (4). Indeed, the rules are normalized in this layer in the scale of [0, 1]

$$O_i^{(4)} = \frac{O_i^{(3)}}{\sum_{i=1}^{m} O_i^{(3)}}, \qquad (4)$$

where m is the number of all rule sets. The output of this layer is also called normalized firing power.

Combination and Defuzzification Layer (Step 5)

After applying the decision matrix to the fuzzified expression levels in step 4, the membership degree of a target to a statement is determined. At step 5, the fuzzy target value is transformed back into a value between 0 and 1 via the process of defuzzification; indeed, the output is the predicted value of the RBNFN.

Equation (5) formulates the output of layer 5, as the overall output of RBNFN. In fact, this equation combines and defuzzifies all the rules from previous steps

$$O_i^{(5)} = \sum_{i=1}^{m} w_i O_i^{(4)}, \tag{5}$$

where w_i is the weight multiplied to the ith rule.

Figure 5 illustrates this process as an example. In this figure, each row indicates a rule calculated from normalized expression values of input genes at a sample time point.

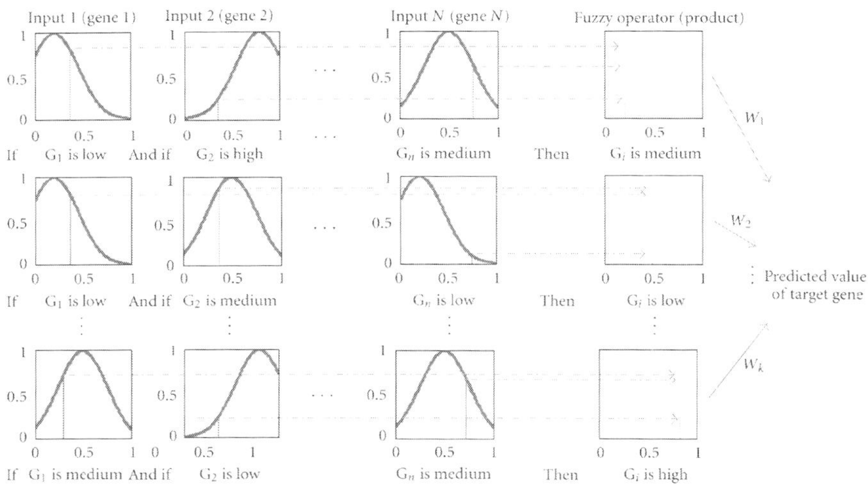

Figure 5: The defuzzification process.

As described above, in our methodology, we have an RBNFN for each gene. In a particular RBNFN, the expression value of the output gene is assumed to be the outcome of all other genes. In other words, all remaining genes are considered to be either activators or repressors for the target gene, and the predicted expression pattern for this gene is deduced from the expression levels of all other ones.

Also we compare the prediction of the RBNFN with the real expression pattern of a target gene in order to calculate the weights of the network. A greater weight specifies that the corresponding input gene has a greater affect on the output, so it can be considered as a greater activator or repressor effect on the target gene.

RBNFNs Learning Procedure

The parameters of proposed RBNFN must be trained to obtain a prediction system for our data. Therefore, in the learning procedure, the related parameters of rules (weights) are trained by LS the least squares (LS) learning algorithm. The simplicity of LS algorithm makes it a common and suitable method for training and tuning parameters of neural or fuzzy networks.

An error criterion is defined for each RBNFN in order to test the accuracy of a given data set. The error criterion is an overall error measurement based on the differences between the predictions and target values. We used mean square error (MSE) as our error criterion which is a very common criterion described by

$$E_i = \frac{1}{N}\sum_{i=1}^{N}\left(y_i^{\text{Predicted}} - y_i^{\text{Measured}}\right)^2, \tag{6}$$

where N is the number of samples in the data set. Intuitively, an increase in the number of rules results in a decrease in the error measure of (6).

In the next step, we attempt to extract the most of available information from provided genetic data in order to make the most possible understanding of behaviour and interaction among genes.

Extracting Fuzzy IF-THEN Rules

As stated above, in the proposed HRBNF algorithm, we construct network models for as many genes as possible from the data available. In each model, one of the genes is considered as the output and the others as the input nodes. The model attempts to describe the behaviour of the output gene based on the possible interactions of input genes.

Finally, it stores the derived knowledge from gene-gene interactions in the fuzzy rules and their corresponding weights.

Table 2 presents the extracted rules of ith network. In this table, i, k run through all the input variables and extracted rules, respectively, for n genes. $L^i_{k,j}$ indicates the fuzzy label of jth gene in kth rule of ith RBNFN.

Table 2: Extracted rules of ith network

Rule	Gene								
	G_1	G_2	...	G_{i-1}	G_{i+1}	...	G_n	W	G_i
Rule 1	$L^i_{1,1}$	$L^i_{1,2}$	...	$L^i_{1,i-1}$	$L^i_{1,i+1}$	...	$L^i_{1,n}$	w^i_1	$L^i_{1,i}$
Rule 2	$L^i_{2,1}$	$L^i_{2,2}$	...	$L^i_{2,i-1}$	$L^i_{2,i+1}$	...	$L^i_{2,n}$	w^i_2	$L^i_{2,i}$
⋮	⋮	⋮	⋮	⋮	⋮	⋮	⋮	⋮	⋮
Rule K	$L^i_{K,1}$	$L^i_{K,2}$	...	$L^i_{K,i-1}$	$L^i_{K,i+1}$	...	$L^i_{K,n}$	w^i_K	$L^i_{K,i}$

Fuzzy Rules Preparation

After extracting fuzzy IF-THEN rules such as those in Table 2, the rules are sorted in a way that the ith gene takes the ith position in the matrix; so we shift it between i − 1th and i + 1th column of the matrix. This leads us to have similar rule matrices in all RBNFNs and a decision agent to find gene interactions. The interactions are delineated according to the weight values and upon comparing them in similar rules of different RBNFNs. As stated in previous sections, to extract more precise relationships between genes, a full capacity of all available data is used in HRBNF algorithm. The proposed analysis of obtained rules and the way interactions among genes occurs are described in the next stage.

Similar Rules Extraction

Assuming that we have N genes, the outputs of previous stages are N matrices with K similar rules in each one. In this stage, we create a new set of matrices from previous tables in order to extract information. To achieve this target, N matrices are defined with K rules in them. A new table in this stage consists of K similar rules gained from all RBNFNs. Regarding the weights of a specified rule in RBNFNs, we can find the effect of that particular rule in every RBNFN; therefore, its effect on every output gene can be evaluated. This guides us to extract the relationships between genes by exploiting these matrices.

In Table 3, an example to the jth rule is illustrated. We call this table as the table of network effects on the similar rule (NESR) matrix.

Table 3: The networks effects on jth rule

Rule	Gene								
	G_1	G_2	$\cdots$	G_{i-1}	G_i	G_{i+1}	$\cdots$	G_n	W
Rule j	$L^1_{j,1}$	$L^1_{j,2}$	$\cdots$	$L^1_{j,i-1}$	$L^1_{j,i}$	$L^1_{j,i+1}$	$\cdots$	$L^1_{j,n}$	w^1_j
Rule j	$L^2_{j,1}$	$L^2_{j,2}$	$\cdots$	$L^2_{j,i-1}$	$L^2_{j,i}$	$L^2_{j,i+1}$	$\cdots$	$L^2_{j,n}$	w^2_j
$\vdots$	$\vdots$	$\vdots$	$\vdots$	$\vdots$	$\vdots$	$\vdots$	$\vdots$	$\vdots$	$\vdots$
Rule j	$L^n_{j,1}$	$L^n_{j,2}$	$\cdots$	$L^n_{j,i-1}$	$L^n_{j,i}$	$L^n_{j,i+1}$	$\cdots$	$L^n_{j,n}$	w^n_j

In order to estimate gene interactions from NESR matrix, the weights have to be analyzed. Therefore, we define some thresholds to identify the effect of a rule on all genes.

It is not easy to determine an optimal threshold value since a large value may result in the elimination of true connections with small effects during the procedure. On the other hand, a small threshold is not able to retrieve true connections from false connections in an inherently sparse genetic dataset. Our proposed criterion to obtain an appropriate threshold is called "effectivity threshold (ET)." This threshold is different for each RBNFN and is defined as the average of all weights calculated in that RBNFN, as shown in (7). We defined different thresholds for RBNFNs because the range of weights in each RBNFN is different from others; thus, by having different thresholds, a more realistic view of the effect of similar rules in different networks is provided

$$T_i = \frac{1}{k}\sum_{p=1}^{k} w^i_p. \tag{7}$$

In (7) i, k denote ith RBNFN and the number of rules in ith RBNFN, respectively. Also, w^i_p is the calculated weight value for pth rule in ith RBNFN.

After defining the thresholds, ETs are compared with thresholds. We also define a parameter named effectivity symbol (ES), which describes the state of each ET. According to (8), an ES equals to "+1" when the related weight is positive with absolute value greater than the defined threshold, or equals to "−1" when the related weight is negative with absolute value bigger than the defined threshold. An ES is considered to be "0" if the absolute value of related weight is less than the threshold

$$\text{if } \left|w_j^i\right| < T_i \quad \text{then } \mathrm{ES}_j^i = 0,$$

$$\text{if } \left|w_j^i\right| > T_i,\ w_j^i > 0 \quad \text{then } \mathrm{ES}_j^i = 1,$$

$$\text{if } \left|w_j^i\right| > T_i,\ w_j^i < 0 \quad \text{then } \mathrm{ES}_j^i = -1. \tag{8}$$

In the above equations, i, j denotes ith RBNFN for jth rule; and ES_j^i shows whether the jth rule has an effect more than the threshold on the ith gene or not.

By defining ESs, new matrices are obtained from NESR matrices with ES values instead of weights. We named these matrices as "ES matrices." Table 4 is an instance of an ES matrix related to jth rule. This matrix contains similar jth rule extracted from different RBNFNs.

Table 4: Decision-making table related to jth rule for depicting graph

Rule	Gene								
	G_1	G_2	...	G_{i-1}	G_i	G_{i+1}	...	G_n	ES
Rule j	$L_{j,1}^1$	$L_{j,2}^1$	...	$L_{j,i-1}^1$	$L_{j,i}^1$	$L_{j,i+1}^1$	...	$L_{j,n}^1$	ES_j^1
Rule j	$L_{j,1}^2$	$L_{j,2}^2$	...	$L_{j,i-1}^2$	$L_{j,i}^2$	$L_{j,i+1}^2$	...	$L_{j,n}^2$	ES_j^2
⋮	⋮	⋮	⋮	⋮	⋮	⋮	⋮	⋮	⋮
Rule j	$L_{j,1}^n$	$L_{j,2}^n$	...	$L_{j,i-1}^n$	$L_{j,i}^n$	$L_{j,i+1}^n$	...	$L_{j,n}^n$	ES_j^n

This stage is finalized by a decision-making process, in which we extract the connections from ES matrices. The connection of two genes is displayed by a directed graph in which an edge from one gene is directed to the other one, showing that the former gene has effect on the latter one. Our proposed procedure of extracting the connections among genes from the ES matrix is outlined in the following.

As stated before, a rule with ES equals to "0" implies that the weight of this rule in the related RBNFN is less than defined threshold; this means that the input genes of the RBNFN have effects less than the threshold on the output gene. Thus, we can consider that the output gene is not affected by the input genes. In other words, a rule with ES = "0" in an RBNFN means there is no inferred connection from the input gene to output gene. On the other hand, a rule with ES = "1" in an RBNFN means that the input genes have effects on the output gene as activators. Finally, a rule with ES = "−1" in an RBNFN indicates that the input genes are repressors of the output gene.

The decision-making process is formalized by (9):

$$\begin{aligned}
&\text{In Rule } j\text{: if } ES_j^i = 0 \;\&\; ES_j^p \neq 0,\ p = 1,\ldots,n,\ p \neq i \\
&\Rightarrow \\
&\quad G(i) \text{ connect to } G(p) \\
&\quad \text{if } ES_j^p = -1 \quad \text{then connection is repressor} \\
&\quad \text{if } ES_j^p = 1 \quad \text{then connection is activator}
\end{aligned} \tag{9}$$

Final Genetic Interaction Network Extraction

In the former section, we presented a procedure for extracting the connections among genes from ES tables. This procedure leads us to create two connection networks (CNs) for each ES matrix: one for activator connections and the other for repressor connections. As previously discussed, ES matrices are obtained from the same rules extracted from all RBNFNs. Thus, the number of CNs attained from a genetic dataset is two times that of the extracted rules. In this section, we attempt to achieve a final genetic interaction network from these CNs. This goal is acquired by analyzing and interpreting the activator and repressor connections of all rules.

We define two criteria for all connections of different CNs: activator criterion (AC) and repressor criterion (RC). Activator criterion, as shown in (10), has been employed to survey the activator interactions; as a result of which, Table 6 is obtained from all connections in CNs

$$\begin{aligned}
&\text{For Rule } j \\
&\Rightarrow \text{if } G(i) \text{ connect to } G(p)\ (p = 1,\ldots,n,\ p \neq i) \\
&\quad \text{and connection is activator} \\
&\qquad AC_{i\to p}^j = 1 \\
&\quad \text{else } AC_{i\to p}^j = 0
\end{aligned} \tag{10}$$

In (10), AC = 1 denotes the existence of an activator connection and AC = 0 does not indicate whether an activator connection exists or not. The interaction is estimated by summation of ACs as shown in Table 5. In this estimation, first we calculate the summation of ACs that confirm an interaction from ith gene to pth gene $(\sum_{j=1}^{k} AC_{i\to p}^j)$, and the summation of ACs that confirm an interaction from pth gene to ith gene $(\sum_{j=1}^{k} AC_{p\to i}^j)$; then, the interaction between ith gene and pth gene is estimated as the subtraction of these two values. By estimating the interactions between all pairs of genes, the final activator interaction network can be ascertained.

Table 5: Resultant of activator interactions

Gene	Gene					
	G_1	G_2	$\cdots$	G_i	$\cdots$	G_n
G_1	0	$\sum_{j=1}^{k} AC_{1\to 2}^{j}$	$\cdots$	$\sum_{j=1}^{k} AC_{1\to i}^{j}$	$\cdots$	$\sum_{j=1}^{k} AC_{1\to n}^{j}$
G_2	$\sum_{j=1}^{k} AC_{2\to 1}^{j}$	0	$\cdots$	$\sum_{j=1}^{k} AC_{2\to i}^{j}$	$\cdots$	$\sum_{j=1}^{k} AC_{2\to n}^{j}$
$\vdots$	$\vdots$	$\vdots$	$\vdots$	$\vdots$	$\vdots$	$\vdots$
G_i	$\sum_{j=1}^{k} AC_{i\to 1}^{j}$	$\sum_{j=1}^{k} AC_{i\to 2}^{j}$	$\cdots$	0	$\cdots$	$\sum_{j=1}^{k} AC_{i\to n}^{j}$
$\vdots$	$\vdots$	$\vdots$	$\vdots$	$\vdots$	$\vdots$	$\vdots$
G_n	$\sum_{j=1}^{k} AC_{n\to 1}^{j}$	$\sum_{j=1}^{k} AC_{n\to 2}^{j}$	$\cdots$	$\sum_{j=1}^{k} AC_{n\to i}^{j}$	$\cdots$	0

Table 6: Resultant of repressor interactions

Gene	Gene					
	G_1	G_2	$\cdots$	G_i	$\cdots$	G_n
G_1	0	$\sum_{j=1}^{k} RC_{1\to 2}^{j}$	$\cdots$	$\sum_{j=1}^{k} RC_{1\to i}^{j}$	$\cdots$	$\sum_{j=1}^{k} RC_{1\to n}^{j}$
G_2	$\sum_{j=1}^{k} RC_{2\to 1}^{j}$	0	$\cdots$	$\sum_{j=1}^{k} RC_{2\to i}^{j}$	$\cdots$	$\sum_{j=1}^{k} RC_{2\to n}^{j}$
$\vdots$	$\vdots$	$\vdots$	$\vdots$	$\vdots$	$\vdots$	$\vdots$
G_i	$\sum_{j=1}^{k} RC_{i\to 1}^{j}$	$\sum_{j=1}^{k} RC_{i\to 2}^{j}$	$\cdots$	0	$\cdots$	$\sum_{j=1}^{k} RC_{i\to n}^{j}$
$\vdots$	$\vdots$	$\vdots$	$\vdots$	$\vdots$	$\vdots$	$\vdots$
G_n	$\sum_{j=1}^{k} RC_{n\to 1}^{j}$	$\sum_{j=1}^{k} RC_{n\to 2}^{j}$	$\cdots$	$\sum_{j=1}^{k} RC_{n\to i}^{j}$	$\cdots$	0

Similar to the procedure described above, we define a repressor criterion in order to obtain the repressor connections and to achieve the final repression interactions network. The repressor criterion is presented in (11) and the repressor interactions are illustrated in Table 6

$$\begin{aligned}&\overset{\text{For Rule } j}{\Rightarrow} \text{if } G(i) \text{ connect to } G(p)\,(p = 1,\ldots,n,\ p \neq i)\\ &\quad \text{and connection is repressor}\\ &\quad\quad RC_{i\to p}^{j} = 1\\ &\quad \text{else } RC_{i\to p}^{j} = 0,\end{aligned} \tag{11}$$

where RC denotes the repressor connections.

In Table 6, similar to Table 5, the repressor interaction between ith gene and pth gene is estimated as the subtraction of $\sum_{j=1}^{k} RC_{i\to p}^{j}$ and $\sum_{j=1}^{k} RC_{p\to i}^{j}$. The final repressor interaction network will be provided by estimating the repressor interactions between all pairs of genes.

RESULTS

In this section, we present the results of HRBNF algorithm for an experimental data. In Section 3.1, we will introduce yeast (Saccharomyces cerevisiae) cell cycle microarray time series data sets presented in [47, 48]. This data has been extensively exploited for both practical and academic applications. Researchers frequently use these data sets to demonstrate and validate statistical and clustering analysis (e.g., [49–52]), mathematical modelling [37, 38], and reverse engineering methods [36, 53].

In Section 3.2, we present the results of RBNFNs performance in predicting the time series data sets. Finally, in Section 3.3, the designed interaction network, the comparison of the results with experimentally evaluated network and previous works are presented.

Data: Yeast Cell Cycle Dataset

In order to evaluate HRBNF algorithm, we generated fuzzy gene networks based on yeast cell cycle microarray time series data sets. We focused on twelve yeast genes playing key roles in the control of cell cycle as listed in Table 7 with descriptions taken from the Yeast Proteome Database [46]. The protein-protein and the regulatory interaction of the coded proteins for these genes which are involved in yeast cell-cycle are well-studied. The high-throughput techniques such as microarray provided us with the time-series expression of these genes reflecting the dynamic behaviour of these genes in cell cycle. Although the new techniques such as RNA-seq are already out there which give us better resolution of expression profile, the main pattern of genes expression profile can be extracted from microarray expression data. Consequently, HRBNF algorithm is used as a "reverse engineering" method to find all possible genetic interaction network models that fit the data for the set of twelve genes and to demonstrate its ability to handle other similarly noisy data sets.

Table 7: List of a subset of genes involved in the yeast cell cycle selected for our demonstration network. Descriptions were adopted from the Yeast Protein Database [46]

Gene Name	ORF	Description
SIC01	YLR079W	inhibitor of the Cdc28-Clb protein kinase complex
CLB05	YPR120C	B-type cyclin
CDC20	YGL116W	cell division control protein
CLN03	YAL040C	G1/S-specific cyclin
SWI06	YLR182W	transcription factor, subunit of SBF and MBF
CLN01	YMR199W	G1/S-spec fic cyclin
CLN02	YPL256C	G1/S-spec fic cyclin
CLB06	YGR109C	B-type cyclin
CDC28	YBR160W	cyclin-dependent protein kinase
MBP01	YDL056W	transcription factor, subunit of MBF
CDC06	YJL194W	initiates DNA replication, active late G1/S
SWI04	YER111C	transcription factor, subunit of the SBF factor

In addition to the vast studies on the yeast data sets specially in systems biology that allow us to have a better assessment on the acquired results, these data sets are advantageous because of having adequate number of genes and time points for testing results.

As described by Cho et al. [48], gene expression profiles for yeast cell cycle have been studied through four microarray time series data sets: Alpha, Cdc15, Cdc28, and Elu with 18, 24, 17, and 14 time points, respectively. We have used Alpha, Cdc15, and Cdc28 data sets, in which still some samples were missed. The missing values are computed using an estimation method based on the K-Nearest Neighbourhood (KNN) algorithm [54].

S. cerevisiae cell cycle regulatory protein-DNA interactions were also the subject of a recent extensive experimental study [55] for which a great deal of information has been compiled in KEGG pathway database [44, 45]. Figure 6, shows the interactions of the cell cycle regulatory protein subset shown experimentally so far [56]. We applied HRBNF algorithm on gene expression values corresponding to these twelve genes, and compared to the estimated directed network with the pathway depicted in KEGG.

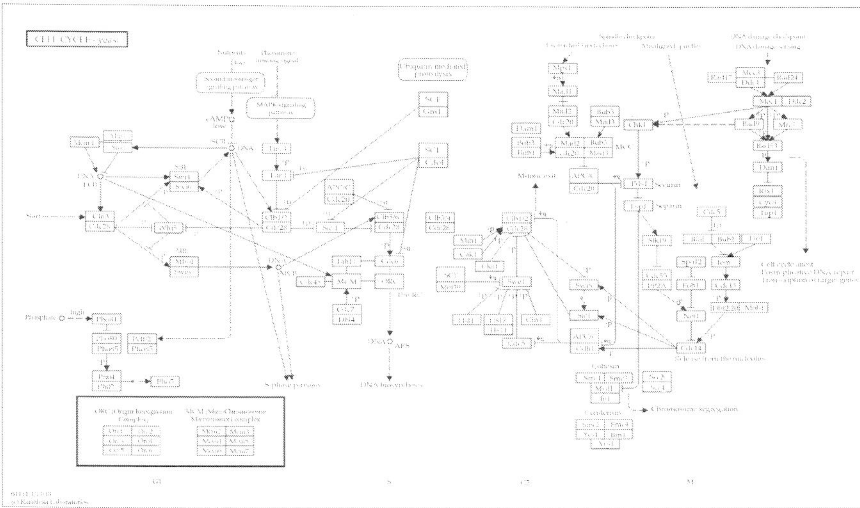

Figure 6: KEGG yeast cell cycle interactions—Schematic of known yeast cell cycle interactions between protein products of twelve studied genes regulatory (Table 7). An arrow indicates a positive interaction and a closed circle indicates negative interaction [44, 45].

Prediction of Gene Expression Level

In order to constrain the number of parameters, we have to restrict the structure of our RBNFNs according to the size of employed data set. As a result, we consider the following hypothesis for the RBNFNs: (i) focusing on 12-key yeast cell cycle genes as the input/output nodes of the RBNFNs; (ii) setting the number of MFs to three, as the minimum meaningful resolution. The centers and variances of search space with three MFs are considered as follows (identified empirically):

$$\text{Center} = [0.2, 0.5, 0.8]; \qquad \text{Variance} = 0.25. \tag{12}$$

The centers are selected according to the considered three MFs and linguistic values of "Low," "Medium," and "High"; the variances are obtained from results of Table 8. This table presents the accuracy of extracted interactions compared with real interactions, and shows that variance of 0.25 can extract the best result. Also this variance represents a distribution over the entire range of our search space.

Table 8: The accuracy of extracted interactions after considering different variances

Correct Interactions				Variances			
	0.05	0.1	0.15	0.2	0.25	0.3	0.35
%	24.2%	27.3%	27.3%	33.3%	39.4%	27.3%	24.2%

The RBNFNs are trained using Cdc15 dataset, and tested by Cdc28 and Alpha datasets. The test results are shown in Table 9.

Table 9: MSE errors of HRBNF models for testing data

Genes Name	MSE error	
	Data Set (Cdc28)	Data Set (Alpha)
SIC01	0.047	0.125
CLB05	0.096	0.020
CDC20	0.051	0.082
CLN03	0.059	0.163
SWI06	0.184	0.102
CLN01	0.108	0.107
CLN02	0.044	0.016
CLB06	0.046	0.024
CDC28	0.112	0.123
MBP01	0.146	0.138
CDC06	0.123	0.094
SWI04	0.050	0.054

Genetic Interaction Network Reconstruction for Yeast Cell Cycle Data

To test the capability of our proposed method, we used the RBNFNs to predict time-series gene expression values and then, extract gene regulatory network (GRN) structures from inferred rules. The results of GRN reconstruction were evaluated by a part of yeast cell cycle regulatory network extracted from KEGG database [44, 45]. Figure 7 illustrates the extracted genetic network of activator and repressor interactions.

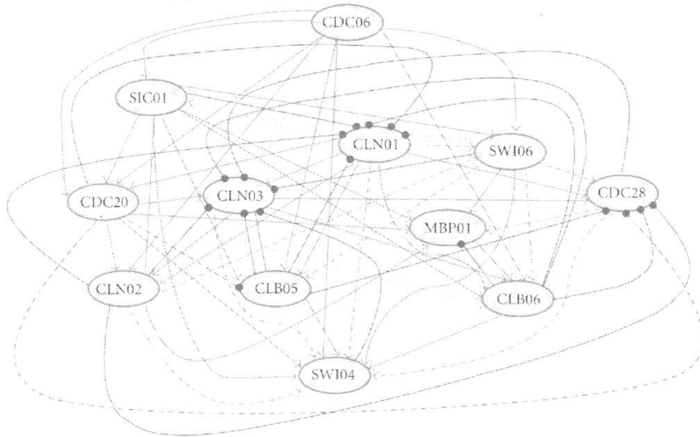

Figure 7: Inferred genetic interaction network for twelve yeast cell cycle regulatory genes. Each node represents a gene and the presence of an edge between the two nodes represents the existence of interaction between the two genes. Symbols " → " and "—○", shown by blue and red edges, illustrate activator and repressor interactions, respectively. Dashed edges represent interactions that have been verified. In contrast, dotted edges are incorrect extracted interactions.

DISCUSSION

We compared genetic interaction networks constructed based on five algorithms from the same dataset. The results show that the inferred network from our proposed algorithm has the best performance.

As shown in Table 10, the extracted network from HRBNF algorithm demonstrates a more complete match with the KEGG pathway than proposed methods from literature. Indeed, we observe that our HRBNF algorithm is capable to extract 13 true connections out of 33 available experimentally illustrated connections, while only 4, 5, 10, and 7 true connections are captured by DBN [57], VBEM [58], Time delay-ARACNE [59], and PF subjected to LASSO [30], respectively. It is clear that we reduced the misdirected edges.

Table 10: Comparison of our algorithm performance by other methods in literature in detecting interactions among experimentally known gene interactions

No	Interactions	DBN [57]	VBEM [58]	Time delay-AR-ACNE [59]	PF subjected to LASSO [30]	Propose Algorithm
1	CLN03 → SWI04	Consistent	Consistent	Consistent	Not Found	Consistent
2	CLN03 → SWI06	Not Found	Not Found	Consistent	Inconsistent	Not Found
3	CLN03 → MBP01	Not Found	Not Found	Consistent	Not Found	Not Found
4	CDC28 → MBP01	Not Found	Consistent	Consistent	Not Found	Consistent
5	CDC28 → SWI04	Not Found	Not Found	Not Found	Not Found	Consistent
6	CDC28 → SWI06	Not Found	Not Found	Consistent	Inconsistent	Not Found
7	CDC28 → CDC06	Not Found	Consistent	Not Found	Not Found	Not Found
8	CDC28 —○ SIC01	Not Found	Not Found	Not Found	Not Found	Not Found
9	SWI04 → CLN01	Consistent	Consistent	Consistent	Not Found	Not Found
10	SWI04 → CLN02	Consistent	Not Found	Not Found	Not Found	Not Found
11	SWI04 → CDC28	Not Found	Not Found	Not Found	Not Found	Not Found
12	SWI06 → CLB05	Not Found	Not Found	Not Found	Consistent	Consistent
13	SWI06 → CLB06	Not Found	Consistent	Consistent	Not Found	Consistent
14	SWI06 → CLN1	Inconsistent	Inconsistent	Consistent	Consistent	Inconsistent
15	SWI06 → CLN2	Not Found	Not Found	Not Found	Consistent	Consistent
16	SWI06 → CDC28	Not Found	Not Found	Not Found	Not Found	Consistent
17	MBP01 → CLB05	Not Found	Not Found	Not Found	Not Found	Not Found
18	MBP01 → CLB06	Not Found	Not Found	Not Found	Not Found	Consistent
19	MBP01 → CDC28	Not Found	Not Found	Not Found	Not Found	Not Found
20	CLN01 → SWI06	Not Found	Not Found	Not Found	Consistent	Consistent
21	CLN01 → SWI04	Not Found	Not Found	Not Found	Not Found	Consistent
22	CLN01 —○ SIC01	Consistent	Inconsistent	Inconsistent	Consistent	Inconsistent
23	CLN02 —○ SIC01	Not Found	Not Found	Inconsistent	Not Found	Not Found
24	CLN02 → SWI04	Not Found	Not Found	Consistent	Not Found	Consistent
25	CLN02 → SWI06	Not Found	Not Found	Not Found	Consistent	Not Found
26	SIC01 —○ CDC28	Not Found	Not Found	Not Found	Inconsistent	Not Found
27	SIC01 —○ CLB05	Not Found	Not Found	Not Found	Not Found	Not Found
28	SIC01 —○ CLB06	Not Found	Not Found	Not Found	Not Found	Not Found
29	CDC20 —○ CLB05	Not Found	Not Found	Not Found	Not Found	Consistent
30	CDC20 —○ CLB06	Not Found	Not Found	Not Found	Not Found	Inconsistent
31	CDC20 —○ CDC28	Not Found	Not Found	Not Found	Consistent	Consistent
32	CLB05 → CDC06	Not Found	Not Found	Not Found	Not Found	Not Found
33	CLB06 → CDC06	Not Found	Inconsistent	Consistent	Not Found	Not Found

* Symbols " → " and "—o" illustrate activator and repressor interactions, respectively

Some criteria are useful to evaluate the goodness of fit of the inferred network [60]: the proportion of recovered true edges in the target network that is called Recall and precision corresponds to the expected success rate in the experimental validation of the predicted interactions.

We can compute sensitivity and precision from following equations:

$$\text{Sensitivity} = \frac{TP}{TP + FN},$$

$$\text{Precision} = \frac{TP}{TP + FP}, \tag{13}$$

where true positive (TP) is the inferred number of edges identified correctly, false negative (FN) is the number of edges that were not identified, and false positives (FP) is the number of edges identified incorrectly.

Also, we calculated the -score by using (14) to further quantify the performance of the algorithms [60]:

$$F\text{-score} = \frac{1}{\alpha(1/\text{Precision}) + (1-\alpha)(1/\text{Sensitivity})}, \tag{14}$$

where α is a weighting factor and here we consider $\alpha = 0.5$ that is called the harmonic mean of precision and sensitivity because the importance of precision and sensitivity is even.

Consequently, the goodness of fit of the results based on HRBNF and the other structure learning approaches in predicting the connectivity network of the KEGG pathway were compared using estimates of above criteria.

The results, presented in Table 11, show that evaluation criteria, sensitivity, and F-score, are distinctively higher for our approach compared to previous methods indicating the efficiency of our approach. In a good system, precision decreases as sensitivity increases.

Table 11: Comparison of the proposed algorithm with other methods using statistical criteria

	TP	FP	FN	Sensitivity	Precision	*F*-Score
DBN [57]	4	1	29	12.1%	80%	21%
VBEM [58]	5	3	28	15.2%	62.5%	23.5%
Time delay-ARACNE [59]	10	2	23	30.3%	83.3%	44.4%
PF subjected to LASSO [30]	7	3	26	21.2%	70%	32.5%
Proposed Algorithm	13	3	20	39.4%	81.3%	53.1%

Considering the results obtained from proposed method, it can be concluded that there are considerable agreements between the findings of

HRBNF algorithm and experimental results reported in the literature. The main advantage of proposed method is that HRBNF algorithm searches for the relationships that fit to a logical understanding of how a set of genes should interact. By using the same criteria that biologists would use to describe the gene regulatory function (i.e., framed as an "IF-THEN-ELSE" relationship in terms of expression level), HRBNF algorithm based on fuzzy logic approximates the thought process that an expert would use to analyze these kinds of data; however, in contrast to an expert, HRBNF algorithm is automated and unbiased. If gene expression data is not analyzed properly, it can be difficult to interpret and can easily be misconstrued.

In comparison to other methods, our computational algorithm analyses the data more efficient, unbiased, and fast. Our proposed model and Time delay-ARACNE [59] used less than one second against PF subjected to LASSO [30] which in turn used 23 minutes and 37 seconds on a Core i7 PC with 8 GB main memory, respectively. We speculate that this is due to the reduction in search space that results from by applying relevant rule selection (Section 2.1.3) prior to neurofuzzy network learning.

Algorithm based on neurofuzzy networks found a disproportionately large number of interactions for the roles of activators and repressors due to available datasets with small size samples. Of course, if expression profiling technologies become more sensitive and faster, our HRBNF algorithm would detect a more precise effect of gene expression on the activator and repressor roles, providing more information to decrease the number of interactions, which leads to a better understanding.

CONCLUSIONS

In this paper, we described a novel algorithm based on RBNFNs for gene regulatory network reconstruction. We demonstrated our approach by developing RBNFN models that more accurately predict gene expression data from typically noisy microarray experiments. Looking at the gene regulatory networks provides us a systemic view at the "gene interaction" level.

Our HRBNF algorithm was successfully validated using yeast cell cycle data. The rules of inferred networks reflect most interactions previously identified by genome-scale analysis and match the existing literature. At the network level, the inferred rules provide more detailed information about genes and the interactions among them. Potential new interesting interactions were identified, which provide novel hypotheses for new lines of further research. As a consequence, common rules among all the RBNFNs and the plausible model were identified, giving rise to better understanding of the system.

RBNFNs can simultaneously extract both quantitative and qualitative information. However, the insufficiency in both data and biological understanding of gene interactions limit the results obtained from RBNFN models.

We arrive to the conclusion that the presented HRBNF algorithm provides a comprehensive method with the capability of capturing meaningful results. It should be noted that the goal of HRBNF algorithm is not only to provide quantitative predictions, but also to extract knowledge for interactions among genes.

Finally, among these five algorithms, our proposed algorithm has the increased performance in terms of execution time, sensitivity, precision, and F-score. This study found that proposed method is a good strategy to more readily infer gene networks due to its better performance and short execution time.

Despite great progress made with different algorithms, many problems remain unsolved, and much improvement is still required.

REFERENCES

1. P. Du, J. Gong, E. S. Wurtele, and J. A. Dickerson, "Modeling gene expression networks using fuzzy logic," IEEE Transactions on Systems, Man, and Cybernetics B, vol. 35, no. 6, pp. 1351–1359, 2005.
2. M. K. Kerr and G. A. Churchill, "Experimental design for gene expression microarrays," Biostatistics, vol. 2, no. 2, pp. 183–202, 2001.
3. R. Schmid, P. Baum, C. Ittrich et al., "Comparison of normalization methods for Illumina BeadChip HumanHT-12 v3," BMC Genomics, vol. 11, no. 1, article 349, 2010.
4. J. Rudy and F. Valafar, "Empirical comparison of cross-platform normalization methods for gene expression data," BMC Bioinformatics, vol. 12, no. 1, article 467, 2011.
5. A. W. C. Liew, N. F. Law, and H. Yan, "Missing value imputation for gene expression data: computational techniques to recover missing data from available information," Briefings in Bioinformatics, vol. 12, no. 5, pp. 498–513, 2011.
6. M. B. Eisen, P. T. Spellman, P. O. Brown, and D. Botstein, "Cluster analysis and display of genome-wide expression patterns," Proceedings of the National Academy of Sciences of the United States of America, vol. 95, no. 25, pp. 14863–14868, 1998.
7. P. D'Haeseleer, "How does gene expression clustering work?" Nature Biotechnology, vol. 23, no. 12, pp. 1499–1501, 2005.
8. P. Xu, G. N. Brock, and R. S. Parrish, "Modified linear discriminant analysis approaches for classification of high-dimensional microarray data," Computational Statistics and Data Analysis, vol. 53, no. 5, pp. 1674–1687, 2009.
9. Y. Tan and Y. Liu, "Comparison of methods for identifying differentially expressed genes across multiple conditions from microarray data," Bioinformation, vol. 7, no. 8, pp. 400–404, 2011.
10. Y. Huang, I. Tienda Luna, and Y. Wang, "Reverse engineering gene regulatory networks: a survey of statistical models," IEEE Signal Processing, vol. 59, no. 2, pp. 113–125, 2009.
11. X. Cai and X. Wang, "Stochastic modelling and simulation of gene networks—a review of the state-of-the-art research on stochastic simulations," IEEE Signal Processing Transactions, vol. 24, no. 1, pp. 27–36, 2007.

12. J. T. Trevors, J. D. Elsas, and A. K. Bej, "The molecularly crowded cytoplasm of bacterial cells: dividing cells contrasted with viable but non-culturable (VBNC) bacterial cells," Current Issues in Molecular Biology, vol. 15, no. 1, pp. 1–6, 2012.
13. M. Ding, S. Cui, C. Li et al., "Loss of the tumor suppressor Vhlh leads to upregulation of Cxcr4 and rapidly progressive glomerulonephritis in mice," Nature Medicine, vol. 12, no. 9, pp. 1081–1087, 2006.
14. M. V. Karpuj, M. W. Becher, J. E. Springer et al., "Prolonged survival and decreased abnormal movements in transgenic model of Huntington disease, with administration of the transglutaminase inhibitor cystamine," Nature Medicine, vol. 8, no. 2, pp. 143–149, 2002.
15. U. M. Braga-Neto, "Fads and fallacies in the name of small-sample microarray classification—a highlight of misunderstanding and erroneous usage in the applications of genomic signal processing," IEEE Signal Processing Magazine, vol. 24, no. 1, pp. 91–99, 2007.
16. J. Ernst, G. J. Nau, and Z. Bar-Joseph, "Clustering short time series gene expression data," Bioinformatics, vol. 21, no. 1, pp. I159–I168, 2005.
17. S. Liang, S. Fuhrman, and R. Somogyi, "Reveal, a general reverse engineering algorithm for inference of genetic network architectures," Pacific Symposium on Biocomputing, pp. 18–29, 1998.
18. T. Akutsu, S. Miyano, and S. Kuhara, "Identification of genetic networks from a small number of gene expression patterns under the Boolean network model," Pacific Symposium on Biocomputing, pp. 17–28, 1999.
19. M. Chaves, E. D. Sontag, and R. Albert, "Methods of robustness analysis for Boolean models of gene control networks," IEE Proceedings: Systems Biology, vol. 153, no. 4, pp. 154–167, 2006.
20. N. Friedman, M. Linial, I. Nachman, and D. Pe'er, "Using Bayesian networks to analyze expression data," Journal of Computational Biology, vol. 7, no. 3-4, pp. 601–620, 2000.
21. S. Y. Kim, S. Imoto, and S. Miyano, "Inferring gene networks from time series microarray data using dynamic Bayesian networks," Briefings in Bioinformatics, vol. 4, no. 3, pp. 228–235, 2003.
22. M. Zou and S. D. Conzen, "A new dynamic Bayesian network (DBN) approach for identifying gene regulatory networks from time course microarray data," Bioinformatics, vol. 21, no. 1, pp. 71–79, 2005.

23. A. V. Werhli and D. Husmeier, "Reconstructing gene regulatory networks with Bayesian networks by combining expression data with multiple sources of prior knowledge," Statistical Applications in Genetics and Molecular Biology, vol. 6, no. 1, article 15, 2007.
24. A. Schliep, A. Schönhuth, and C. Steinhoff, "Using hidden Markov models to analyze gene expression time course data," Bioinformatics, vol. 19, supplement 1, pp. i255–i263, 2003.
25. H. Toh and K. Horimoto, "Inference of a genetic network by a combined approach of cluster analysis and graphical Gaussian modeling," Bioinformatics, vol. 18, no. 2, pp. 287–297, 2002.
26. M. Quach, N. Brunel, and F. D'alché-Buc, "Estimating parameters and hidden variables in non-linear state-space models based on ODEs for biological networks inference," Bioinformatics, vol. 23, no. 23, pp. 3209–3216, 2007.
27. Z. Wang, F. Yang, D. W. C. Ho, S. Swift, A. Tucker, and X. Liu, "Stochastic dynamic modeling of short gene expression time-series data," IEEE Transactions on Nanobioscience, vol. 7, no. 1, pp. 44–55, 2008.
28. L. Qian, H. Wang, and E. R. Dougherty, "Inference of noisy nonlinear differential equation models for gene regulatory networks using genetic programming and Kalman filtering," IEEE Transactions on Signal Processing, vol. 56, no. 7, pp. 3327–3339, 2008.
29. Z. Wang, X. Liu, Y. Liu, J. Liang, and V. Vinciotti, "An extended kalman filtering approach to modeling nonlinear dynamic gene regulatory networks via short gene expression time series," IEEE/ACM Transactions on Computational Biology and Bioinformatics, vol. 6, no. 3, pp. 410–419, 2009.
30. A. Noor, E. Serpedin, M. Nounou, and H. Nounou, "Inferring gene regulatory networks via nonlinear state-space models and exploiting sparsity," IEEE/ACM Transactions on Computational Biology and Bioinformatics, vol. 9, no. 4, pp. 1203–1211, 2012.
31. N. H. Lee, "Genomic approaches for reconstructing gene networks," Pharmacogenomics, vol. 6, no. 3, pp. 245–258, 2005.
32. S. Mehra, W. S. Hu, and G. Karypis, "A Boolean algorithm for reconstructing the structure of regulatory networks," Metabolic Engineering, vol. 6, no. 4, pp. 326–339, 2004.

33. L. A. Soinov, M. A. Krestyaninova, and A. Brazma, "Towards reconstruction of gene networks from expression data by supervised learning," Genome Biology, vol. 4, no. 1, article R6, 2003.
34. G. J. Hickman and T. C. Hodgman, "Inference of gene regulatory networks using Boolean-network inference methods," Journal of Bioinformatics and Computational Biology, vol. 7, no. 6, pp. 1013–1029, 2009.
35. S. C. Madeira, M. C. Teixeira, I. Sá-Correia, and A. L. Oliveira, "Identification of regulatory modules in time series gene expression data using a linear time biclustering algorithm," IEEE/ACM Transactions on Computational Biology and Bioinformatics, vol. 7, no. 1, pp. 153–165, 2010.
36. Y. Xiao, "A tutorial on analysis and simulation of Boolean gene regulatory network models," Current Genomics, vol. 10, no. 7, pp. 511–525, 2009.
37. H. Kim, J. K. Lee, and T. Park, "Boolean networks using the chi-square test for inferring large-scale gene regulatory networks," BMC Bioinformatics, vol. 8, article 37, 2007.
38. K. P. Murphy, Dynamic Bayesian Networks: Representation, Inference and Learning, Computer Science, University of California, Berkeley, Calif, USA, 2002.
39. L. A. Zadeh, "Fuzzy sets," Information and Control, vol. 8, no. 3, pp. 338–353, 1965.
40. E. Ruspini, P. Bonissone, and W. Pedrycz, Handbook of Fuzzy Computation, Iop Pub/Inst of Physics, 1998.
41. E. Cox, The Fuzzy Systems Handbook, AP Professional, New York, NY, USA, 1994.
42. S. Haykin, Neural Networks: A Comprehensive Foundation, Prentice Hall, New York, NY, USA, 1999.
43. K. Mehrotra, C. K. Mohan, and S. Ranka, Elements of Artificial Neural Networks, MIT Press, Boston, Mass, USA, 1997.
44. M. Kanehisa, S. Goto, M. Furumichi, M. Tanabe, and M. Hirakawa, "KEGG for representation and analysis of molecular networks involving diseases and drugs," Nucleic Acids Research, vol. 38, no. 1, pp. D355–D360, 2009.
45. M. Kanehisa, S. Goto, Y. Sato, M. Furumichi, and M. Tanabe, "KEGG for integration and interpretation of large-scale molecular

datasets," Nucleic Acids Research, vol. 40, no. 1, pp. D109–D114, 2012.

46. M. C. Costanzo, J. D. Hogan, M. E. Cusick et al., "The yeast proteome database (YPD) and Caenorhabditis elegans proteome database (WormPD): comprehensive resources for the organization and comparison of model organism protein information," Nucleic Acids Research, vol. 28, no. 1, pp. 73–76, 2000.
47. P. T. Spellman, G. Sherlock, M. Q. Zhang et al., "Comprehensive identification of cell cycle-regulated genes of the yeast Saccharomyces cerevisiae by microarray hybridization," Molecular Biology of the Cell, vol. 9, no. 12, pp. 3273–3297, 1998.
48. R. J. Cho, M. J. Campbell, E. A. Winzeler et al., "A genome-wide transcriptional analysis of the mitotic cell cycle," Molecular Cell, vol. 2, no. 1, pp. 65–73, 1998.
49. L. Zeng, J. Wu, and J. Xie, "Statistical methods in integrative analysis for gene regulatory modules," Statistical Applications in Genetics and Molecular Biology, vol. 7, no. 1, article 28, 2008.
50. M. S. Shotwell and E. H. Slate, "Bayesian outlier detectionwith dirichlet process mixtures," Bayesian Analysis, vol. 6, no. 4, pp. 665–690, 2011.
51. B. Grün, T. Scharl, and F. Leisch, "Modelling time course gene expression data with finite mixtures of linear additive models," Bioinformatics, vol. 28, no. 2, pp. 222–228, 2012.
52. L. P. Tian, L. Z. Liu, Q. W. Zhang, and F. X. Wu, "Nonlinear model-based method for clustering periodically expressed genes," The Scientific World Journal, vol. 11, pp. 2051–2061, 2011.
53. M. Böck, S. Ogishima, H. Tanaka, S. Kramer, and L. Kaderali, "Hub-centered gene network reconstruction using automatic relevance determination," PLoS ONE, vol. 7, no. 5, Article ID e35077, 2012.
54. O. Troyanskaya, M. Cantor, G. Sherlock et al., "Missing value estimation methods for DNA microarrays," Bioinformatics, vol. 17, no. 6, pp. 520–525, 2001.
55. T. I. Lee, N. J. Rinaldi, F. Robert et al., "Transcriptional regulatory networks in Saccharomyces cerevisiae," Science, vol. 298, no. 5594, pp. 763–764, 2002.
56. http://www.genome.jp/kegg/.
57. S. Kim, S. Imoto, and S. Miyano, "Dynamic Bayesian network and nonparametric regression for nonlinear modeling of gene networks

from time series gene expression data," BioSystems, vol. 75, no. 1–3, pp. 57–65, 2004.

58. I. M. Tienda-Luna, M. C. C. Perez, D. P. R. Padillo, Y. Yin, and Y. Huang, "Sensitivity and specificity of inferring genetic regulatory interactions with the VBEM algorithm," IADIS International Journal on Computer Science and Information Systems, vol. 4, no. 1, pp. 54–63, 2009.
59. P. Zoppoli, S. Morganella, and M. Ceccarelli, "TimeDelay-ARACNE: reverse engineering of gene networks from time-course data by an information theoretic approach," BMC Bioinformatics, vol. 11, article 154, 2010.
60. F. Emmert-Streib, "Statistical inference and reverse engineering of gene regulatory networks from observational expression data," Frontiers in Genetics, vol. 3, article 8, 2012.

SECTION IV
BAYESIAN NETWORKS

CHAPTER 10

BAYESIAN NETWORKS ANALYSIS OF MALOCCLUSION DATA

Marco Scutari[1], Pietro Auconi[2], Guido Caldarelli[3,4,5] & Lorenzo Franchi[6,7]

[1]Department of Statistics, University of Oxford, 24-29 St Giles', Oxford, OX1 3LB, UK

[2]Private Practice of Orthodontics, Roma, Italy

[3] IMT School for Advanced Studies, Piazza San Francesco 19, 55100, Lucca, Italy

[4] Istituto dei Sistemi Complessi CNR, Unita Sapienza, Dip. Fisica, P.le A. Moro 2, 00185, Rome, Italy

[5]London Institute for Mathematical Sciences, 35a South St, Mayfair, London, W1K 2XF, UK

CITATION (APA): Scutari, M., Auconi, P., Caldarelli, G., & Franchi, L. (2017). Bayesian Networks Analysis of Malocclusion Data. Scientific Reports, 7(1), 15236. (11 pages).

[6]Dipartimento di Chirurgia e Medicina Traslazionale, Universita degli Studi di Firenze, Firenze, Italy

[7]Thomas M. Graber Visiting Scholar, Department of Orthodontics and Pediatric Dentistry, School of Dentistry, The University of Michigan, Ann Arbor, MI, USA

ABSTRACT

In this paper we use Bayesian networks to determine and visualise the interactions among various Class III malocclusion maxillofacial features during growth and treatment. We start from a sample of 143 patients characterised through a series of a maximum of 21 different craniofacial features. We estimate a network model from these data and we test its consistency by verifying some commonly accepted hypotheses on the evolution of these disharmonies by means of Bayesian statistics. We show that untreated subjects develop different Class III craniofacial growth patterns as compared to patients submitted to orthodontic treatment with rapid maxillary expansion and facemask therapy. Among treated patients the *CoA* segment (the maxillary length) and the *ANB*angle (the antero-posterior relation of the maxilla to the mandible) seem to be the skeletal subspaces that receive the main effect of the treatment.

INTRODUCTION

The use of statistical methods in medicine is crucial to overcome the large individual variability in the pathological features of different patients1. In the orthodontic discipline, the variability of craniofacial disharmonies is especially relevant due to important differences between individuals in the amount and direction of facial growth due to heredity, gender, ethnic background, and functional characteristics2. In this paper we introduce an approach based on modern techniques from Bayesian statistics for Complex Network analysis to estimate and describe the evolution of orthodontic features measured simultaneously on a set of patients. An incredibly large amount of integrations of the various components of the craniomaxillary and mandibular combinations are possible during the growth process: the integration of these features determines the ultimate dentofacial harmony or disharmony3. An in-depth understanding of the resulting large amounts of interrelated data obtained from clinical, radiographic, and functional analyses is required to establish a solid knowledge basis for orthodontic diagnoses. Malocclusions are isoforms of disharmony: they express a form of organic integrity during the growth process by assimilating existing elements in a new synthesis. These isoforms incur costs in terms of weakness of mechanotrasduction, cumulative occlusal trauma, adaptability, local optimisation, competition between tooth elements for space, and outcome uncertainty about the ultimate facial appearance[4]. These conditions are rarely a consequence of an abnormality in a single craniofacial component, so individual clinical and radiological measurements are likely to be less indicative than the interplay between the measurements themselves. In the case of patients affected by Class III malocclusion (characterised by the protrusion of lower dental arch), skeletal imbalance is established early in life, becomes more pronounced during puberty, and continues to increase until skeletal maturation is complete[4]. Therefore, predicting treatment success or failure early in a single Class III patient based on a small number of morphometric determinants is problematic[5].

Here we present a methodology that makes use of longitudinal data collected from a sample of orthodontic patients to evaluate possible causal paths linking orthodontical features during the growth process and the changes in those paths induced by the treatment. Practising orthodontists often perform clinical reasoning under uncertainty about facial growth, with incomplete information, receiving far more inputs than they can consciously consider; and as a result they are forced to distil clinical and/or radiological evidence into regularities and patterns[6]. Modern techniques in computational statistics build on fundamental principles of probability theory[7] to provide a better understanding and visualisation of complex data by learning those regularities and patterns directly from the data, thus producing rigorous yet tractable models of domains in which expensive computations are required for quantitative reasoning[8]. In particular, Bayesian statistics[9,10] develops the idea of combining the information contained in experimental data with prior knowledge available from the literature and from previous experiments to evaluate the probability of specific hypotheses; an approach that is natural and especially useful in biological and medical research[11,12]. These computational tools can summarise a biological system involving multiple interacting components into a simplified representation that captures the interplay between those components; and that can provide insights on how those components influence (and possibly are causal for) each other[13]. A convenient device to represent such complex patterns of relationships are complex networks[14], which provide a high-level, abstracted view of the interplay between the variables of interest by representing them as nodes and by linking them with arcs that show how those variables interact with each other. A popular choice for this kind of representation are Directed Acyclic Graphs (DAGs), in which links represent direct probabilistic dependencies and have arrows indicating the direction of the dependence. In this paper we intend to develop such a model in the context of orthodontics, combining DAGs with the joint probability distribution of the craniofacial variables of interest15. The contribution of the DAG in this case is to visualise the set of relationships between these variables and to determine how they may be grouped into communities. Networks have already been used in the literature to describe the evolution of patients with malocclusions16 and to help in the formulation of diagnosis17. Indeed, the exact focal morphological areas of the treatment effect and treatment priorities are both still under discussion in the clinical orthodontics community.

Unfortunately, current clinical evidence has been unable to fully elucidate the network of causalities that link the relevant skeletal components, the starting point of the treatment, treatment priorities, and the best way to channel and disseminate the effects of the treatment17. Here we will try to address such questions for Class III malocclusion, a dysmorphosis characterised by growth excess of the mandible and/or a defective growth of the maxilla, with protrusion of the lower dental arch. To this end we will use a set of 147 longitudinal measurements of various craniofacial features on 143 Class III growing patients evaluated at least twice between the ages of 6 and 19. Sixty-six of these subjects were undergoing orthodontic treatment by early rapid maxillary expansion and facemask therapy followed by fixed appliances, while the remaining 77 were not subject to any treatment. We will estimate Bayesian networks from these data and we will use resampling techniques from modern statistics to produce a consensus network model that describes the relationships between treatment and craniofacial figures and to evaluate its predictive accuracy. We find the resulting network to be consistent with a number of key characteristics of Class III malocclusion as known from current clinical evidence and literature, which we use to validate the relationships we learn from the longitudinal data. Furthermore, the network displays good predictive accuracy for the dynamics of Class III malocclusion in new patients. Finally, we use the network to identify the focal morphological areas of the treatment effect on the basis of the causal relationships captured by the network structure.

METHODS

The Data

The data contain longitudinal measurements on a set of 147 Class III growing patients (83 female, 60 male) evaluated at least twice between the ages of 6 and 19. Two sets of simultaneous measurements at ages T_1(6 to 19 years, average 8±1 years) and T_2 (5 to 19 years, average 15±1years) are available for all patients, in addition to a *Treatment* variable identifying treated from untreated patients. For each untreated subject, a *Growth* variable indicating the prognosis as positive or negative in comparison with the normal craniofacial progression was reported. The complete list and details for the 8 variables for this data set can be found in the Supplementary Information.

Correlation Networks

We represent the entire craniofacial system as an aggregate structure of a variety of agents where the clinical (e.g., radiographic, functional) features are the vertices of a network whose edges are the relationships between them. To build the network we start from a measure of correlation among the cephalometric variables X_a,X_b and in particular we compute the Pearson correlation coefficient r defined as

$$r_{ab} = \frac{\sum_{i=1}^{n}(X_a(i) - \overline{X}_a)(X_b(i) - \overline{X}_b)}{\sqrt{\sum_{i=1}^{n}(X_a(i) - \overline{X}_a)^2}\sqrt{\sum_{i=1}^{n}(X_b(i) - \overline{X}_b)^2}} \quad (1)$$

where X_a(i) is the *i*-th value of the feature X_a as observed in the data and $\overline{X}_a$ is the arithmetic mean of the *n* values observed for X_a, *e.g.* $\overline{X}_a = (\sum_{i=1}^{n} X_a(i))/n$. Pairs of variables with values of $|r_{ab}|$ above the threshold of 0.4 are linked by edges in the correlation network.

Bayesian Statistics

The field of statistics provides several approaches to estimate the probability of particular events of interest and to model the laws that govern the phenomena under investigation. For instance, we can estimate the former with its observed, empirical frequency (*frequentist statistics*); and the latter by making assumptions on the distribution of the data and estimating the values of the parameters of the model as those that are best supported by the data (that is, having the *maximum likelihood*)18. A second approach is given by *Bayesian statistics* 19, in which we also assume *a priori* a distribution for the parameters of the model. That distribution is then updated based on the observed data to reflect the current understanding of the phenomenon; the result is the *posterior distribution* of the parameters given the data. For instance, consider the probability $p(e|c_i)$ of the occurrence of an event *e* we observe under one of several possible conditions c_i, $i=1,\ldots,k$. A classic approach in statistics is to estimate $p(e|c_i)$ by means of its frequency (the ratio of how many times the event is observed over the total number of measurements), and then to diagnose the condition as that that has the largest $p(e|c_i)$. On the other hand, Bayesian statistics answers a different question, what is the probability of each condition c_i given the event e? Using Bayes theorem, we can write

$$p(c_i|e) = \frac{p(e, c_i)}{p(e)} = \frac{p(c_i)p(e|c_i)}{p(e)} = \frac{p(c_i)p(e|c_i)}{\sum_{j=1}^{k} p(c_j)p(e|c_j)} \quad (2)$$

to express the *a posteriori* probability of c_i as a function of the *a priori* probability $p(c_i)$ of the condition and $p(e|c_i)$ relative that of the complete set of conditions. As a result, we obtain a completely specified probabilistic model we can use to test experimental hypotheses and that can easily incorporate additional information available from external sources via the $p(c_i)$ terms. Importantly, this makes it possible to iteratively update $p(c_i|e)$ as new data becomes available by taking the current estimates of $p(c_i|e)$ as the *a priori* $p(c_i)$ for the new data to compute new, up-to-date estimates of $p(c_i|e)$.

Differential Equations Models

Since we are interested in modelling the evolution of malocclusion and its response to treatment over time, we will model the data using the differences of the craniofacial features between different time points instead of the raw point measurements. We assume that each difference can be modelled with a linear regression20 of the form

$$\Delta Y = \mu + \Delta T\beta_1 + \Delta X_1\beta_2 + \ldots + \varepsilon_{\Delta Y} \quad (3)$$

where $\varepsilon_{\Delta Y} \sim N(0, \sigma^2_{\Delta Y})$, $\Delta T = T_2 - T_1$ and $\Delta Y = Y_{T_2} - Y_{T_1}$ and so forth for the other regressors. We can then rewrite Eq. 3 as

$$\frac{\Delta Y}{\Delta T} = \mu^* + \frac{\Delta X_1}{\Delta T}\beta_2^* + \ldots + \varepsilon_{\frac{\Delta Y}{\Delta T}} \quad (4)$$

which in the limit of $\Delta T \to 0$ can be considered as a set of differential equations that models the rates of change. (This is a particular case of structural equation models21, which are widely used in statistical genetics and systems biology22·23.) The relationships between the differences are assumed to be well approximated by a linear behaviour. This constraint is intrinsically enforced by the data: only 120 out of 147 patients have been measured only twice, making it impossible to estimate any trend more complex than linear. Note that Eqs 3 and 4imply that craniofacial features change linearly over time, because each rate of change ΔY/ΔT depends on the rates of change of other variables but not on time itself. To have a nonlinear trend we would need

$$\Delta Y = \mu + \Delta T \beta_1 + (\Delta T)^2 \beta_2 + \ldots \Rightarrow \frac{\Delta Y}{\Delta T} = \mu^* + \Delta T \beta_2^* + \ldots \Rightarrow \frac{\Delta Y}{\Delta T^2} = \beta_2^* \neq 0. \quad (5)$$

Furthermore, including the *Growth* and *Treatment* in the differential equations makes it possible to have regression models of the form

$$\frac{\Delta Y}{\Delta T} = \mu^* + \frac{\text{Growth}}{\Delta T} \beta_G^* + \frac{\text{Treatment}}{\Delta T} \beta_{TR}^* + \frac{\Delta X_1}{\Delta T} \beta_2^* + \ldots + \varepsilon_{\frac{\Delta Y}{\Delta T}} \quad (6)$$

thus allowing for different rates of change depending on whether the patient shows positive developments or not in the malocclusion and whether he is being treated or not. Conversely, we do not allow the treatment level to depend on ΔT, since patients are either treated or untreated for the whole period of observation; and we do not assign a regression model to ΔT because we assume that it does depend on any measured variables.

Bayesian Networks

A Bayesian network 24· 25 is a statistical model to describe probabilistic relationships among a set of variables using a directed acyclic graph (DAG). The global distribution of the variables $X=\{X_1,\ldots,X_N\}$, where N is the number of different features (in this case N=8), is decomposed into a the local distributions of the individual variables X_i as

$$p(\mathbf{X}) = \prod_{i=1}^{N} p(X_i | Pa(X_i)) \quad (7)$$

where $Pa(X_i)$ are the variables that correspond to the parents of X_i in the DAG (i.e. the nodes with an arc pointing towards X_i). The process of estimating such model is called *learning*, and consists in two steps:

- "learning" which arcs are present in the graph (i.e. which probabilistic relationships are supported by the data);

- "learning" the parameters that regulate the strength of those dependencies.

The former is known as *structure learning*, and the latter as *parameter learning*. In the context of the differential equations described above in Eq. 3, in structure learning we determine which regressors (if any) are present in each differential, while in parameter learning we estimate the values of the corresponding regression coefficients. In order to do that, we assume that the errors in each differential equation (represented by the $\varepsilon\Delta Y$ term) are normally distributed, independent, homoscedastic and with mean zero. Under these assumptions, each differential equation can be treated as a classic linear regression model and estimated by ordinary least squares26; and the regressors correspond to the variables associated to the nodes that are parents of ΔY in the DAG.

Structure learning is similarly based on model selection procedures for classic regression models. Since we operate in a Bayesian setting, we select which variables are statistically significant regressors in each differential equation as those that maximise the posterior probability of the Bayesian network, which we approximate with the *Bayesian Information Criterion* 27. Those regressors are the parents of the node corresponding to the response variable in the DAG, and are chosen using hill-climbing, a greedy search algorithm based on step-wise selection28. The only restriction imposed by Bayesian networks is that, once the probabilistic relationships are represented as a directed graph, the graph should be acyclic.

While it is possible in principle to learn all dependencies from the data, Bayesian networks can easily include prior knowledge available from the literature and the practice of the discipline to produce more informative models and to overcome the inherent noisiness of orthodontic data. This can be done by encoding the available prior knowledge in sets of *whitelisted arcs* (which we know represent real dependencies and thus should be forced to be present in the graph) and *blacklisted arcs* (which correspond to relationships we know to be impossible). In particular:

- Craniofacial features do not determine ΔT or *Treatment*, so we blacklist any arc from the former to the latter. We also blacklist any arc from the craniofacial features to *Growth*, as we interpret them to be determined by the overall evolution of the malocclusion (including unobserved factors) as expressed by *Growth*. This also leads to a more intuitive parameterisation of the differential equations, with different regimes for the craniofacial features depending on the prognosis.

- We blacklist any arc from ΔT and *Treatment* as discussed above.
- We whitelist the dependence structure $\Delta ANB \rightarrow \Delta IMPA \leftarrow \Delta PPPM$ 1·4·5.
- We whitelist the arc from ΔT to *Growth* to allow the prognosis to change over time.

We want to clarify the meaning of the variable *Growth* we use in our analysis: *Growth* reflects the expected prognosis of the patient at the time of the visit. In such respect this is a "static" variable and does not reflect a measure of the effective growth. Quality of growth has been evaluated by considering the normal evolution of the maxillomandibular sagittal imbalance (CoGn-CoA) with respect to average population. Patients near the average values were diagnosed as "good growers" while the others were indicated as "bad growers"29. Furthermore, to reduce the impact of the noise present in the data, we use a second Bayesian technique called *model averaging* to improve the reliability of structure learning30. Typically, to examine the phenomenon under investigation we estimate a single model from the data, and we draw our conclusions from that model treating it as a "fixed" quantity. In doing so we underestimate the degree of uncertainty present in those statistical conclusions by ignoring the fact that the estimated model is not "fixed", but carries its own uncertainty from the selection procedure used to learn it from the data31. Intuitively, we can imagine that adding or removing a few observations from the data may result in a different model being identified, in turn leading to different conclusions. To reduce this model uncertainty, we re-sample the data 200 times using bootstrap32 and we perform structure learning separately on each of the resulting samples, thus collecting 200 DAGs. We then compute the frequency with which each appears in those 200 graphs, known as the *arc strength*, and we compute an "average", consensus DAG by selecting those arcs that have a frequency above a certain threshold. (The threshold can either be estimated from the data or set to an arbitrary value, such as 0.85 below, for the purpose of obtaining a sparse DAG that is easier to interpret.) The averaged Bayesian network model has a number of favourable statistical properties; in particular, it is less sensitive to noisy data and typically produces more accurate predictions for new observations.

Once we have estimated the average Bayesian network and the values of the regression coefficients in the differential equations it describes, we evaluate its predictive accuracy using 10-fold cross-validation33. 10-fold cross-validation is a model validation technique that assesses how well a statistical model generalises to independent data or, in other words, how accurately it will predict the behaviour of new observations. It is implemented as follows.

1. We split the data into 10 subsets (called *folds*) of the same size (or as close as possible).
2. For each fold in turn:
   (a)we take that fold as the *test set*;
   (b)we take the rest of the data as the *training set*;
   (c)we learn the Bayesian network model on the training set, both the structure and the parameters;
   (d)we predict each variable in turn for the observations in the test set, from the model we learned from the training set and from all the other variables in the test set;
3. We collect the pairs of (observed, predicted) values for all the observations and:
   (a)for each continuous variable, we compute the correlation between the observed and predicted pairs (this quantity is called *predictive correlation*);
   (b)for Growth, we compute the number of misclassified predicted values using the observed values as the true values (this is the *predictive classification error*, which the complement of predictive accuracy).

In addition, we use the averaged Bayesian network for inference to answer a number of crucial questions and to check whether it reflects the available knowledge of how the measured variables interact with each other and with the treatment. In the context of Bayesian networks, this is typically done using a technique called *belief updating*, in which we estimate the posterior probability of a certain event or the posterior estimate of some parameter conditional on some evidence on the values of one or more variables. Several exact and approximate approaches are available in the literature25; in this paper we use logic sampling for its simplicity. Logic sampling is defined as follows:

1. Define a counter n_E=0 for the evidence and a counter n_q=0 for the event of interest;
2. For a suitably large number of times (10^4–10^6):
   (a)generate a random sample from the Bayesian network.
   (b)if the random sample matches the evidence, set n_E=n_E+1;
   (c)if the random sample matched both the evidence and the

event, set $n_{E,q}=n_{E,q}+1$.

3. Estimate the conditional probability of the event given the evidence with $n_{E,q}/n_E$; or compute the posterior estimates of the parameters of interest using those random observations that match the event.

By applying this approach we can answer arbitrary questions, which are called *conditional probability queries*, from a Bayesian network. We perform the whole analysis using the bnlearn package15 for R34.

RESULTS

Raw Data

As discussed in the literature16·17, craniofacial features evolve and respond to external stimuli as a system and therefore they form clusters of variables with high (greater than 0.40 in absolute value) Pearson's correlation. We observe this phenomenon in Fig. 1 (right panel), in which all the craniofacial features are connected with the exception of ΔIMPA and ΔPPPM. *Treatment* is also connected to four of these features, and presents similar correlations values with all of them (Fig. 1, left panel), making it difficult to establish the focal point of its action.

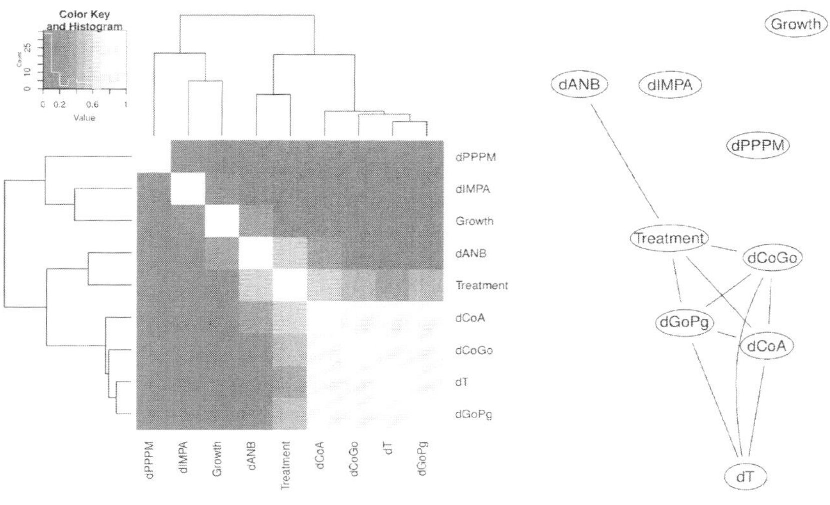

Figure 1: (left) A heatmap of the Pearson's correlation between variables measured on the patients. (right) The correlation network displaying Pearson's correlations greater than 0.40 in absolute value.

The Bayesian network consensus model constructed by learning 200 networks from the data and keeping the arcs that appear at least 50% of the time (threshold estimated from the data) is shown in Fig. 2. All the directions of the arcs seem to be well established; this can probably be attributed to the use of a whitelist and a blacklist, as they force the directions of nearby arcs to cascade into place. Furthermore, a cursory examination of the arc strengths above the threshold confirms that 15 out of 18 arcs in the consensus network appear in fact with a frequency of at least 0.85. All arc directions are also clearly established (all frequencies are equal to 1). This allows to further simplify the consensus network as shown in Fig. 3 while losing little information in the process. While the skeletal growth process influences the evolution of sagittal maxillomandibular imbalance (ΔANB) and mandible ramps height (ΔCoGo), the treatment effects mainly influence the maxillary length growth (ΔCoA), and the progression of maxillomandibular imbalance (ΔANB).

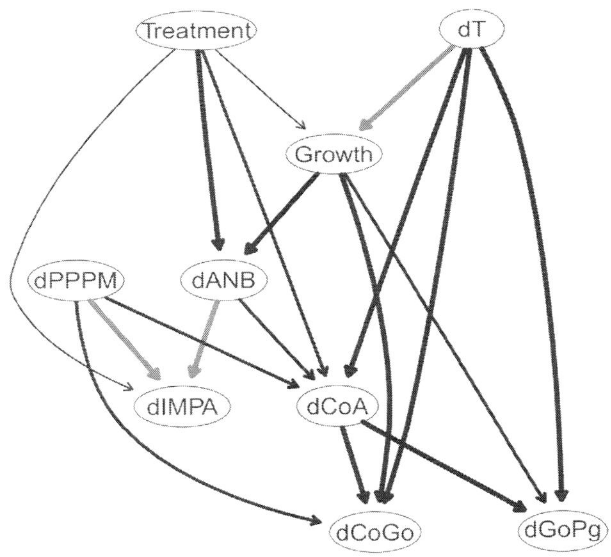

Figure 2: The DAG underlying the consensus Bayesian network learned from the variables measured on all 143 patients. Arcs in red are constrained to be present in the network by the whitelist. The thickness of the arcs is in the proportion to their strength; only arcs with a strength greater than 0.5 are included in the consensus network.

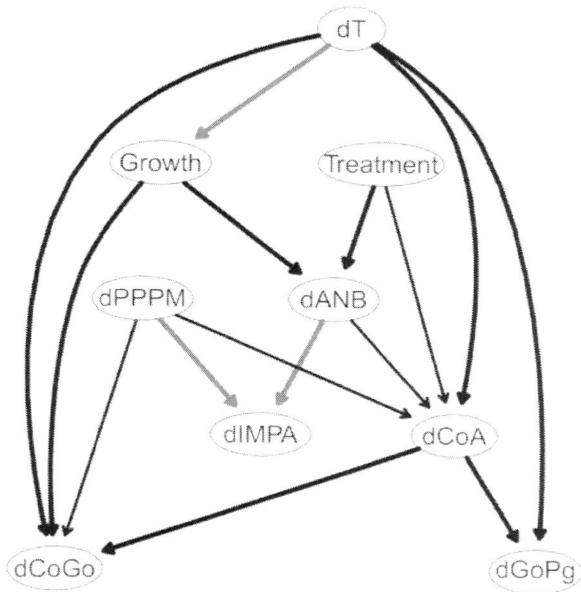

Figure 3: A simplified DAG derived from that in Fig. 2 after removing arcs with a strength smaller than 0.85.

To further validate the Bayesian network, we check whether it is consistent with prior information on Class III malocclusion that has not been used in the construction of the model. We formalise this prior information into four hypotheses, and we use conditional probability queries as described in the Methods (with 10^4 random samples) to test them.

- In Class III growing subjects an excessive growth of *CoGo* induces a reduction in *PPPM*, assuming no treatment is taking place. In the differential equations in the network, we have that as ΔCoGo increases (which indicates an increasingly rapid growth) ΔPPPM becomes increasingly negative (which indicates a reduction in the angle). This is shown in Fig. 4.

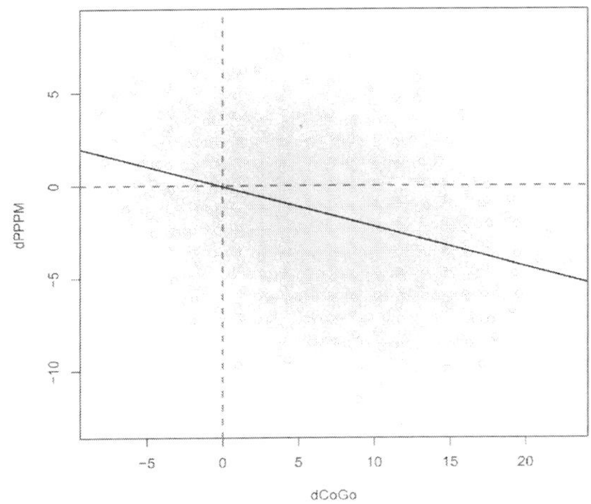

Figure 4: Values simulated from the Bayesian network for ΔPPPM and ΔCoGo. The black line represents the regression line of ΔPPPM against ΔCoGo; its negative slope confirms that as ΔCoGo increases (which indicates an increasingly rapid growth) ΔPPPM becomes increasingly negative (which indicates a reduction in the angle).

- In Class III growing subjects, if *ANB* decreases *IMPA* decreases to compensate; ΔANB is proportional to ΔIMPA, as shown in Fig. 5, so a decrease in one suggests a decrease in the other.

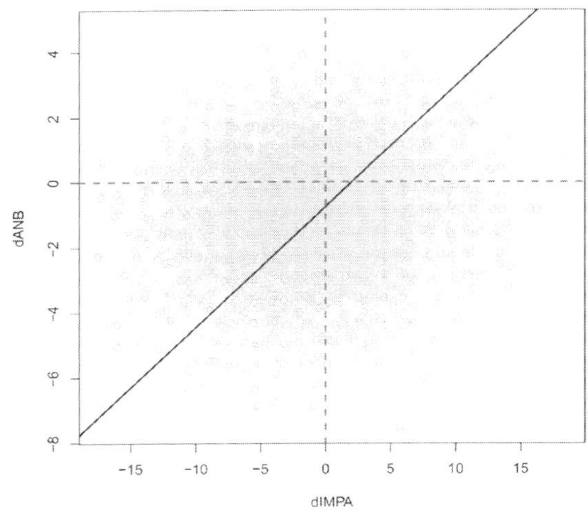

Figure 5: Values simulated from the Bayesian network for ΔANB and ΔIMPA. The black line represents the regression line of ΔANB against ΔIMPA; its positive slope suggests that ΔANB is proportional to ΔIMPA, so a decrease in one suggests a decrease in the other.

- Since Class III orthodontic treatment is aimed at stopping the decrease of *ANB* (ΔANB≈0), we expect to observe different dynamics for *ANB* in treated and untreated patients. First, we note that the Bayesian network correctly assigns a higher probability of a favourable prognosis to treated (0.63) compared to untreated (0.51) patients. The unfavourable prognosis of treated patients was defined as the concurred presence of Class III permanent molar relationship and negative overjet35. If we simulate the treatment effect and fix ΔANB≈0 (thus making it independent from its parents and removing the corresponding arcs), we have that the probability of a favourable prognosis is the same (0.58) for both treated and untreated patients and thus it does not depend on the treatment. This suggests that a favourable prognosis of a Class III malocclusion is determined mainly by preventing changes in *ANB*.
- If we use *GoPg* as a proxy for point B, the treatment does not affect point B after controlling for point A: if we keep *GoPg* fixed (ΔGoPg≈0) the angle between point A and point B (ΔANB) evolves differently for treated and untreated patients. On average, ΔANB increases for treated patients (+0.37 degrees; strongly negative values denote horizontal imbalance, so a positive rate of changes indicate a reduction in imbalance) and decreases for untreated patients (−1.13 degrees; the imbalance slowly worsens over time).

Finally, we also consider the predictive accuracy of the consensus Bayesian network. Using cross-validation as described in the Methods, we find that the prognosis is accurately predicted with probability 0.73. The predictive correlations for the craniofacial features are 0.86 for ΔCoGo, 0.91 for ΔGoPg, 0.92 for ΔCoA, 0.23 for ΔIMPA, 0.42 for ΔPPPM and 0.65 for ΔANB.

We also learn two separate consensus Bayesian networks from treated and untreated patients, which are shown Fig. 6. There are significant differences between the influence networks pertaining to treated and untreated subjects. The treatment effects on the craniofacial subspaces are channelled through the maxillary node ΔCoA to the mandibulary nodes ΔCoGo (mandibular ramus) and ΔGoPg (mandibular body). The adaptive ΔIMPA node (which aims to maintain unchanged the sagittal relationship between the maxilla and the mandible) is influenced by both horizontal and vertical skeletal imbalances during the treatment process (i.e., by ΔPPPM and ΔANB).

On the contrary, among Class III untreated subjects the progression of the horizontal skeletal imbalance (ΔANB) strictly influences the maxillary node ΔCoA, which in turn influences the progression of mandibulary nodes ΔGoPg and ΔCoGo during the growth process.

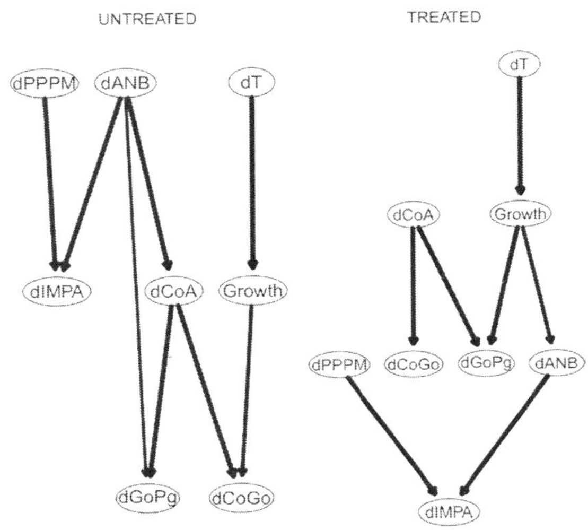

Figure 6: The DAGs underlying the consensus Bayesian networks for treated and untreated patients on the 8 variables measured for both, after adjusting them using the population reference values from Bathia and Leighton41.

ADJUSTED DATA

With the aim of confirming the methodological approach we introduced in this paper, we considered another case study. This is obtained by adjusting the data by subtracting the corresponding reference values from an Atlas of normal cephalometric features2 for infancy and childhood. A consensus Bayesian networks built in the same way as that for the raw data is shown in Fig. 7. The threshold for the significance of the arcs is about the same as before (0.5) and the number of arcs is also similar. Again we can simplify the network by retaining only the arcs with an arc strength of at least 0.85. The most striking feature of this new network, shown in Fig. 8, is the absence of arcs between ΔT and the orthodontic variables; the only arc from ΔT points to *Growth* and is only included because of the whitelist. (Note that the same happens when we exclude the individuals with the most extreme ΔT, and the resulting networks in the two cases are very similar.) This seems to suggest that much of the dependence on ΔT observed in the raw data is not a consequence of the evolution of the malocclusion but a result of ageing; and it is consistent with the fact that if we reduce the spread in the observed ages by removing the most extreme ΔT, most of the dependencies on ΔT vanish. We conjecture that the nonlinear trend of the raw values of the orthodontic variables (that is, the fact that that their rate of change is a function of time) can be decomposed into two components: a general population average and a deviation from that average given the malocclusion. The former effectively changes with time (that is, the trend of the population mean over time is not constant) while the latter does not (that is, the rate of change of the deviation from the population mean depends only on other orthodontic values and on the treatment). In other words, the population average evolves with age for all orthodontic variables, which is expected as the patients are not yet fully grown adults. However, the deviations from the populations averages do not seem to evolve with age, or to be a function of the passage of time. This would imply that the effects applying a change to one of the orthodontic variables propagate to related orthodontic variables and cause them to respond them in the same way regardless of how quickly the change is applied. (e.g. a one-degree shift in *ANB* influences neighbouring variables such as *CoA* and *IMPA* in the same way regardless of how quickly that one-degree change happens; it can be one year, it can be two years, etc. but those neighbouring variables will have the same value at the end).

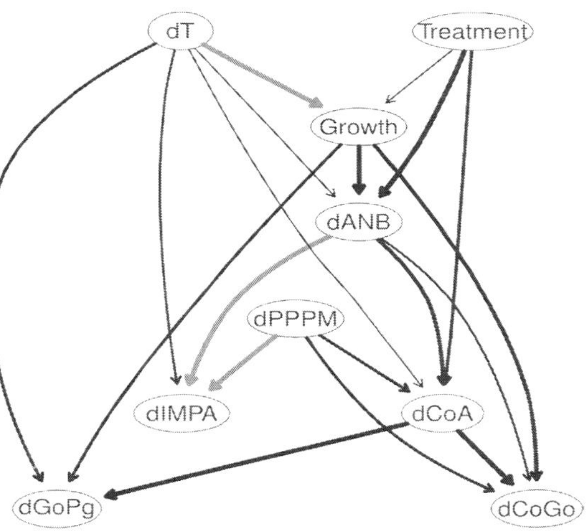

Figure 7: The DAG underlying the Bayesian network learned from the 9 variables measured on all 143 patients after adjusting them using the population reference values from Bathia and Leighton41.

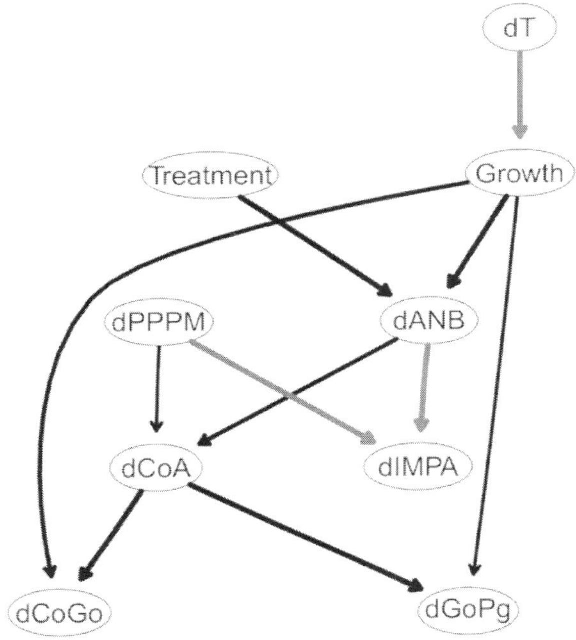

Figure 8: A simplified DAG derived from that in Fig. 7 after removing arcs with a strength smaller than 0.85.

We consider the predictive accuracy of this new consensus Bayesian network, using cross-validation as we did for the network we learned from the raw data. We find that the probability of correctly predicting the prognosis is similar (0.74 vs 0.73). The predictive correlations for most craniofacial features, however, are smaller: 0.64 (−0.22) for ΔCoGo, 0.68 (−0.23) for ΔGoPg, 0.81 (−0.11) for ΔCoA, 0.28 (+0.05) for ΔIMPA, 0.39 (−0.03) for ΔPPPM and 0.71 (+0.06) for ΔANB. This is expected since we are now modelling deviations from the population average, which are intrinsically more difficult to analyse.

We also learn two separate consensus Bayesian networks from treated and untreated patients, which are shown Fig. 9. Among both untreated and treated subjects the progression of skeletal imbalance is influenced by the evolution of the maxillary sagittal dimensions (ΔCoA); however, only treated patients exhibit the strong dependency of the mandibulary corpus from ΔCoA.

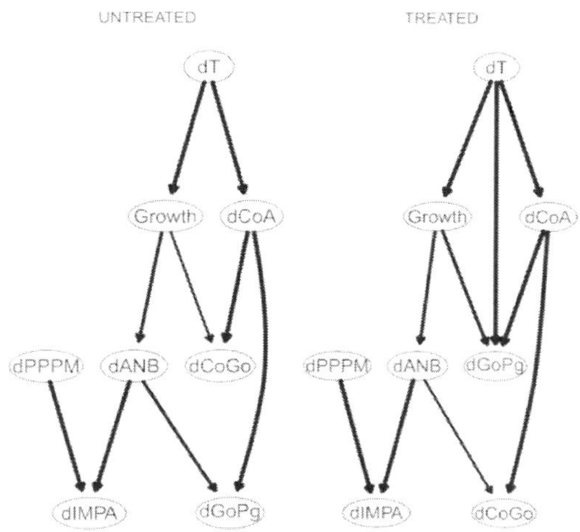

Figure 9: The DAGs underlying the consensus Bayesian networks for treated and untreated patients on the 8 variables measured for both.

DISCUSSION

Previous studies have proposed different cephalometric models to determine specific facial parameters related to abnormal growth patterns in Class III untreated and treated patients2·4·5. Multilevel, nonparametric, and predictive function algorithms have provided growth predictions and treatment outcomes based on a variety of facial characteristics4·5. Recently, network approaches to understanding morphological and functional relationships among orthodontic data have been proposed16·17. While these approaches improved the interpretation of quantitative, patient-specific information, networks were unable to elucidate the effects of influences (possibly, causal influences) between craniofacial variables16. Craniofacial features change and adapt as a system in response to both natural stimuli such as growth and external stimuli such as clinical treatments. This implies that efforts towards understanding diseases such as malocclusion must reflect this interplay in the choice of statistical models and in how clinically relevant hypotheses are tested. This has motivated the use of network approaches16·17 which represent features as nodes in an undirected graph and explicitly groups them into clusters based on their pair-wise correlation. These clusters can describe particular regions of the craniomaxillary and mandibular complexes and other broad features (symmetry, proportions, etc.) that are impacted by malocclusion and that must be targeted by treatment.

This modelling approach, however, has three important limitations. The first is that the use of pair-wise correlations makes it impossible to distinguish direct relationships between two features from indirect ones that are mediated by other features, thus making it difficult to get a clear picture of system as a whole and to identify the best target for the treatment36. Furthermore, the direction of the relationships is not taken into consideration by the model nor it is represented in the graph, making it impossible to infer cause-effect relationships even in the presence of data systematically collected from a clinical trial13. Finally, limiting the model to a representation of pair-wise dependencies falls short of characterising the full probability distribution of the features, and that in turn makes it impossible to use it to test the complex hypotheses required for model validation and treatment evaluation.

Bayesian networks suffer from none of these limitations12. Thanks to their modular model structure and to the availability of efficient software implementations15, they can be used to simultaneously explore a large number of features. The number of features does not impact the interpretability of the network: focusing on direct dependencies means that each feature is described by a local distribution that depends only those features (the "parents") for which the corresponding nodes have an arc with an arrow pointing to that particular feature. Therefore, complex models are divided into a collection of simpler problems which are mathematically tractable and computationally simpler. Furthermore, the DAG can always be used as a high level abstraction for qualitative reasoning in the context of exploratory analysis and to investigate hypotheses on the whether various sets of features are related to each other24. Finally, a Bayesian network can also be interpreted as a causal network in the absence of confounding factors13 and used to examine or generate novel clinical hypotheses37. The inherently Bayesian nature of a Bayesian network facilitates such reasoning by incorporating both prior knowledge about the variables of interest and the uncertainty present in the data38; and by not defining model estimation and inference around a single response at the expense of the ability to reason about other variables (unlike, e.g., linear regression models).

However, information becomes knowledge only when it is placed in context: without it, the orthodontist has no way to differentiate signal from noise, so the research for better diagnosis and treatment might be swamped by false positives and false assumptions. Often in everyday practice the orthodontist's efforts are aimed (or perhaps compelled) to the optimisation of therapy more than the optimisation of diagnosis. The result is that the therapy is effective, sometimes extraordinarily effective, but the price of a hasty diagnosis is paid in terms of relapse of the pretreatment craniofacial features. The aim of this work is therefore to obtain an integrated view of the craniofacial features, the treatment and the prognosis to allow systematic reasoning in the diagnostic process.

The use of Bayesian networks allows us to achieve this aim. We identify the focal morphological areas of the treatment for Class III malocclusion as the *CoA* segment (the maxillary length) and the *ANB*angle (the antero-posterior relation of the maxilla to the mandible); therefore, any apparent effect of the treatment on other cranial features can be disregarded as noise since it is actually mediated by these two features. Furthermore, by modelling the putative causal relationships we can study how the effect of an intervention on one feature propagates by identifying neighbouring features in the DAG and by studying how their distribution changes in response to various stimuli. We performed such an exercise to validate the consensus Bayesian network with respect to prior knowledge on malocclusion, with promising results.

To our knowledge, this is the first time a complex system such as craniofacial features has been modelled in this way with a formal statistical and causal Bayesian network. The usefulness of such a model is two-fold: it provides an intuitive qualitative description (in the form of a DAG) of the relationships that link the craniofacial features beyond mere physical proximity; and it also provides a quantitative description of their behaviour that can be used to validate the model and to test novel hypotheses by simulation. Collecting clinical data on either treated or untreated patients in the context of clinical trials is expensive, time consuming and subject to many practical, legal and deontological problems. In this context, Bayesian networks provide a way to perform a preliminary verification of the hypotheses that would be targeted by those trials to prioritise the trials and allocate resources efficiently39. For instance, in this paper we identified the focal point of the effect of the facemask therapy. To perform the same task experimentally without the help of the Bayesian network would require us to check many different locations; but with the indications provided by the Bayesian networks we can concentrate on *CoA* and *ANB* first and possibly avoid further experiments involving the remaining features. This sequential approach to experimental design and planning is becoming increasingly common in systems biology to reduce the cost of *in vitro* and *in vivo* research programmes40, and by the pharmaceutical industry to reduce the costs and risk of clinical trials11.

While Bayesian networks can deal with the uncertainty in the data, their main limitations lie in the impact of confounding variables and in the assumptions they make about the distribution of the features. With a larger number of measurements per patients, for instance, we expect that assuming linear relationships between the features would be a significant limitation, since we would have enough statistical power to detect nonlinear relationships.

The results of this study show that the Bayesian networks applied to a growing craniofacial complex are a useful tool to define a more detailed individualised prognosis for patients affected by the Class III malocclusion, and to mitigate an unpredictable ultimate outcome of this dysmorphosis.

ACKNOWLEDGEMENTS

GC acknowledges support from EU Projects MULTIPLEX (317532).

REFERENCES

1. Scotch, M., Duggal, M., Brandt, C., Lin, Z. & Shiffman, R. Use of Statistical Analysis in the Biomedical Informatics Literature. *Journal of the American Medical Informatics Association* 17, 3–5 (2010).
2. Ravindra, N. & Uribe, R. A. *Atlas of Complex Orthodontics* (Elsevier, 2016).
3. Zaidel, D. W. & Cohen, J. A. The Face, Beauty, and Symmetry: Perceiving Asymmetry in Beautiful Faces. *The International Journal of Neuroscience* 115, 1165–73 (2005).
4. Baccetti, T., Franchi, L. & McNamara, J. A. Growth in the Untreated Class III Subject. *Seminars in Orthodontics* 13, 130–142 (2007).
5. Fudalej, P., Dragan, M. & Wedrychowska-Szulc, B. Prediction of the Outcome of Orthodontic Treatment of Class III Malocclusions–a Systematic Revie. *w. European Journal of Orthodontics* 33, 190–197 (2011).
6. Silver, N. *The Signal and the Noise: Why So Many Predictions Fail–but Some Don't* (Penguin, 2015).
7. Cumming, G. The New Statistics: Why and How. *Psychological Science* 25, 7–29 (2014).
8. Pearl, J. *Probabilistic Reasoning in Intelligent Systems: Networks of Plausible Inference* (Morgan Kaufmann Publishers Inc., 1988).
9. Bayes, T. & Price, M. An Essay Towards Solving a Problem in the Doctrine of Chances. *Philosophical Transactions* 1683–1775, 370–418 (1763).
10. Laplace, P. S. Memoir on the Probability of the causes of Events (1774) Translated from French. *Statistical Science* 1, 364–378 (1986).
11. Berry, S. M., Carlin, B. P., Lee, J. J. & Müller, P. *Bayesian Adaptive Methods for Clinical Trials* (Chapman & Hall, 2011).
12. Nagarajan, R., Scutari, M. & Lèbre, S. *Bayesian Networks in R with Applications in Systems Biology*. Use R! series (Springer, 2013).
13. Pearl, J. *Causality: Models, Reasoning and Inference*, 2nd edn (Cambridge University Press, 2009).
14. Caldarelli, G. *Scale-Free Networks: Complex Webs in Nature and Technology* (Oxford University Press, 2007).
15. Scutari, M. Learning Bayesian Networks with the bnlearn R Package. *Journal of Statistical Software* 35, 1–22 (2010).

16. Scala, A. *et al*. Complex Networks for Data-Driven Medicine: the Case of Class III Dentoskeletal Disharmony. *New Journal of Physics* 16, 115017 (2014).
17. Auconi, P., Caldarelli, G., Scala, A., Ierardo, G. & Polimeni, A. A Network Approach to Orthodontic Diagnosis. *Orthodontics & Craniofacial Research* 14, 189–197 (2011).
18. Young, G. A. & Smith, R. L. *Essentials of Statistical Inference* (Cambridge University Press, 2005).
19. Lee, P. M. *Bayesian Statistics: an Introduction*, 2nd edn (Wiley, 2012).
20. Weisberg, S. *Applied Linear Regression*, 4th edn (Wiley, 2013).
21. Hoyle, R. H. *Structural Equation Modeling: Concepts, Issues, and Applications* (SAGE, 1995).
22. Li, R. *et al*. Structural Model Analysis of Multiple Quantitative Traits. *PLoS Genetics* 2, e114 (2006).
23. Rosa, G. J. M. *et al*. Inferring Causal Phenotype Networks Using Structural Equation Models. *Genomic Selection Evolution* 43, 6 (2011).
24. Koller, D. & Friedman, N. *Probabilistic Graphical Models: Principles and Techniques* (MIT Press, 2009).
25. Scutari, M. & Denis, J.-B. *Bayesian Networks with Examples in R*(Chapman & Hall, 2014).
26. Draper, N. R. & Smith, H. *Applied Regression Analysis*, 3rd edn(Wiley, 1998).
27. Schwarz, G. Estimating the Dimension of a Model. *The Annals of Statistics* 6, 461–464 (1978).
28. Russell, S. J. & Norvig, P. *Artificial Intelligence: A Modern Approach*, 3rd edn (Prentice Hall, 2009).
29. Auconi, P. *et al*. Understanding Interactions among Cephalometrics Variables During Growth in Untreated Class III Subjects. *European Journal of Orthodontics* (2017).
30. Scutari, M. & Nagarajan, R. On Identifying Significant Edges in Graphical Models of Molecular Networks. *Artificial Intelligence in Medicine* 57, 207–217 (2013).
31. Claeskens, G. & Hjort, N. L. *Model Selection and Model Averaging*(Cambridge University Press, 2008).
32. Efron, B. & Tibshirani, R. *An Introduction to the Bootstrap*(Chapman & Hall, New York, 1993).

33. Hastie, T., Tibshirani, R. & Friedman, J. *The Elements of Statistical Learning: Data Mining, Inference, and Prediction* (Springer, 2009), 2nd edn.

34. R Development Core Team. *R: A Language and Environment for Statistical Computing*. R Foundation for Statistical Computing, Vienna, Austria (2016).

35. Auconi, P., Scazzocchio, M., Cozza, P., McNamara, J. A. & Franchi, L. Prediction of Class III Treatment Outcomes Through Orthodontic Data Mining. *European Journal of Orthodontics* 37, 257–267 (2015).

36. Opgen-Rhein, R. & Strimmer, K. Learning Causal Networks from Systems Biology Time Course Data: an Effective Model Selection Procedure for the Vector Autoregressive Process. *BMC Bioinformatics* 8(Suppl. 2), S3 (2007).

37. Lucas, P. J. F., van der Gaag, L. C. & Abu-Hanna, A. Bayesian Networks in Biomedicine and Health-Care. *Artificial Intelligence in Medicine* 30, 201–214 (2004).

38. Mukherjee, S. & Speed, T. P. Network Inference Using Informative Priors. *Proceedings of the National Academy of Sciences* 105, 14313–14318 (2008).

39. Ness, R. O., Sachs, K., Mallick, P. & Vitek, O. A Bayesian Active Learning Experimental Design for Inferring Signaling Networks. In *Proceedings of the 21st Annual International Conference on Research in Computational Molecular Biology*, 134–156 (Springer, 2017).

40. Pauwels, E., Lajaunie, C. & Vert, J.-P. A Bayesian Active Learning Strategy for Sequential Experimental Design in Systems Biology. *BMC Systems Biology* 8, 102 (2014).

41. Bathia, S. N. & Leighton, B. C. *A Manual of Facial Growth* (Oxford University Press, 1993).

CHAPTER 11

A BAYESIAN NETWORK APPROACH TO CAUSATION ANALYSIS OF ROAD ACCIDENTS USING NETICA

Xin Zou and Wen Long Yue

School of Natural and Built Environments, University of South Australia, Adelaide, SA 5095, Australia

ABSTRACT

Based on an overall consideration of factors affecting road safety evaluations, the Bayesian network theory based on probability risk analysis was applied to the causation analysis of road accidents. By taking Adelaide Central Business District (CBD) in South Australia as a case, the Bayesian network structure was established by integrating K2 algorithm with experts' knowledge, and Expectation-Maximization algorithm that could process missing data was

CITATION (APA): Zou, X., & Yue, W. L. (2017). A Bayesian Network Approach to Causation Analysis of Road Accidents Using Netica. Journal of Advanced Transportation, 2017. (18 pages)

adopted to conduct the parameter learning in Netica, thereby establishing the Bayesian network model for the causation analysis of road accidents. Then Netica was used to carry out posterior probability reasoning, the most probable explanation, and inferential analysis. The results showed that the Bayesian network model could effectively explore the complex logical relation in road accidents and express the uncertain relation among related variables. The model not only can quantitatively predict the probability of an accident in certain road traffic condition but also can find the key reasons and the most unfavorable state combination which leads to the occurrence of an accident. The results of the study can provide theoretical support for urban road management authorities to thoroughly analyse the induction factors of road accidents and then establish basis in improving the safety performance of the urban road traffic system.

INTRODUCTION

With the expansion of urban development and the surging of vehicle ownership, urban travel becomes vulnerable to three "chronic diseases," which are congestion, accident, and pollution. Among the above three, accident has been recognised as the most negative aspect, in particular in and around Central Business District (CBD). According to Global Plan for the Decade of Action for Road Safety 2011–2020 developed by the UN Road Safety Collaboration in 2011, nearly 1.3 million people die as a result of road traffic collisions per annum, which means more than 3,000 fatalities per day. And 20 to 50 million more people sustained nonfatal injuries from collisions, and these injuries were an important cause of disability worldwide. The case in Australia is also at an alarming level; there were around 25 deaths and 700 serious injuries per week, and cost to tax payers was more than 32 billion dollars a year [1]. Unless immediate and effective action is taken, road traffic injuries are predicted to become the fifth leading cause of death in the world. Therefore, the analysis and evaluation on the influencing factors on traffic accident, estimation of the potential safety hazards, and selection of appropriate measures in advance, so as to reduce the frequency and severity of traffic accidents, are important research topics in road safety engineering.

Previous studies showed that there are many reasons behind road accidents; these causes may be coherent to each other, and, for instance, poor road alignments and unexpected vehicle compositions or behaviours may result in the confusion of road users, which may lead to traffic accidents.

However, many official records of road accidents indicated that most of the crashes are only pointed to single causes, especially human errors. For example, according to a crash causation survey released by the US National Highway Traffic Safety Administration (NHTSA) in 2015 [2, 3], drivers are to be criticised for 94% of crash cases. Apparently, the 94% of such accidents are also related to other causes from common experience, such as road alignment [4–6], traffic sign [7–9], and weather condition [10–12]. Therefore, the existing road accident statistics cannot fully reveal the causes, and traffic engineers and road infrastructure designers are provided with limited information for the accident mechanism and the formulation of improvement plans. It is of great importance to take full advantage of the traffic accident statistics and mine potential information so as to provide a basis for the analysis of accident mechanism and the improvement of road safety.

Bayesian network is one of the effective methods in the field of artificial intelligence to express uncertainty analysis and probability reasoning of a system. It can exploit the dependence relationships based on local conditions in a model to conduct bidirectional uncertainty investigation for prediction, classification, and diagnostic analyses. At present, there are some software platforms available for the construction of a Bayesian network, such as Bayes Net Toolbox (BNT), BayesBuilder, and JavaBayes, of which the MATLAB-based BNT developed by Murphy [13] is extensively used. This toolbox provides a lot of underlying basic function libraries for Bayesian network learning, but it does not integrate the basic functions for Bayesian network learning into a system. Moreover, BNT does not have Graphical User Interface (GUI), which is not user-friendly, nor can it be well generalized. Netica is a Bayesian network learning software developed by Norsys Software Corporation in Canada, which has been extensively applied in uncertainty management such as business, engineering, medicine, and ecology [14–16] due to its powerful functions, friendly GUI, reliable computation, and good performance. In this paper, a model is formulated using Bayesian network for road accident studies, and then a Bayesian network learning process, posterior probability reasoning, most probable explanation, and inferential analysis are conducted by using Netica.

This paper is organized as follows: Section 2 reviews the related literature on causation analysis of road accidents; Section 3 describes the construction of a Bayesian network model; Section 4 presents a case study on Bayesian network model application for Adelaide Central Business District (CBD) in South Australia; the findings of this study are summarized in Section 5.

LITERATURE REVIEW

The use of causation theory for road accident analysis aims to extract the accident mechanisms and accident models from a large number of typical accidents so as to provide theoretical basis for the qualitative and quantitative analyses, the predication and prevention of accidents, and the improvement of safety management. Scholars across the world have done some researches in road accident causation analysis and various data sources, variables, sample sizes, and analytical models, such as aggregated models which include Frequency Analysis [17–19] and χ^2 Test [20, 21].

In terms of disaggregated models, as the frequency of road accidents is in a form of nonnegative, discrete, and abnormal distribution and based on experience the frequency of accidents follows Poisson distribution, the Poisson regression model can be applied to analyse the influence of each risk factor on the frequency of accidents [22]. The negative binomial distribution regression is based on Poisson distribution, but its specification error follows Gamma distribution. The negative binomial regression model has been extensively applied in traffic safety analysis model [23–27]. However, the assumption that the mean value of Poisson distribution is equal to the variance is often inconsistent with realities. And in the analysis of longitudinal data samples, the adoptions of Poisson regression model and negative binomial regression model are likely to generate biased estimate and even incorrect results. When the explained variables only take a limited number of multiple discrete values, the established regression model is a discrete choice model, in which Logit model is the earliest discrete choice model and is one of the widely used models [28–31]. For an applicable statistical model, research object is required to be in independent distribution; while the safety data has a complex spatial distribution, the accuracy and robustness of safety level estimation will be greatly affected if the spatial feature is neglected.

Through the review of the existing literature, it has been discovered that past researches on the causation analysis of traffic accidents are gradually evolving from the descriptive simple analysis based on aggregated models to the multivariable complex modeling analysis based on disaggregated models. However, the deficiencies of the existing studies are the following: the influencing factors on accidents are not fully considered; most are based on specific, isolated, superficial single-factor analysis, considering only the main influence factors. These studies revealed the inherent rules of the occurrence of accidents in one aspect or case but ignored the multidimensionality of accident relationships and their correlations, so that the complex logical relationship between causes, accident occurrence, and accident consequence was not reflected. Therefore, research methods and analysis technologies are not generally applicable. Although some scholars used Decision Tree [26, 32, 33], Bayesian network [34–36], and other complex systems to research the correlation between accident causes, the theoretical systems and related supporting technologies have not been systematically established.

CONSTRUCTION OF BAYESIAN NETWORK MODEL

Basic Principles of Bayesian Network

Bayesian network, also referred to as belief network, is considered as one of the most effective theoretical models in the fields of uncertainty knowledge representation and reasoning. It is a directed acyclic network topology consisting of node set and directed edge, and each node denotes one variable state, while directed edge denotes the dependence between variables. The correlation intension or confidence coefficient between variables is described by using Conditional Probability Table (CPT). Prediction, diagnosis, classification, and other tasks can be achieved by using learning and statistical inference functions of Bayes theorem. Bayesian network uses probability to denote the uncertainty of all forms and uses the probabilistic rules to achieve learning and reasoning process. It has the following relationship:

$$p(X) = \prod_{i=1}^{n} p(X_i \mid pa_i). \tag{1}$$

A set of variables $X = \{X_1, X_2, \ldots, X_n\}$ of Bayesian network consists of the following components [37] S is a network structure which denotes the conditional independent assertion in variable set X, P is a set of local probability distributions associated with each variable, X_i denotes the variable node, and pai denotes the father node of X_i in S.

S and P define the joint probability distribution of X. S is a directed acyclic graph (DAG), and each node in S corresponds to a variable in X (Figure 1). The default arc between nodes of S denotes conditional independence.

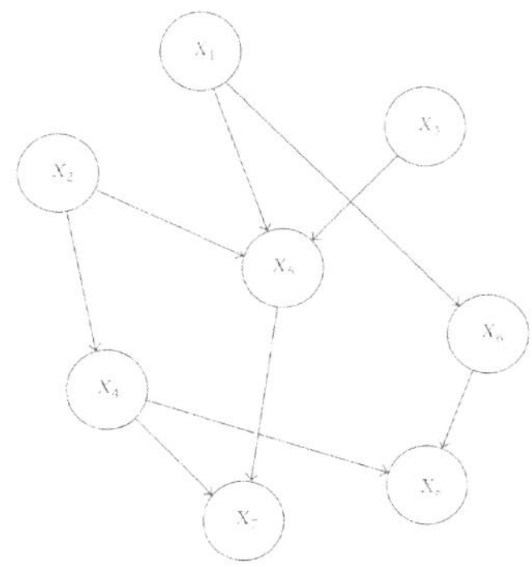

Figure 1: Graph of a valid Bayesian network (no cycle exists).

Use P to denote the local probability distribution in (1), namely, the product term $(X_i \mid pa_i)$ $(i = 1, 2, \ldots, n)$; then the binary group (S, P) denotes the joint probability distribution $p(X)$.

The construction of a Bayesian network mainly involves the following steps:(1)Structure learning: determine the factor variables (nodes) related to the study object, and then determine the dependent or independent relationship between the nodes so as to construct a directed acyclic network structure(2)Parameter learning: based on the given Bayesian network structure, learn the Conditional Probability Table (CPT) at each node of the Bayesian network model

Structure Learning

As the network structure and data set can be used to determine the parameters, structure learning is the basis of Bayesian network learning, and the effective structure learning is the key to constructing the optimal network structure.

The construction of Bayesian network structure includes the following three points:(1)Based on expert experience and prior knowledge, determine the variable nodes of Bayesian network so as to determine the structure of Bayesian network.(2)Through the learning of sample data, automatically acquire the Bayesian network structure by using machine learning algorithm. (3)Based on expert experience and machine learning of data, acquire the Bayesian network structure by using data fusion method.

As the third point combines the advantages of expert experience and machine learning and avoids the disadvantage of using one method to determine the Bayesian network structure only, in this paper, the third method to determine the Bayesian network structure for the causation analysis of road accidents will be used. Common machine learning methods include K2 algorithm, MCMC algorithm, and hill-climbing algorithm. K2 algorithm is based on the scoring function and hill-climbing algorithm, which lies in the basic principle: from an empty network, according to the predefined order of nodes, select the node with the most posterior probability as the father node of this node, sequentially traverse all nodes, and gradually add the optimal father node to each variable. K2 algorithm uses posterior probabilities as the scoring function, which is described as follows:

$$P(D \mid B_S) = \prod_{i=1}^{n} \text{score}(i, pa_i), \tag{2}$$

Where

$$\text{score}(i, pa_i) = \prod_{j=1}^{q_i} \left[\frac{\Gamma(\partial_{ij})}{\Gamma(\partial_{ij} + N_{ij})} \prod_{k=1}^{r_i} \frac{\Gamma(\partial_{ijk} + N_{ijk})}{\Gamma(\partial_{ijk})} \right]. \tag{3}$$

D is a set of variables.

B_S is the network structure.

n are the numbers of nodes in the graph.

q_i are configurations (states) of the parents of the ith node.

r_i are mutual exclusive states of the ith node.

N_{ijk} are instances of the ith node being in the kth state when its parents are in their jth configuration, and $N_{ij} = \sum_{k=1}^{r_i} N_{ijk}$.

∂_{ijk} are the hyperparameters of the Dirichlet distribution and correspond to the a priori probability distribution of X_i taking on its kth state while its parents are in their jth configuration. $\partial_{ij} = \sum_{k=1}^{r_i} \partial_{ijk}$. The gamma function $\Gamma(X) = \int_0^{+\infty} t^{X-1} e^{-t} dt$ satisfies $\Gamma(X+1) = X\Gamma(X)$ and $\Gamma(1) = 1$.

K2 algorithm uses a variable order ρ and a positive integer u to limit the search space, which seeks the optimal model $\wp$ that meets the following two conditions: (1) the number of father nodes of any variable in $\wp$ should not be greater than u and (2) ρ is a topological order of $\wp$. However, as K2 algorithm adopts greedy search strategy, which may easily fall into the local optimal solution and cannot guarantee that the network acquired is the optimal network, the knowledge and experience of experts need to be integrated so as to acquire the optimal network structure. In this paper, the combination of expert experience and K2 algorithm will perform the Bayesian network structure learning for the causation analysis of road accidents.

Parameter Learning

After determining the topological structure of Bayesian network, the parameter learning of Bayesian network can be performed. In the process of collecting road accidents information, missing data often occurs due to various reasons, for instance, recording instruments malfunction and confusion of respondents in answering questions. Most of statistical models cannot directly analyse the data with missing values, and in the case of any missing values, the record with missing values is generally eliminated directly to ensure that the statistical model can be properly fitted. If the missing values are less, this will not greatly affect the results if the record with missing values is directly eliminated. However, if the multivariate analysis is performed, more variables will be studied, which means that more records will be eliminated; it may cause a loss of information, reduce the power of test, and cause some bias to research results [38].

The Expectation-Maximization (EM) algorithm is an asymptotic deterministic estimation method for the unknown parameter θ with missing data. It can be used to perform maximum likelihood estimation on the parameters from incomplete data set, which is a practical learning algorithm [39]. EM algorithm can be widely used to deal with incomplete data, such as missing data and censored data. EM algorithm mainly involves two steps: Expectation Step (E-Step) and Maximization Step (M-Step). The algorithm is described as follows.

(1) Initialize $^{(0)}$. Set accuracy ε and correction value $\widehat{\theta}'$ of estimated value $\widehat{\theta}$.

While $\left|\widehat{\theta} - \widehat{\theta}'\right| > \varepsilon$,

$$\text{do } \widehat{\theta} \longleftarrow \widehat{\theta}'. \tag{4}$$

(2) E-Step. Calculate the expected sufficient statistic of missing value e^*.

The probability distribution of e* is

$$P\left(e^{*} \mid e, \hat{\theta}\right) = \frac{P\left(e \mid e^{*}, \hat{\theta}\right) P\left(e^{*} \mid \hat{\theta}\right)}{\sum_{e^{*}} P\left(e \mid e^{*}, \hat{\theta}\right) P\left(e^{*} \mid \hat{\theta}\right)}, \tag{5}$$

Where

$$P\left(e \mid e^{*}, \hat{\theta}\right) = \frac{P\left(e, e^{*}, \hat{\theta}\right)}{P\left(e^{*}, \hat{\theta}\right)},$$

$$P\left(e^{*} \mid \hat{\theta}\right) = \frac{P\left(e^{*}, \hat{\theta}\right)}{P\left(\hat{\theta}\right)}. \tag{6}$$

The sufficient statistic is

$$E_{P(X \mid e, \hat{\theta})} N_{ijk} = \sum_{j,k} P\left(X_{ij}, \pi\left(X_{ik}\right) \mid \hat{\theta}\right), \tag{7}$$

where $P(e \mid e^{*}, \hat{\theta})$ is the probability distribution of e under the condition that e^{*} and $\hat{\theta}$ are known, $P(e^{*}, \hat{\theta})$ is the joint distribution of $e*$ and $\hat{\theta}$, X_i is the ith variable, N_{ijk} is the count of all possible joint instantiations between Xi and $\pi(X_i)$ denoted by j and k, respectively.

(3) *M*-Step. Calculate the new maximum likelihood (ML) or maximum a posteriori (MAP) values of $\hat{\theta}'$ in the given condition $P(e^{*} \mid e, \hat{\theta})$.

In Expectation-Maximization, we have the following:

ML:

$$\hat{\theta}'_{ijk} = \frac{E_{P(X \mid e, \hat{\theta})} N_{ijk}}{\sum_{k'} E_{P(X \mid e, \hat{\theta})} N_{ijk'}}. \tag{8}$$

MAP:

$$\hat{\theta}'_{ijk} = \frac{\alpha_{ijk} + E_{P(X \mid e, \hat{\theta})} N_{ijk}}{\sum_{k'} \left(\alpha_{ijk'} + E_{P(X \mid e, \hat{\theta})} N_{ijk'}\right)}, \tag{9}$$

where α_{ijk} is the Dirichlet parameter that can be obtained through the iteration process of E-Step and M-Step.

E-Step is used to calculate the expected sufficient statistic of e^{*}, and M-Step is used to conduct new estimation of learning parameter by using the statistic obtained in E-Step. In this paper, the Bayesian network parameter learning of road accidents is performed by using EM algorithm in Netica.

CASE STUDIES

Study Area and Data Source

Adelaide Central Business District (CBD) in South Australia is selected as the study case, as it attracts 22% of metropolitan Adelaide's work trips [40] and has the first and the second most dangerous accident concentration areas which are North Terrace and West Terrace in the CBD [41].

The crash data of South Australia from 2006 to 2008 were provided by the Department of Planning, Transport and Infrastructure (DPTI), and ArcGIS 10.5 software was used to locate the precise crash sites, as shown in Figure 2.

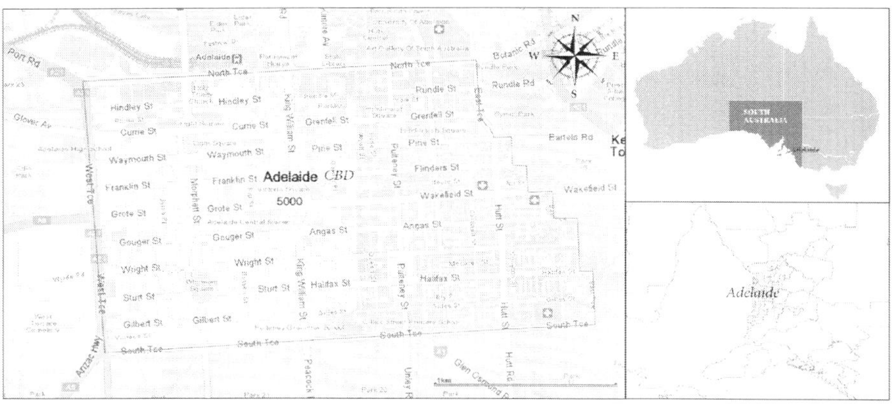

Figure 2: The location and region of Adelaide CBD.

Variable Selection and Data Preprocessing

By using ArcGIS 10.5, 1558 and 756 data sets of road accidents in Adelaide CBD from 2006-2007 and 2008 are obtained, respectively. The statistical data from 2006 to 2007 will be used for the construction of Bayesian network model and calibration, and the statistical data in 2008 will be used for the model validation process.

Previous studies [33, 42–45] provided some in-depth insights to guide the variable selection, discretization, and classification in this research. As a result, fourteen variables are selected from the data sets as having "significant influence," that is, "crash type," "driver's apparent error," "road geometry," "vehicle type," and others, as shown in Table 1. However, according to the statistical result in Table 1, it can be seen that the percentage of "inattention" reaches up to 39.84%, which is the biggest contributing factor in "driver's apparent error" category. In our daily routine, "inattention" is explained as "failure to give attention, or negligence." Generally, such usage is quite convenient for record purposes; however, from the perspective of psychology and physiology, the usage is not clear and definite. There are lots of other reasons that may be behind traffic accidents, such as human factors (driver's physical and mental state, knowledge and skill, and the operational approach), objective factors (vehicles, roads, and road facilities), and safety management. If all above factors are simply summarized as "inattention," then the causes of traffic accidents are to be extremely simplified. And the prevention measures will be hardly developed. Therefore, in this research, the factor of "inattention" will be excluded, and all variables used for modeling are shown in Table 2.

Table 1: Variables of road accidents, Adelaide CBD, 2006–2008

Variable class	Variable name	Discretization value	Value description	Frequency	Percentage
Driver	Apparent error (X_1)	1	Fail to stand	307	13.25%
		2	Change lanes to endanger	221	9.54%
		3	Incorrect turn	31	1.34%
		4	Reverse without due care	92	3.97%
		5	Follow too closely	173	7.47%
		6	Overtake without due care	52	2.24%
		7	Disobey traffic lights	171	7.38%
		8	Disobey stop sign	23	0.99%
		9	Disobey give way sign	47	2.03%
		10	Inattention	923	39.84%
		11	DUI	23	0.99%
		12	Fail to give way	254	10.96%

Road	Road geometry (X_2)	1	Cross road	1116	48.17%
		2	Y junction	57	2.46%
		3	T junction	450	19.42%
		4	Multiple	33	1.42%
		5	Divided road	349	15.06%
		6	Not divided	294	12.69%
		7	Pedestrian crossing	18	0.78%
	Road moisture condition (X_3)	1	Wet	237	10.23%
		2	Dry	2080	89.77%
	Traffic control (X_4)	1	Traffic signals	1282	55.33%
		2	Stop sign	42	1.81%
		3	Give way sign	118	5.09%
		4	No control	875	37.76%
Environment	Weather condition (X_5)	1	Raining	151	6.52%
		2	Not raining	2166	93.48%
	Light condition (X_6)	1	Daylight	1716	74.06%
		2	Night	601	25.94%
Vehicle	Vehicle type (X_7)	1	Heavy	172	7.42%
		2	Medium	901	38.89%
		3	Light	1244	53.69%
	Vehicle movement (X_8)	1	Right turn	426	18.39%
		2	Left turn	95	4.10%
		3	U turn	122	5.27%
		4	Swerving	242	10.44%
		5	Reversing	77	3.32%
		6	Straight ahead	1238	53.43%
		7	Entering private driveway	15	0.65%
		8	Leaving private driveway	50	2.16%
		9	Overtaking on right	39	1.68%
		10	Overtaking on left	13	0.56%

Road crash	Crash type (Y_1)	1	Rear end	1010	43.59%
		2	Hit fixed object	59	2.55%
		3	Side swipe	392	16.92%
		4	Right angle	377	16.27%
		5	Head on	5	0.22%
		6	Hit pedestrian	50	2.16%
		7	Right turn	333	14.37%
		8	Hit parked vehicle	91	3.93%
	Crash severity (Y_2)	1	PDO (property damage only)	1748	75.44%
		2	Injury	569	24.56%
	Total units (involved in a road crash) (Y_3)	1	Two units	2012	86.84%
		2	Three units	258	11.14%
		3	Four units	40	1.73%
		4	Five units	7	0.30%
	Total casualties (fatalities and treated injuries) (Y_4)	1	None	1748	75.44%
		2	One casualty	495	21.36%
		3	Two casualties	62	2.68%
		4	Three casualties	12	0.52%
	Total serious injuries (Y_5)	1	None	2267	97.84%
		2	One serious injury	50	2.16%
	Total estimated damage (A$) ($Y_6$)	1	[0, 5000)	1139	49.16%
		2	[5000, 10000)	795	34.31%
		3	[10000, +∞)	383	16.53%

Table 2: Variables used for the Construction of Bayesian Network

Variable class	Variable name	Discretization value	Value description	Frequency	Percentage
Driver	Apparent error (X_1)	1	Fail to stand	199	21.47%
		2	Change lanes to endanger	141	15.21%
		3	Incorrect turn	24	2.59%
		4	Reverse without due care	61	6.58%
		5	Follow too closely	119	12.84%
		6	Overtake without due care	35	3.78%
		7	Disobey traffic lights	115	12.41%
		8	Disobey stop sign	18	1.94%
		9	Disobey give way sign	27	2.91%
		10	DUI	9	0.97%
		11	Fail to give way	179	19.31%
Road	Road geometry (X_2)	1	Cross road	454	48.98%
		2	Y junction	11	1.19%
		3	T junction	196	21.14%
		4	Multiple	13	1.40%
		5	Divided road	132	14.24%
		6	Not divided	121	13.05%
	Road moisture condition (X_3)	1	Wet	100	10.79%
		2	Dry	827	89.21%
	Traffic control (X_4)	1	Traffic signals	477	51.46%
		2	Stop sign	27	2.91%
		3	Give way sign	56	6.04%
		4	No control	367	39.59%
Environment	Weather condition (X_5)	1	Raining	65	7.01%
		2	Not raining	862	92.99%
	Light condition (X_6)	1	Daylight	677	73.03%
		2	Night	250	26.97%

Vehicle	Vehicle type (X_7)	1	Heavy	63	6.80%
		2	Medium	379	40.88%
		3	Light	485	52.32%
	Vehicle movement (X_8)	1	Right turn	267	28.80%
		2	Left turn	41	4.42%
		3	U turn	79	8.52%
		4	Swerving	137	14.78%
		5	Reversing	52	5.61%
		6	Straight ahead	275	29.67%
		7	Entering private driveway	10	1.08%
		8	Leaving private driveway	31	3.34%
		9	Overtaking on right	25	2.70%
		10	Overtaking on left	10	1.08%
Road crash	Crash type (Y_1)	1	Rear end	155	16.72%
		2	Hit fixed object	4	0.43%
		3	Side swipe	255	27.51%
		4	Right angle	258	27.83%
		5	Head on	3	0.32%
		6	Hit pedestrian	22	2.37%
		7	Right turn	216	23.30%
		8	Hit parked vehicle	14	1.51%
	Crash severity (Y_2)	1	PDO (property damage only)	689	74.33%
		2	Injury	238	25.67%
	Total units (involved in a road crash) (Y_3)	1	Two units	860	92.77%
		2	Three units	54	5.83%
		3	Four units	9	0.97%
		4	Five units	4	0.43%
	Total casualties (fatalities and treated injuries) (Y_4)	1	None	689	74.33%
		2	One casualty	213	22.98%
		3	Two casualties	21	2.27%
		4	Three casualties	4	0.43%
	Total serious injuries (Y_5)	1	None	904	97.52%
		2	One serious injury	23	2.48%
	Total estimated damage (A$) ($Y_6$)	1	[0, 5000)	423	45.63%
		2	[5000, 10000)	342	36.89%
		3	[10000, +∞)	162	17.48%

Bayesian network can be used to process continuous variables and discrete variables. As the classification result of traffic accident variables obviously has the discrete characteristic, discrete variables are adopted for Bayesian network analysis of road accidents. Before structure learning, discretization processing has to be conducted for road accident variable. The discretization values and value descriptions of processed variables are shown in Table 2.

Structure Learning

In this paper, the method combining K2 algorithm and experts' knowledge is used to formulate the Bayesian network structure. Based on K2 algorithm, FullBNT-1.0.7 is utilized to conduct structure learning via MATLAB. Through repeated selection and sequencing of variables by experts, the Bayesian network structure is finally developed, as shown in Figure 3. The network is composed by 14 nodes and several lines. The 14 nodes refer to 14 variables, and lines between these nodes indicate the relationships among the variables.

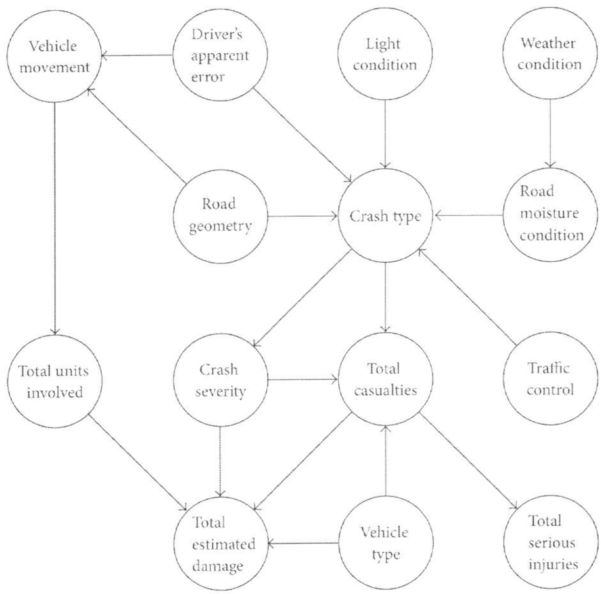

Figure 3: The Bayesian network structure for the road accident analysis.

It can be seen from Figure 3 that some road accident variables have demonstrated clear hierarchical relations of affecting others and being affected by others. Road accidents result from the interaction of variables from "traffic participant, vehicle, road, and environment," which is fully reflected by the Bayesian network structure as well. For instance, "vehicle movement" is affected by "road geometry" and "driver's apparent error," but it can also affect "total units involved" at the same time. The actual situation of road accidents can be fully embodied by the interactional hierarchical relationship of variables in Bayesian network.

Parameter Learning

Once a Bayesian network structure is formed, parameter learning can be carried out. The Bayesian network structure can be created in Netica, and then parameter learning can be conducted, thus obtaining the conditional probability distribution of nodes. Finally, the Bayesian network model for the road accident causation analysis can be determined, as shown in Figure 4.

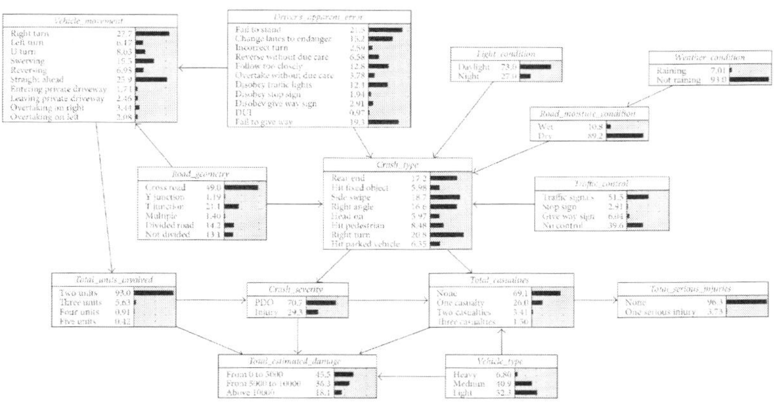

Figure 4: The Bayesian network model after parameter learning in Netica 6.02.

Model Calibration and Validation

To validate the parameter learning accuracy and prediction accuracy of the Bayesian network model, sensitivity analysis is used to identify the sensitive factors with a significant impact on the target node from a number of uncertain factors, and then the target node is set as the evidence variable to conduct model fitting and prediction with these sensitive factors.

Sensitivity Analysis

In Bayesian network, the sensitivity analysis refers to the analysis of the influence and influence degrees of multiple causes (node states) on result (target node). Based on sensitivity analysis, the elementary events with relatively greater contribution to the probabilities of the consequential events can be determined to facilitate the reduction of probabilities of these elementary events by taking effective measures, so that the probabilities of the consequential events will be reduced.

The sensitivity analysis function of Netica can be used to identify which factors have more important safety management values in analyzing traffic accidents. In Netica, select the target node and then analyse the impact degrees of other nodes on the target node in a descending order. Taking the node "crash type" as an example, make sensitivity analysis of it and the result is as shown in Table 3.

Table 3: Sensitivity analysis result of the node "crash type"

Node	Mutual info	Percent	Variance of beliefs
Crash type	2.82978	100	0.7195390
Driver's apparent error	0.57217	20.2	0.0869963
Vehicle movement	0.33948	12	0.0351281
Total casualties	0.21563	7.62	0.0057073
Crash severity	0.17897	6.32	0.0037534
Road geometry	0.04044	1.43	0.0008944
Total serious injuries	0.01888	0.667	0.0004707
Traffic control	0.01125	0.398	0.0001573
Road moisture condition	0.00503	0.178	0.0001027
Light condition	0.00417	0.147	0.0000915
Total estimated damage	0.00407	0.144	0.0000533
Total units involved	0.00394	0.139	0.0002397
Weather condition	0.00313	0.111	0.0000638
Vehicle type	0.00000	0.000	0.0000000

The mutual information refers to the direct or indirect information flow rate and measures the degree of dependence between nodes. In other words, the mutual information between two nodes can indicate if the two nodes are dependent on each other and if so how close their relationship is [46]. As shown in Table 3, it can be seen that the mutual info (=0.57217) of node "driver's apparent error" is the largest, which means that it has the strongest impact on "crash type," followed by "vehicle movement" and "road geometry" which have mutual info = 0.33948 and 0.04044, respectively.

Model Fitting and Prediction

Based on the sensitivity analysis result of "crash type," the posterior probabilities of "driver's apparent error," "vehicle movement," and "road geometry" obtained from the Bayesian network are compared with the actual calculations from 2006 to 2007 and from 2008, respectively. Due to the large amount of data, "rear end" from the parameter learning results of "crash type" is used for exemplificative explanation.

Figures 5, 6, and 7 show the posterior and actual probability distributions of "driver's apparent error," "vehicle movement," and "road geometry," respectively, when the evidence variable is "rear end." Compared with the actual computational results from 2006 to 2007, the maximum mean absolute error (MAE) of Bayesian network model is 5.58%. Similarly, compared with the actual computational results of 2008, the maximum MAE is 6.14%, which suggests that the Bayesian network model has both high fitting accuracy and high prediction accuracy. Therefore, it is feasible to use the Bayesian network model to conduct result prediction and inferential analysis of each variable of road accidents accordingly.

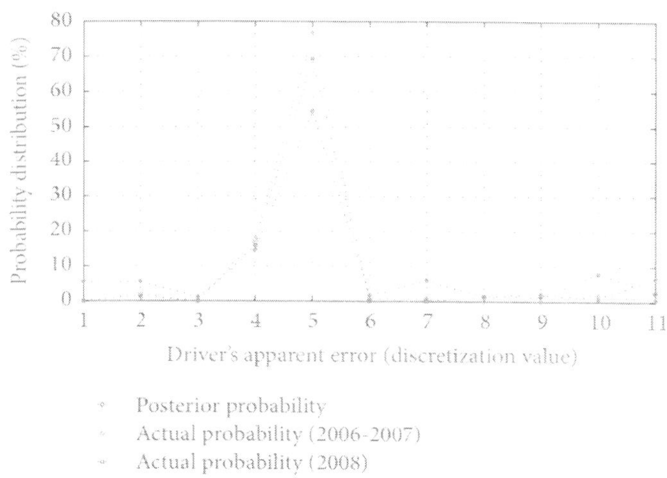

Figure 5: The comparison of posterior and actual probability curves of driver's apparent error when the evidence variable is "rear end."

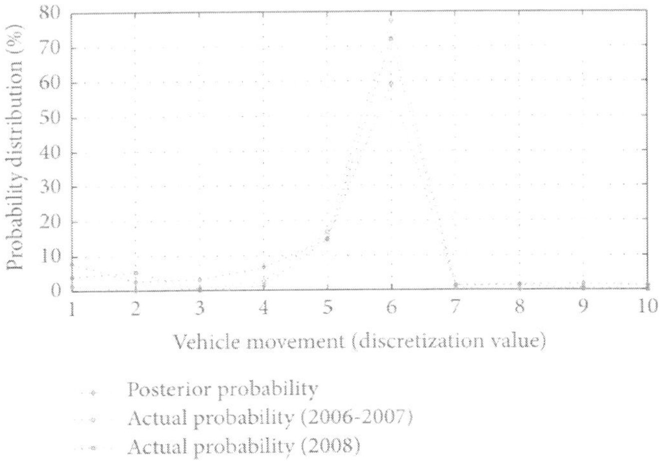

Figure 6: The comparison of posterior and actual probability curves of vehicle movement when the evidence variable is "rear end."

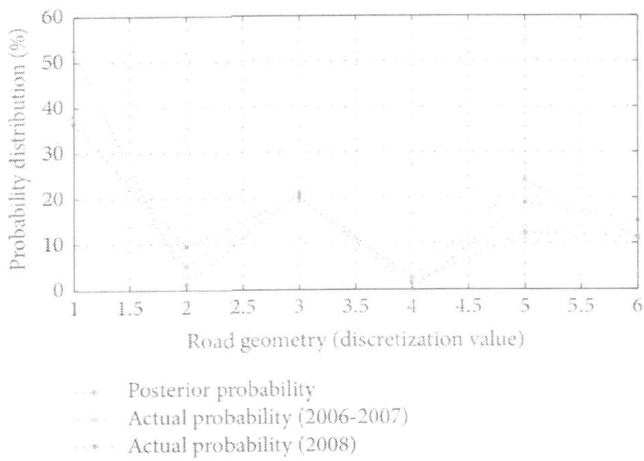

Figure 7: The comparison of posterior and actual probability curves of road geometry when the evidence variable is "rear end."

Bayesian Network Model Application

Posterior Probability Reasoning

Bayesian network model can be used to perform probability reasoning, including posterior probability calculations. Precisely, it aims to calculate the posterior probabilities of some targeted nodes, control the influence degrees of determined specific nodes on the nodes of interest, predict the possibility of accident occurrence, and analyse the major accident sources under the condition that states of specific nodes are determined. In brief, the posterior probabilities in inferring the result from cause and inferring the cause from result are referred to as accident prediction and causal inference, respectively.

(1) Accident Prediction. Figure 8 is the accident prediction on the assumption that a driver "disobeys traffic lights" when driving in Adelaide CBD. Input the evidence variables ("traffic signals" and "disobey traffic lights") emerging from this circumstance into the Bayesian network, so it becomes a problem to solve the posterior probabilities of other nodes, with the known status of some evidence variables.

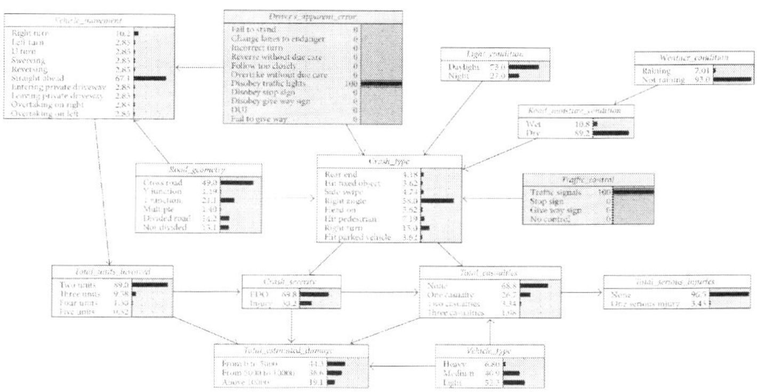

Figure 8: Road accident prediction when the evidence variables are "traffic signals" and "disobey traffic lights."

In Netica, set both the statuses of "traffic signals" and "disobey traffic lights" as 100%; that is, the statuses of the evidence variables are determined. Then update the probabilities of the whole network; the probability change of relevant nodes, namely, the probability change of "crash type" and other nodes, can be observed. In this case, the probability of "right angle" in "crash type" is found to increase from the initial 16.6% to 58.0%. This suggests that if the driver "disobeys traffic lights," the probability of "right angle" will significantly increase. As shown in Figure 9, in addition to "disobey traffic lights," assume that the driving time is at night; namely, the status of "night" in "light condition" is set as 100%. After automatically updating the probabilities of the whole network, the probability of "right angle" is found to further increase from 58.0% to 59.4%, which means that the probability of "right-angle" traffic accident is higher. Go one step further and assume that it is also a rainy night (namely, set the status of "raining" in "weather condition" as 100%). According to Figure 10, once again, it can be found that the probability of "right angle" further increases from 59.4% to 63.3%. This suggests that "driver's apparent error," "light condition," and "weather condition" will all affect the probability of "right angle" to various degrees. Therefore, it can be found that the status change of evidence node variables will affect the probabilities of query nodes, which is consistent with the engineering practice.

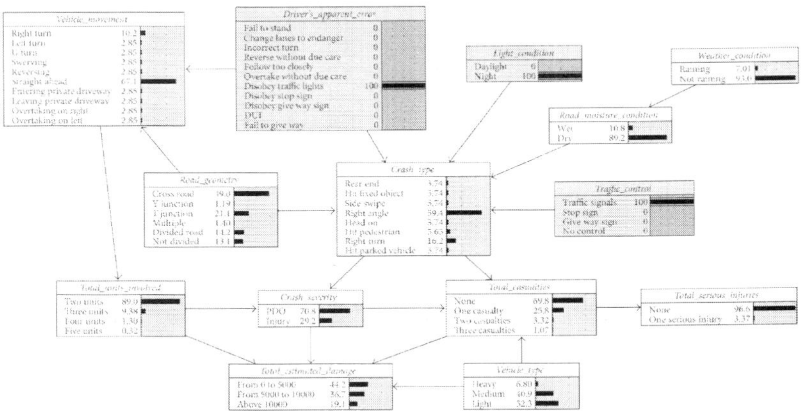

Figure 9: Road accident prediction when the evidence variables are "traffic signals," "disobey traffic lights," and "night."

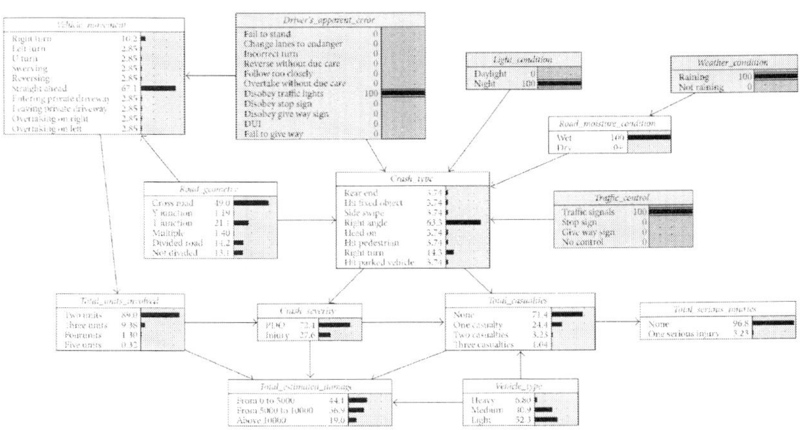

Figure 10: Road accident prediction when the evidence variables are "traffic signals," "disobey traffic lights," "night," and "raining."

(2) Causal Inference. Another important application of the Bayesian network is fault diagnosis of the system. The bidirectional reasoning technology of the Bayesian network can calculate not only the probability of a system failure under combined fault conditions but also the posterior probabilities of various components under the system fault condition and easily find out the most likely combination that caused system failure, thereby making the computational analysis more intuitive and flexible.

Conduct causal inference by taking the "side swipe" in "crash type" as an example. In this case, the evidence variable is "side swipe," so its status probability is 100%. As shown in Figure 11, after inputting the evidence, the probability of "change lanes to endanger" in "driver's apparent error" increases greatly from 15.2% to 46.7% through the automatic updating function of Netica. And the probability of "swerving" in "vehicle movement" also increases from 15.5% to 44.9%, which reaches the maximum probability. This suggests that, in the absence of other evidences, the most probable cause to "side swipe" is "swerving" (vehicle) caused by "change lanes to endanger" (driver).

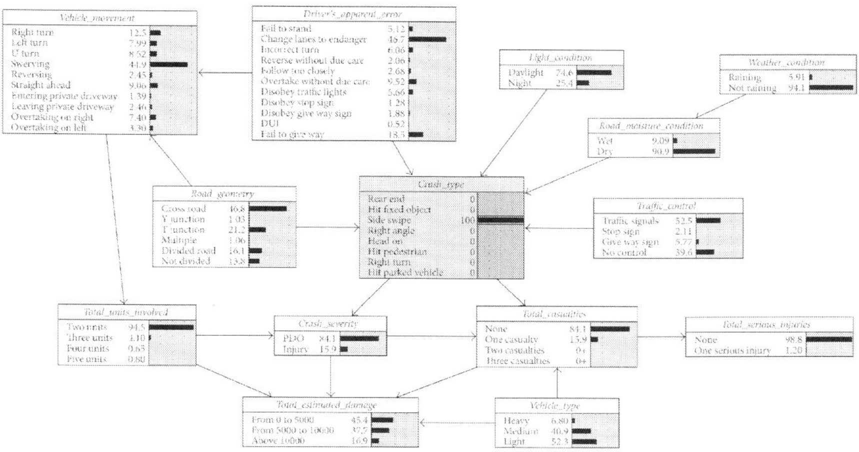

Figure 11: The posterior probability when the evidence variable is "side swipe."

Most Probable Explanation

Bayesian network model can be used to make the most probable explanations, precisely, from sets of multiple causes (node states) which are likely to lead to a conclusion; use Netica to identify the set that is most likely to lead to the result, and this set with the maximum likelihood will be the most probable explanation.

In the example of "side swipe" as illustrated in Figure 12, use "Most Probable Explanation" function in Netica to find out the most probable cause set. As shown in Figure 12, the most probable explanation cause (node state) set of "side swipe" is [cross road, change lanes to endanger, swerving, daylight, not raining, dry, traffic signals]. It explicitly shows that most probable explanation and causal inference are highly consistent when the evidence variable is "side swipe," and the set is also consistent with the engineering practice.

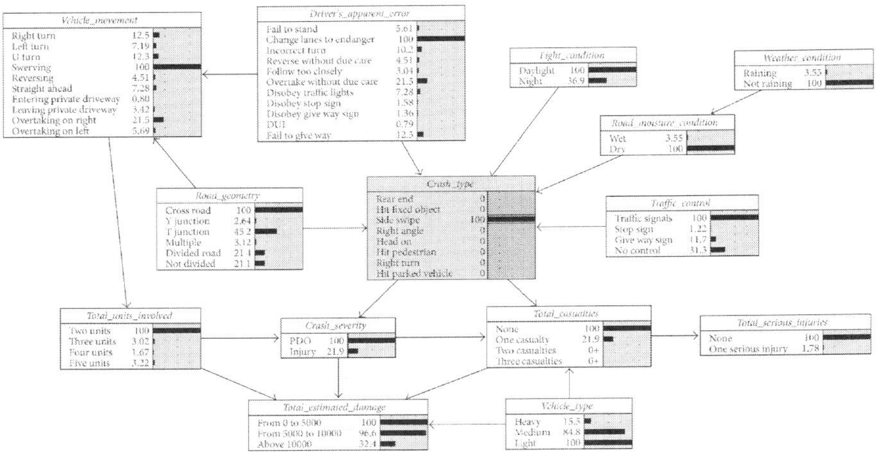

Figure 12: The most probable explanation when the evidence variable is "side swipe."

Inferential Analysis of Accidents Based on "Serious Injuries" and "Total Estimated Damage"

The application of the Bayesian network model in Netica to solve the posterior probability reasoning problem, maximum posterior hypothesis problem, and most probable explanation problem highlighted the inferential capability of the Bayesian network model. To further analyse the factors contributing to traffic accidents, especially serious traffic accidents, the Bayesian network model was used to calculate the probabilities of "serious injuries" and "total estimated damage over 10,000 AUD" under the influence of "driver's apparent error," "road geometry," "weather condition," "light condition," and "crash type," respectively. The results are shown in Table 4.

Table 4: Inference results for variables that are associated with "serious injuries" and "total estimated damage" in serious traffic accidents

Variable class	Variable name	Serious injuries/%	Total estimated damage (≥10,000 AUD)/%
Driver's apparent error	Fail to stand	3.54	17.5
	Change lanes to endanger	2.83	17.5
	Incorrect turn	3.61	17.1
	Reverse without due care	4.03	17.9
	Follow too closely	3.91	20.0
	Overtake without due care	3.36	18.7
	Disobey traffic lights	4.15	19.4
	Disobey stop sign	4.81	19.5
	Disobey give way sign	4.68	18.9
	DUI	5.17	20.4
	Fail to give way	3.91	17.2
Road geometry	Cross road	4.47	19.6
	Y junction	3.69	18.2
	T junction	3.27	18.0
	Multiple	4.17	18.4
	Divided road	4.05	18.1
	Not divided	4.12	18.3
Weather condition	Raining	4.22	18.4
	Not raining	3.69	18.1
Light condition	Daylight	3.65	18.1
	Night	3.93	18.2
Crash type	Rear end	3.40	19.0
	Hit fixed object	2.76	17.4
	Side swipe	1.20	16.9
	Right angle	2.48	17.6
	Head on	15.4	26.0
	Hit pedestrian	9.37	18.6
	Right turn	2.93	17.2
	Hit parked vehicle	0.34	16.9

Driver's Apparent Error

The results presented in Table 4 indicate that "driving under the influence" (DUI) will most likely cause serious injuries and heavy property damage, as DUI is more easily to lead to dangerous behaviours including speeding, not wearing a safety belt, and reckless or erratic driving. According to the inference results, DUI is most likely to cause traffic accidents on cross roads with the inference probability of 49.0%. As for the crash types, the inference probabilities of rear-end and head-on crashes are the two highest (resp., 15.6% and 15.4%). In previous studies, among all traffic accidents caused by DUI, unrestrained occupants were 4.70 times more likely to die or 4.66 times more likely to be injured than restrained occupants [47]. Besides, drunk drivers show weak control ability of vehicles, and the higher ethanol content in their blood, the higher probability that they will have illegal actions mentioned previously [48], which are more likely to cause serious traffic accidents with heavy casualties and property damage.

Road Geometry

Intersections are an important part of road system and are potentially the most dangerous locations in a network as well. Previous studies have shown that intersections, especially cross roads, have higher crash rates and greater crash severity, particularly in urban areas [49–51]. As shown in Table 4, the most dangerous "road geometry" in Adelaide CBD also is "cross road," which means that the probabilities of causing serious injuries and heavy property damage at "cross roads" are both the biggest. According to the accident records in Adelaide CBD, among serious traffic accidents caused at cross roads, there are two main crash types, "right angle" and "right turn," respectively, accounting for 44.35% and 42.61%, while "fail to stand" and "disobey traffic lights" become the two main reasons of traffic accidents caused at cross roads, respectively, accounting for 38.26% and 34.78%.

Weather Condition

Rainfall will not only decrease the effectiveness of drivers' visual search [12] but also lower the friction coefficient of roads, which makes roads slippery, increases braking distance greatly, and thus results in the possibility of traffic accidents. A study for Melbourne, Australia, by Keay and Simmonds [10] found that rainfall was the strongest factor that correlated to weather parameter and it had the greatest impact in winter and spring. Keay and Simmonds [52] also found a contributing parameter, which is the lagged effect of rain. Symons and Perry [53] found that wet roads or raining is increasing the probability of traffic accidents which can reach up to 70 percent. Similarly, Qiu and Nixon [11] found that rain can increase the crash rate by 71% and the injury rate by 49%. This coincides with the results found in this research, which indicate that rainfall is associated with traffic accidents that had serious injuries and heavy property damage.

Light Condition (Urban Heat Island)

Traditional bituminous pavement can absorb and store large amounts of heat during the day and continuously output the heat to the external environment at night, which will result in an increase in external environment temperature and lead to urban heat island (UHI). UHI might have some impacts on road durability and safety, such as accelerating bituminous pavement aging and exacerbating road high-temperature rutting may lead to road accidents. With the continuous expansion of city size, the comprehensive phenomenon of such microclimatic variation will become increasingly obvious [54]. As one of the major Australian cities, Adelaide is also affected by UHI [55]. And as waste heat from vehicles and temperature regulation of buildings is an important determinant of UHI magnitudes, Adelaide CBD, which has the largest density of traffic network and the largest number of buildings, has become the center of UHI. As shown in Table 4, the number of road accidents that occurred at night demonstrates a larger proportion of 55.56% than by daylight. Under the influence of UHI, Jusuf et al. [56] found that, at nighttime, commercial area has the highest ambient temperature among the four land use types (commercial, residential, industrial, and airport). Similarly, Parker [57] also found that the urban heat island is strongest at night in high-rise city centers. Therefore, the correlation analysis between road accidents and nighttime UHI is a potential research direction.

Crash Type

Head-on crashes are among the most severe collision types and are of great concern to road safety authorities [58]. For instance, according to an annual report presented by NHTSA in 2015, head-on crashes occupied only 2.3% of total crashes; however, they accounted for 9.6% of fatal crashes. As shown in Table 4, no matter "serious injuries" or "property damage over 10,000 AUD," the possibility of "head-on" crashes always takes the first place and is largely higher than other types of traffic accidents. As for the road geometry, the head-on crashes are most likely to happen on cross roads with the inference probability of 39.6%, while the inference probability of "disobey traffic lights" is the biggest in drivers' apparent errors which cause head-on crashes. These results agree with Bham et al. [59] who found that head-on collisions were at a higher risk for severe injuries compared with other collision types. And Rizzi et al. [60] concluded that 31% and 21% of crashes at intersections could have been avoided entirely or influenced by anti-lock braking system (ABS); however, the head-on crashes were the only crash type for which ABS seemed to be ineffective.

CONCLUSION AND FUTURE WORK

As road accidents are unexpected, random, complex, and latent, it is necessary to conduct investigations on accident mechanism and accurately identify the exact causes. The Bayesian network combining the probability theory with graph theory not only has a rigorous mathematical consistency but also has the structure chart that can intuitively identify problems. Therefore, it is one of the most powerful and effective tools to deal with uncertainties.

The occurrence of road accident results from the interactions of "traffic participant, vehicle, road, and environment," and there is a potential hierarchical relation (impacting and impacted) among the variables. In Bayesian network, the directed acyclic graph is a visual expression form that is closer to the characteristics of thought and reasoning mode of human. In the study, the Bayesian network structure for the road accident causation analysis was achieved by using K2 algorithm and experts' knowledge which combines the advantages of machine learning and experts' knowledge. The structure learning result of the Bayesian network fully reflects the hierarchical relations among the accident related variables and allows for better prediction and analysis of the characteristics of road accidents.

In this study, the Bayesian network model for the road accident causation analyses was established by using Netica, Bayesian network-based software with friendly GUI. The Expectation-Maximization algorithm that can deal with missing data was adopted to process the parameter learning, and then the calibration and validation, posterior probability reasoning, most probable explanation, and inferential analysis were carried out after the construction of the Bayesian network model. The results showed that the Bayesian network model is feasible and effective for road accident causation analyses; in particular the use of posterior probability of the Bayesian network can not only more precisely and quickly find the key causes for traffic accidents but also identify the most likely cause (state) combination. The result can be used as an important theoretical basis in developing road traffic management strategies so as to improve road traffic safety.

Follow-up studies will consider the rationality of the Bayesian model and other factors that may lead to traffic accidents and establish a more accurate and comprehensive model. As the Bayesian model is a probabilistic model, more comprehensive and extensive basic data are needed to enhance its reliability, in which some data can be obtained only by carrying out experiments, despite the ability of the Bayesian mode to make up for missing data. Moreover, as the influencing road accident factors in reality are more than the factors used in this study and as Netica is also suitable for the establishment of a larger and more complex accident analysis model, the model can be expanded to a more sophisticated model that can consider more factors.

REFERENCES

1. Australasian College of Road Safety, 2017 ACRS Submission to Federal Parliamentarians, Retrieved from http://acrs.org.au/wp-content/uploads/2017-ACRS-Submission-to-Federal-Parliamentarians-FINAL.pdf.
2. National Highway Traffic Safety Administration (NHTSA)., "Critical reasons for crashes investigated in the national motor vehicle crash causation survey," vol. 85, 101 pages, 2015.
3. National Highway Traffic Safety Administration (NHTSA), "Traffic safety facts 2014: a compilation of motor vehicle crash data from the fatality analysis reporting system and the general estimates system," 2015.
4. M. G. Karlaftis and I. Golias, "Effects of road geometry and traffic volumes on rural roadway accident rates," Accident Analysis Prevention, vol. 34, no. 3, pp. 357–365, 2002.
5. R. Fu, Y. Guo, W. Yuan, H. Feng, and Y. Ma, "The correlation between gradients of descending roads and accident rates," Safety Science, vol. 49, no. 3, pp. 416–423, 2011.
6. J. P. Gooch, V. V. Gayah, and E. T. Donnell, "Quantifying the safety effects of horizontal curves on two-way, two-lane rural roads," Accident Analysis Prevention, vol. 92, pp. 71–81, 2016.
7. H. Al-Madani and A. R. Al-Janahi, "Role of drivers personal characteristics in understanding traffic sign symbols," Accident Analysis Prevention, vol. 34, no. 2, pp. 185–196, 2002.
8. D. de Waard, F. J. J. M. Steyvers, and K. A. Brookhuis, "How much visual road information is needed to drive safely and comfortably?" Safety Science, vol. 42, no. 7, pp. 639–655, 2004.
9. T. Ben-Bassat and D. Shinar, Ergonomic Guidelines for Traffic Sign Design Increase Sign Comprehension. Human Factors, vol. 48, no. 1, pp. 182–195, 2006.
10. K. Keay and I. Simmonds, "The association of rainfall and other weather variables with road traffic volume in melbourne, australia," Accident Analysis Prevention, vol. 37, no. 1, pp. 109–124, 2005.
11. L. Qiu and W. Nixon, "Effects of adverse weather on traffic crashes: systematic review and meta-analysis," Transportation Research Record: Journal of the Transportation Research Board, pp. 139–146, 2008.

12. P. Konstantopoulos, P. Chapman, and D. Crundall, "Driver's visual attention as a function of driving experience and visibility. Using a driving simulator to explore drivers' eye movements in day, night and rain driving," Accident Analysis & Prevention, vol. 42, no. 3, pp. 827–834, 2010.
13. K. P. Murphy, "The bayes net toolbox for MATLAB," Computing Science and Statistics, vol. 33, no. 2, pp. 1024–1034, 2001, http://www.interfacesymposia.org/I01/I2001Proceedings/KMurphy/KMurphy.pdf.
14. O. Arsene, I. Dumitrache, and I. Mihu, "Medicine expert system dynamic Bayesian Network and ontology based," Expert Systems with Applications, vol. 38, no. 12, 2011.
15. G. Büyüközkan, G. Kayakutlu, and İ. S. Karakadılar, "Assessment of lean manufacturing effect on business performance using bayesian belief networks," Expert Systems with Applications, vol. 42, no. 19, pp. 6539–6551, 2015.
16. H. Mcheick, H. Nasser, M. Dbouk, and A. Nasser, "Stroke prediction context-aware health care system," in Proceedings of IEEE First International Conference on Connected Health: Applications, Systems and Engineering Technologies (CHASE), pp. 30–35, Washington, Wash, USA, 2016.
17. S. Jensen, "Pedestrian Safety in Denmark," Transportation Research Record: Journal of the Transportation Research Board, vol. 1674, pp. 61–69, 1999.
18. M. Stone and J. Broughton, "Getting off your bike: cycling accidents in great britain in 1990–1999," Accident Analysis Prevention, vol. 35, no. 4, pp. 549–556, 2003.
19. D. E. Lefler and H. C. Gabler, "The fatality and injury risk of light truck impacts with pedestrians in the United States," Accident Analysis Prevention, vol. 36, no. 2, pp. 295–304, 2004.
20. O. T. Holubowycz, "Age, sex, and blood alcohol concentration of killed and injured pedestrians," Accident Analysis & Prevention, vol. 27, no. 3, pp. 417–422, 1995.
21. A. S. Al-Ghamdi, "Pedestrian–vehicle crashes and analytical techniques for stratified contingency tables," Accident Analysis Prevention, vol. 34, no. 2, pp. 205–214, 2002.
22. D. Lord, S. P. Washington, and J. N. Ivan, "Poisson, Poisson-gamma

and zero-inflated regression models of motor vehicle crashes: balancing statistical fit and theory," Accident Analysis & Prevention, vol. 37, no. 1, pp. 35–46, 2005.

23. M. Poch and F. Mannering, "Negative binomial analysis of intersection-accident frequencies," Journal of Transportation Engineering, vol. 122, no. 2, pp. 105–113, 1996.
24. J.-L. Martin, "Relationship between crash rate and hourly traffic flow on interurban motorways," Accident Analysis & Prevention, vol. 34, no. 5, pp. 619–629, 2002.
25. K.-s. Ng, W.-t. Hung, and W.-g. Wong, "An algorithm for assessing the risk of traffic accident," Journal of Safety Research, vol. 33, no. 3, pp. 387–410, 2002.
26. L.-Y. Chang and W.-C. Chen, "Data mining of tree-based models to analyze freeway accident frequency," Journal of Safety Research, vol. 36, no. 4, pp. 365–375, 2005.
27. G. Lovegrove and T. Sayed, "Macrolevel collision prediction models to enhance traditional reactive road safety improvement programs," Transportation Research Record: Journal of the Transportation Research Board, pp. 65–73, 2007.
28. M. J. M. Sullman, M. L. Meadows, and K. B. Pajo, "Aberrant driving behaviours amongst new zealand truck drivers," Transportation Research Part F: Traffic Psychology and Behaviour, vol. 5, no. 3, pp. 217–232, 2002.
29. J.-K. Kim, S. Kim, G. F. Ulfarsson, and L. A. Porrello, "Bicyclist injury severities in bicyclemotor vehicle accidents," Accident Analysis & Prevention, vol. 39, no. 2, pp. 238–251, 2007.
30. P. Savolainen and F. Mannering, "Probabilistic models of motorcyclists injury severities in single- and multi-vehicle crashes," Accident Analysis & Prevention, vol. 39, no. 5, pp. 955–963, 2007.
31. R. K. Young and J. Liesman, "Estimating the relationship between measured wind speed and overturning truck crashes using a binary logit model," Accident Analysis & Prevention, vol. 39, no. 3, pp. 574–580, 2007.
32. P. M. Kuhnert, K.-A. Do, and R. McClure, "Combining non-parametric models with logistic regression: an application to motor vehicle injury data," Computational Statistics & Data Analysis, vol. 34, no. 3, pp. 371–386, 2000.

33. J. Abellán, G. López, and J. de Oña, "Analysis of traffic accident severity using decision rules via decision trees," in Expert Systems with Applications, vol. 40, pp. 6047–6054, 2013.
34. J. de Oña, R. O. Mujalli, and F. J. Calvo, "Analysis of traffic accident injury severity on spanish rural highways using bayesian networks," Accident Analysis Prevention, vol. 43, no. 1, pp. 402–411, 2011.
35. S. Heydari, L. F. Miranda-Moreno, D. Lord, and L. Fu, "Bayesian methodology to estimate and update safety performance functions under limited data conditions: A sensitivity analysis," Accident Analysis Prevention, vol. 64, pp. 41–51, 2014.
36. A. C. Mbakwe, A. A. Saka, K. Choi, and Y.-J. Lee, "Alternative method of highway traffic safety analysis for developing countries using delphi technique and Bayesian network," Accident Analysis & Prevention, vol. 93, pp. 135–146, 2016.
37. K. G. Olesen and A. L. Madsen, "Maximal prime subgraph decomposition of Bayesian networks," IEEE Transactions on Systems, Man, and Cybernetics, Part B (Cybernetics), vol. 32, no. 1, pp. 21–31, 2002.
38. S. Demissie, M. P. LaValley, N. J. Horton, R. J. Glynn, and L. A. Cupples, "Bias due to missing exposure data using complete-case analysis in the proportional hazards regression model," Statistics in Medicine, vol. 22, no. 4, pp. 545–557, 2003.
39. R. S. Pilla and B. G. Lindsay, "Alternative EM Methods for Nonparametric Finite Mixture Models," Biometrika, vol. 88, no. 2, pp. 535–550, 2001, Retrieved from http://www.jstor.org/stable/2673498.
40. A. Allan, "Land Use Planning and its Role in Transforming the Adelaide-Gawler Line into a Transit Corridor of Connected Transit Oriented Developments," in Paper presented at the 34th Australasian Transport Research Forum (ATRF) Proceedings, Adelaide, Australia, 2011, http://www.worldtransitresearch.info/research/4335/.
41. Australian Associated Motor Insurers Limited, "Adelaide's most dangerous accident hotspots revealed," 2015, Retrieved from https://www.aami.com.au/aami-answers/press-releases/adelaide-most-dangerous-accident-hotspots-revealed.html.
42. R. O. Mujalli and J. De Oña, "A method for simplifying the analysis of traffic accidents injury severity on two-lane highways using Bayesian

networks," Journal of Safety Research, vol. 42, no. 5, pp. 317–326, 2011.

43. A. Gregoriades and K. C. Mouskos, "Black spots identification through a bayesian networks quantification of accident risk index," Transportation Research Part C: Emerging Technologies, vol. 28, pp. 28–43, 2013.
44. R. O. Mujalli, G. López, and L. Garach, "Bayes classifiers for imbalanced traffic accidents datasets," in Accident Analysis Prevention, (Supplement C), vol. 88, pp. 37–51, URL https, 2016.
45. A. Iranitalab and A. Khattak, "Comparison of four statistical and machine learning methods for crash severity prediction," in Accident Analysis & Prevention, 108 (Supplement C), pp. 27–36, 2017.
46. J. Cheng, R. Greiner, J. Kelly, D. Bell, and W. Liu, "Learning Bayesian networks from data: an information-theory based approach," Artificial Intelligence, vol. 137, no. 1-2, pp. 43–90, 2002.
47. E. Desapriya, I. Pike, and S. Babul, "Public attitudes, epidemiology and consequences of drinking and driving in british columbia," IATSS Research, vol. 30, no. 1, pp. 101–110, 2006.
48. R. Hingson and M. Winter, "Epidemiology and consequences of drinking and driving," Alcohol Research and Health, vol. 27, no. 1, pp. 63–78, 2003, https://pubs.niaaa.nih.gov/publications/arh27-1/63-78.htm.
49. Transport Canada's Motor Vehicle Safety Directorate, "Road Safety in Canada," 2011, Retrieved from https://www.tc.gc.ca/eng/motorvehiclesafety/tp-tp15145-1201.htm.
50. R. Tay and S. M. Rifaat, "Factors contributing to the severity of intersection crashes," Journal of Advanced Transportation, vol. 41, no. 3, pp. 244–265, 2007.
51. E. Hoareau, N. Candappa, and B. Corben, "Intersection Safety: Meeting Victoria's Intersection Challenge," Stage 1 - Task 1: Problem Definition, vol. 1, 2011, http://www.monash.edu/muarc/research/our-publications/muarc316a.
52. K. Keay and I. Simmonds, "Road accidents and rainfall in a large Australian city," Accident Analysis & Prevention, vol. 38, no. 3, pp. 445–454, 2006.

53. L. Symons and A. Perry, "Predicting road hazards caused by rain, freezing rain and wet surfaces and the role of weather radar," Meteorological Applications, vol. 4, no. 1, pp. 17–21, 1997.
54. A. M. Rizwan, L. Y. C. Dennis, and C. Liu, "A review on the generation, determination and mitigation of Urban Heat Island," Journal of Environmental Sciences, vol. 20, no. 1, pp. 120–128, 2008.
55. N. Earl, I. Simmonds, and N. Tapper, "Weekly cycles in peak time temperatures and urban heat island intensity," Environmental Research Letters, vol. 11, no. 7, Article ID 074003, p. 10, 2016.
56. S. K. Jusuf, N. H. Wong, E. Hagen, R. Anggoro, and Y. Hong, "The influence of land use on the urban heat island in Singapore," Habitat International, vol. 31, no. 2, pp. 232–242, 2007.
57. D. E. Parker, "Urban heat island effects on estimates of observed climate change," Wiley Interdisciplinary Reviews: Climate Change, vol. 1, no. 1, pp. 123–133, 2010.
58. M. Hosseinpour, A. S. Yahaya, and A. F. Sadullah, Exploring the effects of roadway characteristics on the frequency and severity of head-on crashes: Case studies from Malaysian Federal Roads. Accident Analysis Prevention, vol. 62, pp. 209–222, 2014.
59. G. H. Bham, B. S. Javvadi, and U. R. R. Manepalli, "Multinomial logistic regression model for single-vehicle and multivehicle collisions on urban u.s. highways in arkansas," Journal of Transportation Engineering, vol. 138, no. 6, pp. 786–797, 2012.
60. M. Rizzi, J. Strandroth, and C. Tingvall, "The effectiveness of antilock brake systems on motorcycles in reducing real-life crashes and injuries.," Traffic Injury Prevention, vol. 10, no. 5, pp. 479–487, 2009.

SECTION V
COMPLEX NETWORKS

CHAPTER 12

INFORMATION THEORETICAL METHODS FOR COMPLEX NETWORK STRUCTURE RECONSTRUCTION

Enrique Hernández-Lemus[1,2] and Jesús M Siqueiros-García[3]

[1]Computational Genomics Department, National Institute of Genomic Medicine, Mexico City, México

[2]Complexity in Systems Biology, Center for Complexity Sciences, National Autonomous University of México, Mexico City, Mexico

[3]Ethics, Legal and Social Studies Department, National Institute Genomic Medicine, Mexico City, México

CITATION (APA): Hernández-Lemus, E., & Siqueiros-García, J. M. (2013). Information theoretical methods for complex network structure reconstruction. Complex Adaptive Systems Modeling, 1(1), 8. (22 pages)

ABSTRACT

Purpose

Complex networks seem to be ubiquitous objects in contemporary research, both in the natural and social sciences. An important area of research regarding the applicability and modeling of graph- theoretical-oriented approaches to complex systems, is the probabilistic inference of such networks. There exist different methods and algorithms designed for this purpose, most of them are inspired in statistical mechanics and rely on information theoretical grounds. An important shortcoming for most of these methods, when it comes to disentangle the actual structure of complex networks, is that they fail to distinguish between direct and indirect interactions. Here, we suggest a method to discover and assess for such indirect interactions within the framework of information theory.

Methods

Information-theoretical measures (in particular, Mutual Information) are applied for the probabilistic inference of complex networks. Data Processing Inequality is used to find and assess for direct and indirect interactions impact in complex networks.

Results

We outline the mathematical basis of information-theoretical assessment of complex network structure and discuss some examples of application in the fields of biological systems and social networks.

Conclusions

Information theory provides to the field of complex networks analysis with effective means for structural assessment with a computational burden low enough to be useful in both, Biological and Social network analysis.

Keywords: Complex networks structure, Probabilistic network inference, Feature selection, Information theory, 94A15, 62B10, 91D30, 05C82

BACKGROUND

Complex networks, no doubt constitute one of the cornerstones of contemporary research in many branches of science (Barabási 2012; Newman 2003). From systems biology to the study of the role of friendship in the spread of diseases, connections between individuals at different organizational levels are outlined using complex graphs.

Alongside with the statistical analysis of large complex networks, the need for robust methodologies to infer such networks from empirical data has risen. Most of these methods rest on the domain of probabilistic inference and computational learning (Bickel and Doksum 2007; Hernández-Lemus and Rangel-Escareño 2011) and as such, they are subject to expectation errors and asymptotic constraints. Apart from the problem of inferring large networks from noisy data sources (Bansal et al. 2007; de Jong 2002; Hernández-Lemus et al. 2009), complex network researchers in general, are confronted with some subtler structural challenges in network reconstruction. One of such challenges lies in the capacity to assess direct from indirect interactions (Chua et al. 2008; Tresch et al. 2007).

The assessment of direct and indirect interactions may play an important role in understanding network navigability, community structure and information flow, specially when considering the range of influence of given nodes within a network, the strength of the interactions and how these interactions shape the whole correlation structure, as well as the topology and dynamics of the system. For instance, in gene regulatory networks, when it comes to the functional role of subnetworks forming either motifs or pathways, there is an important distinction between genes locally involved in the regulation of a small set of highly specific targets and some other genes that are involved in the transcriptional control of a large number of targets, often by means of a chain of indirect interactions. The first set of genes is responsible for the fine tuning processes involved in environment-specific genomic control, while the second set (known as master regulators) is related with long range, large scale control of genome expression used by the cell mechanisms of growth and proliferation (Baca-López et al. 2012).

In the case of social networks, there is also a growing interest in the role that indirect interactions may play in information and influence flow among nodes (Fowler and Christakis 20072010). In some instances (such as the social epidemiology of obesity) (Fowler and Christakis 2007) it has been shown that indirect connections (i.e. second degree links) within a social network may, under some conditions, exert a greater influence than direct interactions that however shape the global structure of the network. Such structural determination may be one of the keystones to discern between diverse features of influence in social networks such as homophily, social contagion and covariation (Shalizi and Thomas 2011). We may envisage other instances in the realm of complex networks where the distinction between direct and indirect links may be of some relevance, such as ecological or economic-financial networks, etc (Beltrán et al. 2012; Callaway and Howard 2007; Tsatskis 2012).

Methods for direct and indirect interactions assessment found in the reviewed literature were designed *ad hoc* for too dense or too small networks (Nawrath et al. 2010; Yan et al. 2007), and most of them require additional information in order to estimate or tune parameters to differentiate the two kinds of interactions (Chua et al. 2008; Yan et al. 2007). Noteworthy is to say, a great deal of importance was expressed in direct and indirect interactions assessment, regardless the particular field of research (Systems Biology, Economics or Ecology). Most efforts invested in distinguishing between these two kinds of interactions among nodes were either done manually (for instance, see (Beltrán et al. 2012; Callaway and Howard 2007)) or rely on extremely specific issues of the underlying networks (Baldazzi et al. 2010; Nawrath et al. 2010), hence the relevance of the method we introduce in this paper.

We suggest a general method to discover and assess for direct and indirect interactions within the framework of information theory. The method we submit in this paper allows to reconstruct the basic structure of complex networks. Since our method rests on the comparison of Mutual Information (MI) among nodes in a triangle, it is not affected by directionality between links, directionality is detected instead, once the basic structure of the network is already in place. In what follows we will discuss some methods based in probabilistic inference and information theory by means of which researchers may be able to infer and assess complex network structure from the probability distributions of some empirical quantitative features.

METHODS

Information Theoretical Approaches in Network Inference

Information theory (IT) offers a powerful theoretical foundation that is well-fit to contribute to the development of computational methodologies intended to deal with network inference problems as applied to real data in several branches of complex systems theory (Hernández-Lemus and Rangel-Escareño 2011). IT also provides an analogy with statistical mechanics (SM), that can be useful for inferring network interactions (links) from between-node correlation measures, thus enabling to use (although in a quite non-trivial manner) the huge arsenal of tools of this science. There are, however a number of open questions in the application of IT to the probabilistic complex network inference. The applied algorithms may be able to return *intelligible*models relying on scarce *a priori* information while dealing with a potentially large number of variables. IT methods may also detect non-linear dependencies in highly noisy non-independent probability distributions. The best benchmarking options for such kind of complex network inference, for us, seems to be the use of sequential search algorithms (instead of stochastic search, typically involving the assignment of structures for large constrained datasets, since these procedures have a high computational complexity, even NP-hard» -exponentially large search-space-) and performance measures based on IT, since this makes feature selection fast end efficient, and it also provides an easy way to communicate results.

Information theoretical measures have been applied intensively to infer interactions in complex networks, in particular in the field of computational biology (Bansal et al. 2007 de Jong 2002 Fleuret 2004 Hernández-Lemus et al. 2009 Margolin et al. 2006 Peng et al. 2005 van Someren et al. 2002) but also in social network studies (Crowley-Riddey 2009 Dong 2011 Mislove 2009 Mislove et al. 2010 Zhao et al. 2011). A group of correlation measures including mutual information, Markov random fields and Kullback-Liebler divergences, amongst others are considered appropriate to perform probabilistic network inference (Hernández-Lemus and Rangel-Escareño 2011). However, since conditional probabilities obey the so-called *tower property*, a number of *false positives* links may appear as a consequence of indirect correlations (Hernández-Lemus and Rangel-Escareño 2011).

For instance, if node (or agent) A has a high value of conditional correlation (say, mutual information) with node B, and B is also highly correlated with node C, most common algorithms would predict (with a marginal probability p^{ind}) the presence of a (possibly non-existent) link between processes A and C. In order to correct for the presence of indirect links we may implement some methods from IT, such as bounds in the information-theoretical probability measures and the use of the Data Processing Inequality (DPI) (Sehgal et al. 2007). DPI can provide a bound to the extent on which *signal processing* may optimize probabilistic inference.

We will discuss these and other ideas in the framework of network inference and structure assessment. We will also discuss some of their implications, and potential applications in the contemporary complex systems scenario.

Some of the essential notions of IT that will be used in this work are: (information-theoretical) entropy, mutual information and other related measures. To do so, let X and Y denote two discrete random variables having the following features:

- Finite alphabet X and Y respectively
- Joint probability mass distribution $p(X, Y)$
- Marginal probability mass distributions $p(X)$ and $p(Y)$
- Conditional probability mass distributions $p(X|Y)$ and $p(Y|X)$

Following Shannon (1949), it is possible to define the *information theoretical entropy* $H$of such distribution as follows

$$H = -K_s \sum_{\nu} p_{\nu}(X) \log p_{\nu}(X) \tag{1}$$

here H is called Shannon-Weaver's entropy, K_s is a constant, useful the determine the units in which entropy is measured (bits, nats, and so on, depending on the base of the log used) and $p_v(X)$ is the mass probability density for state v of the random variable given by $X=x$. IT entropy is a measure of the amount of uncertainty associated to the value of X, hence relating the *predictability* of an outcome to the probability distribution. Let us now consider two discrete random variables (Y, X) with a Joint Probability Distribution (JPD) $p(Y, X)$. For these random variables the *joint entropy*$H(Y, X)$ is:

$$H(Y,X) = -\sum_{y\in\mathcal{Y}}\sum_{x\in\mathcal{X}} p(y,x)\log p(y,x) \tag{2}$$

The maximal joint entropy corresponds to independence conditions of the random variables Y and X *i.e.* when the JPD is factorized $p(Y, X)=p(Y)p(X)$. The entropy of the JPD is then just the sum of their respective entropies. An inequality theorem could be stated as an upper bound for the joint entropy:

$$H(Y,X) \leq H(Y) + H(X) \tag{3}$$

Equality holds iff X and Y are statistically independent.

Also, given a Conditional Probability Distribution (CPD), the corresponding *conditional entropy* of Y given X is given by:

$$H(Y|X) = -\sum_{y \in \mathcal{Y}} \sum_{x \in \mathcal{X}} p(y,x) \log p(y|x) \tag{4}$$

Conditional entropies measure the uncertainty of a random variable once another one (the conditioner) is known. It can be proved (Cover and Thomas 1991) that:

$$H(Y,X) = H(X) + H(Y|X) \leq H(Y) + H(X) \tag{5}$$

Or:

$$H(Y|X) \leq H(Y) \tag{6}$$

Again, equality holds iff X and Y are statistically independent. Equation 6 is useful in the inference/prediction scenario as follows: if Y is a target variable and X is a predictor, adding variables can only decrease the uncertainty on target Y. As it will be shown later, this is essential for network inference when applying IT methods. *Entropy reduction by conditioning* can be accounted if we consider a measure called the *mutual information*, $I(Y,X)$ which is a symmetrical measure (i.e. $I(Y, X) = I(X, Y)$) that is written as:

$$I(Y,X) = H(Y) - H(Y|X) \quad \text{or} \quad I(X,Y) = H(X) - H(X|Y) \tag{7}$$

If we resort to Shannon's definition of entropy (equation 1) (Shannon and Weaver 1949) and substitute it into equation 7 we get:

$$H(Y,X) = -\sum_{y \in \mathcal{Y}} \sum_{x \in \mathcal{X}} p(x,y) \log \frac{p(x,y)}{p(x)p(y)} \tag{8}$$

A comprehensive catalogue of algorithms to calculate diverse information theoretical measures (including mutual information) has been developed for [R], the statistical scientific computing environment (INFOTHEO 2012). We will analyze the special role that MI has in the field of complex networks inference from quantitative feature data. MI has been applied successfully as a measure to infer 2-way interactions in complex networks (quite specially in the field of Gene Regulatory Networks or GRNs) (Andrecut and Kauffman 2006a; Andrecut and Kauffman 2006b; Madni and Andrecut 2007; Margolin et al. 2006). As we have seen, MI quantifies the degree of statistical dependency between two random variables (say α and β). One can see that $MI(\alpha, \beta) = 0$ iff α and β are statistically independent.

Hence, if we measure some quantitative-feature of interest ϑ (say expression level of genes in GRNs), by studying its profile (and more specifically the mutual correlation profile for a set of nodes) we may find interactions conforming a network. A pair of agents characterized by feature distributions ϑ_i and ϑ_j for which $MI(\vartheta_i, \vartheta_j) \neq 0$ are said to interact with each other. Since MI is *reparametrization invariant*, one usually calculates the normalized mutual information. In this case $MI(\vartheta_i, \vartheta_j) \in [0, 1], \forall i, j$.

Distinguishing between Direct and Indirect Interactions

With these definitions in mind, let us consider two random variables, X and Y, whose mutual information is $MI(X, Y)$. Now consider a third random variable, Z, that is a (probabilistic) function of Y only. It can be shown that $P_{Z|X\,Y} = P_{Z|Y}$, which in turn implies that $P_{X|Y\,Z} = P_{X|Y}$, as follows from Bayes' theorem.

An information-theoretical theorem called the Data Processing Inequality (DPI) states that Z cannot have more information about X than Y has about X; that is $MI(X;Z) \leq MI(X;Y)$. We can see that $MI(X;Z) = H(X) - H(X|Z) \leq H(X) - H(X|Y, \quad Z) = H(X) - H(X|Y) = MI(X;Y)$. Inequality follows because conditioning on an extra variable (in this case Y as well as Z) can only *decrease* entropy (in a similar way to what occurs in statistical physics when adding constraints to a thermal system), A formal definition of such a theorem would be:

Definition 1

Three random variables X, Y and Z are said to form a **Markov chain** (in that order) denoted $X \rightarrow Y \rightarrow Z$, if the conditional distribution of Z depends only on Y and is independent of X. i.e. if we know Y, knowing X doesn't add anything new to what we already know about Z than if we know only Y.

If X, Y and Z form a Markov chain, then the Joint Probability Distribution can be written as follows:

$$P(X, Y, Z) = P(X)P(Y|X)P(Z|Y) \tag{9}$$

Theorem 1

Data Processing Inequality: If X, Y and Z form a Markov chain, then

$$MI(X; Z) \leq MI(X; Y) \tag{10}$$

Proof

By the chain rule for mutual information we can state that:

MI(X;Y,Z)=MI(X;Z)+MI(X;Y|Z)MI(X;Y)+MI(X;Z|Y)

By the Markov property, since X and Z are independent, given Y, $MI(X;Z|Y)=0$, then, since $MI(X;Y,Z)\geq 0$ we have: $MI(X;Z)\leq MI(X;Y)$ c.q.d.

In reference (Margolin et al. 2006), the application of DPI has shown that if nodes ϑ_1 and ϑ_3 interact only through a third node, ϑ_2 within a given network, $MI(\vartheta_1, \vartheta_3)\leq \min[MI(\vartheta_1, \vartheta_2);MI(\vartheta_2, \vartheta_3)]$. Hence, the least of the three MIs values may come from indirect interactions. The proposed algorithm examines each triplet vertex for which all three MIs are measured and compared to some threshold value MI_0. If there is an edge with an MI value below the threshold, then it is removed (see Figure 1). DPI is thus useful to quantify efficiently the dependencies among a large number of nodes. The DPI algorithm is useful in the problem of complex network structure assessment as well, since it eliminates those statistical dependencies that might be of an indirect nature.

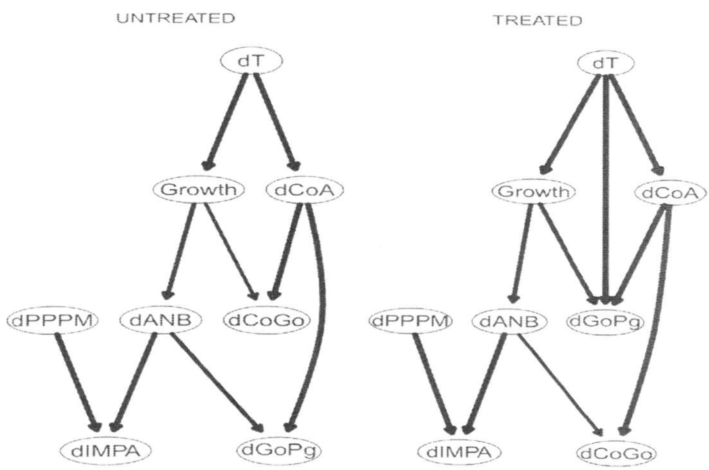

Figure 1: Absolute application of the DPI inequality to assess the direct interaction structure of a three node sub-network. Panel A depicts a three node subnetwork composed by nodes α, β and γ. Interactions were inferred by means of mutual information calculations for a selected quantitative feature. The interaction strength between the nodes i, j is given by the respective MI_{ij} ($i, j=\alpha, \beta, \gamma$) values. Panels B, C, and D result from pruning the subnetwork from the link with the lowest MI_{ij}: $MI_{\alpha,\beta}$ (B), $MI_{\beta,\gamma}$ (C), and $MI_{\alpha,\gamma}$ (D).

In some cases, however, it may happen that the Markov chain structure is not absolutely fulfilled. Say when nodes ϑ_1 and ϑ_3 interact not only through a third node, ϑ_2, but also by means of a direct interaction. Hence ϑ_1 and ϑ_3 may be two-fold connected, in this case pruning-out one of the links may render an inaccurate version of the actual interaction pattern. This scenario can be accounted for by means of establishing a threshold for removing a link, i.e. the link with the lesser *MI* measure would be removed only if its value is below a certain threshold (that we hereon call DPI_{tol}) -to be determined in a case-by-case basis by close examination of the network and also by considering its intended applications- and it stays in the network otherwise (Figure 2).

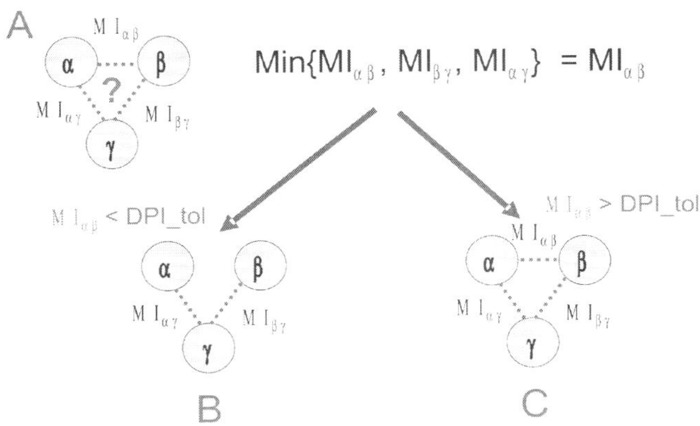

Figure 2: Relative application of the DPI inequality to assess the direct interaction structure of a three node sub-network. Panel A depicts a three node subnetwork composed by nodes α, β and γ. Panels B, and C result from pruning (or not) the subnetwork from the link with the lowest MI_{ij}, in this case $MI_{\alpha,\beta}$. The α- β link is removed only when its respective *MI*-value is lower than a pre-determined threshold *DPI_tol* and left in the network otherwise.

There are similar approaches to the one just presented, for instance the ones in reference (Liang and Wang 2008) and in reference (Zhang et al. 2012). Both approaches are based in *conditional mutual information* (i.e. the degree of information a variable *X* and a variables *Y* share, given a third variable (or group of variables) *Z*). These algorithms try to account for indirect links by means of conditioning the associated mutual information distributions. In reference (Zhang et al. 2012), the authors further improve the performance of the algorithm by using an adaptive computation framework. In addition to these good performance general methods (references (Liang and Wang 2008; Zhang et al. 2012) were developed for gene regulatory networks although with minimal adjustments can ba applied to any other probabilistically inferred networks), there are also more specific approaches based on somehow *Ad Hoc* considerations. We can mention, for instance the MARINa algorithm (Lefebvre et al. 2010; Lefebvre et al. 2007; Mani et al. 2008) developed specifically for the assessment and reconstruction of gene regulatory networks based on *statistical enrichment* of certain signatures (Subramanian et al. 1554), an approach close in philosophy of that of conditioning variables that, however requires for additional information (i.e. the signatures themselves) to be useful, hence is more restricted to its scope and applications as are approaches relying on additional phenotypic information (Wu et al. 2009; Yu et al. 2006).

Data and Algorithms

In order to introduce the importance of telling direct network interactions from indirect ones, we performed a topological analysis on the Gene Network of the fruit fly, *Drosophila melanogaster* (D.m.). The fruit fly Gene Network is a paradigmatic system for genetic studies and one of the best annotated organisms in genomics databases. It also presents a high genomic similarity to that of mammals -humans included- (about 61% of disease-associated genes in humans have a D.m. counterpart) and there is open access to its high-throughput inferred biological network (Costello et al. 2009). By discussing some features of the network structure of this highly studied species we introduce the problem of finding direct and indirect interactions in complex networks inferred from experimental data. Once this problem has been outlined, we proceed to illustrate how the methods of information theory may be appropriate to distinguish between direct and indirect interactions in order to sketch (at least partially), the network structure on a gene regulatory network inferred from experimental data obtained from 1191 whole genome gene expression experiments in breast tissue from breast cancer patients/controls and on a social network inferred from researchers at Mexico's National Institute of Genomic Medicine coauthorship collaborations data, retrieved from the *PubMed* database.

As explained before, the methods of information theory used here correspond to the implementation of MI calculations and DPI to infer and prune respectively such networks. There is a number of different methods for computing this quantities in the literature (Hernández-Lemus and Rangel-Escareño 2011) and most of them are quite functional and almost equivalent in performance. Here we used the C++ implementation of the aracne algorithm (in particular we resort to aracne 1.0 even if there is a new version 2.0 with an improved algorithmic complexity performance, because version 2.0uses a bootstrapping method that we have found to be still a little bit unstable)

(Margolin et al. 2006) for Biological Networks and Python scripts (some customized and others from the NetworkX library) for the Social Networks. The aracne 1.0algorithm is useful for our purposes since it is based on crystal clear MI calculations (Hernández-Lemus and Rangel-Escareño 2011), it is possible to implement DPI thresholds and its algorithmic complexity and performance are quite good (we have benchmarked aracne 1.0 against other information-theoretical methodologies such as Information Based-Similarity (ibs) and linear correlation predictors in the past (Hernández-Lemus et al. 2009) with very acceptable results). Cytoscape and Python's library NetworkX were used to depict and analyze the networks (Assenov et al. 2008).

Microarray pre-processing of the data was performed by using the affy library in BioConductor running under [R] on a 128 Gb RAM 8-Power5+ dual core-processor, symmetric multiprocessing (SMP) unit by IBM. All statistical tests were performed on a Dell Precision Series 16 Gb RAM QuadCore Workstation by using limma package in [R]/BioConductor. Information theoretical measure calculations for biological systems were performed by the aracne v 1.0 program in the IBM SMP machine. Python scripts were used instead for Social network calculations. Graphical depiction and network analyses were performed on a MacBook Pro 8 Gb i7.

The Drosophila melanogaster GRN used to highlight the presence of hierarchical structure was not further used here to demonstrate IT methods of network assessment. The reason for this is that it was inferred (Costello et al. 2009) by using Pearson correlation metric, which is a *linear* measure, thus unable to capture the whole statistical dependency spectrum. Let us recall that for two statistically independent random variables Pearson correlation coefficient is 0. However, the converse is not always true, because Pearson correlation coefficient detects only linear dependencies between two variables. Null Pearson correlation coefficients only implies statistical independence for the special case of *jointly normal* distributions. Since this is not the general case in gene expression distributions, values of linear correlations are not enough to determine statistical dependency (Hernández-Lemus et al. 2009).

RESULTS AND DISCUSSION

Network Assessment in Biological Networks

In order to introduce the need for complex network assessment of indirect interactions, let us consider the case of the gene regulatory network of the fruit fly (*Drosophila melanogaster*, D.m.), which is one of the best curated biological networks that has been constructed ever. By means of computational integration in a probabilistic model of high throughput data sets from DNA, RNA and protein assays, combined with data mining for individual genes to phenotypes treats, such network has been assembled with the goal of providing a *...meaningful, functional gene network and to draw new and unforeseen connections...*(Costello et al. 2009). D.m. is of course a model organism for too many instances in genetics and genomics. One particular feature that is long known for the genetic structure of D.m., is the fact that gene regulation is largely determined by the action of a relatively small set of molecules that are able to exert transcriptional control in the highly active stages of development and proliferation (Baker 2001; Brennecke et al. 2007; Harrison et al. 2011; Wells 2009). For instance, aspects of regeneration are often regulated by complex mechanisms of activation for several growth regulatory pathways in damaged tissue. It has been proved that every pathway involved is being regulated by the *p53* molecule in D.m., (dP53). dP53 is thus a critical master regulator in the GRN of D.m., in particular with regards to cell growth, proliferation and differentiation. Interestingly enough, the human homolog of this very molecule is known to play a quite important role in most human cancers. dP53 is by no means the only master regulator in drosophila; an extremely important gene that acts as a universal master regulator in D.m. is the molecule called eyeless (*ey*). This gene was first study in relation to eye development (hence the name, larvae with knocked-up eye does not develop eyes). However, the role of *ey* is not restricted to eye development. Abnormal expression of the gene is able to convert other tissues into eye-cells, including legs, wings and antennae. *ey* is also involved in controlling the processes that coordinate differentiation of several cell types in a very precise way in order to develop the fly eye. The *ey*-homolog *Pax6* and homologs of other eye determination genes from D.m., are also required for tissue development in vertebrates. It has been shown that *ey* becomes a master regulator of development deep until later stages by means of *regulation by signaling* through the Notch and

EGFR signaling pathways (two outstanding important generalistic signaling pathways) (Baker 2001).

Other master regulator genes are known to play special roles in D.m. Such is the case of the zinc-finger protein *Zelda* (*ZLD*) that plays a key role in transcriptional activation and as a master regulator of genome activation in the earliest stages of D.m. development (Harrison et al. 2011). The ZLD transcription factor bounds to thousands of sites across the genome at all developmental stages of D.m., with relatively small changes in binding between stages. The number and range of ZLD targets - around 2,000 genes have ZLD bound to their promoters and/or enhancers - demonstrate that it plays a major role in maternal-to-zygotic transition activation. Hence Zelda *should* be an important hub in the transcriptional regulation network for D.m. However, probably the most important group of master regulators in the fruit fly GRN is formed by the so-called *piwi* family of transcription factors. The piwi class of genes was identified long ago as encoding regulatory proteins that are responsible for the maintenance of incomplete differentiation in stem cells and also in preserving the stability of cell division rates in germ line cells, a quite important set of processes related again to all stages of development, growth and differentiation (Brennecke et al. 2007). There are plenty of other examples (*Drosophila melanogaster* is known to have some 50 or more master regulators in its GRN (Costello et al. 2009) but we think that these may be sufficient to establish the point that within the GRN for D.m. the presence of such highly connected nodes (hubs) should be evident.

If we refer to Figure 3, in panel A we can see a rendering of the whole D.m., genome GRN (Costello et al. 2009). A clustering algorithm (Bader and Hogue 2003 was used to color-code genes (white-to-red) according to their degree of clustering (red genes are strongly clustered whereas white ones are largely isolated). The relative importance of master regulator genes -that may be measured, for instance through their associated centrality degree- is not evident from the network structure. These genes are also important for biological reasons, since they determine to a large extent the regulation patterns of the entire network. In panel B we present the same network but we color-coded individual nodes according to their degree distribution. Some important *clusters* seem to appear, however it is difficult to asses the relevance of individual nodes. In panels C and D we can see zoomed-in renderings of the rectangular regions in panels A and B respectively. As in panel A, panel C is inadequate to highlight the relative importance in the very same aspect as in panel A of individual nodes. In panel D, highly connected genes are highlighted (big red nodes), however the complex entanglement of direct and indirect interactions makes impossible to detect the actual relative importance of these, since they are surrounded by many medium to medium-high connected nodes. This situation is precisely the one calling for a method to assess for direct and indirect interactions. In what follows we will show a proposal for such method, based in the tenets of information theory as applied to both a biological network (a GRN for primary breast cancer) and (in the next subsection) a social network (a scientific collaboration network based in co-authorship probabilities).

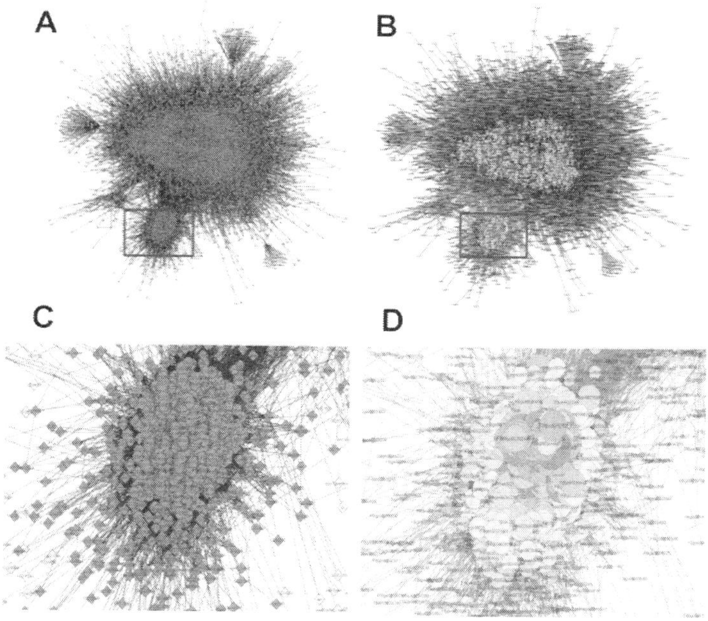

Figure 3: Largest island in the gene regulatory network of drosophila melanogaster. Gene Network constructed from top 200,000 gene pairs (Costello et al. 2009). Panel A: Nodes are color coded (white-to-red) according with their MCode scores (Bader and Hogue 2003) for clustering. Red nodes are genes tightly clustered. It can be seen that there is strong clustering between nearly all genes. However, no definite structural features are evident. Panel B: Same gene network as in panel A but with node-coloring and sizing determined dy connectivity degree: Big red nodes are highly connected genes whereas small green ones are lowly connected genes. Panels C and D present zoomed-in renderings of the boxes in panels A and B respectively. As in panel A, panel C does not permit to observe network structure. Panel D, in the other hand, shows the presence

of some hyper-connected genes (or hubs, here in red) that are densely linked to highly connected nodes (dark orange) passing through less connected ones (light orange) until reaching barely connected genes (yellow and finally green) suggesting some kind of hierarchical structure in the network *wiring*.

Gene Regulatory Network for Primary Breast Cancer

To show the application of IT methods to infer and assess complex biological networks, let us consider the GRN for primary breast cancer as inferred by means of MI-calculations (Margolin et al. 2006) (see Methods) in the whole genome gene expression levels (for differentially expressed genes between biopsy-captured primary breast cancer and healthy breast tissues as controls) (Baca-López et al. 2012). In Figure 4 we present the aforementioned network, where a complex entangled structure reflecting the intricate regulatory relationships driving the cancer phenotype is displayed. As in the D.m. case, in panel B we show a color- and size-coded (see Figure 4 caption) rendering based in individual degree values for every node. The relative importance of a number of genes becomes more evident, but still is not clear. It is now established that there are some (few) genes in such network acting as master regulators (Baca-López et al. 2012). This is still not evident from the topology/visualization in panel B. If we analyze the network topological structure we can see some reasons behind this. In panel C we plot the degree distribution that show an almost homogeneous behavior for two-plus orders of magnitude in the degree: from genes with a few interactions to nodes with more than a hundred connections. This scenario may be consistent with a highly structured hierarchic structure instead of the dominion of a few master regulator genes. The average clustering coefficient distribution in panel D follows a power-law-like behavior, indicating the relative importance of second neighbor and higher order interactions in the structure of the network as opposed to a neighbor dominated by a few hubs. Panel E displays the Shortest-Path length distribution that turned-out to be a short-tailed bimodal (with maxima at distances 2 and 4), indicating high network navigability.

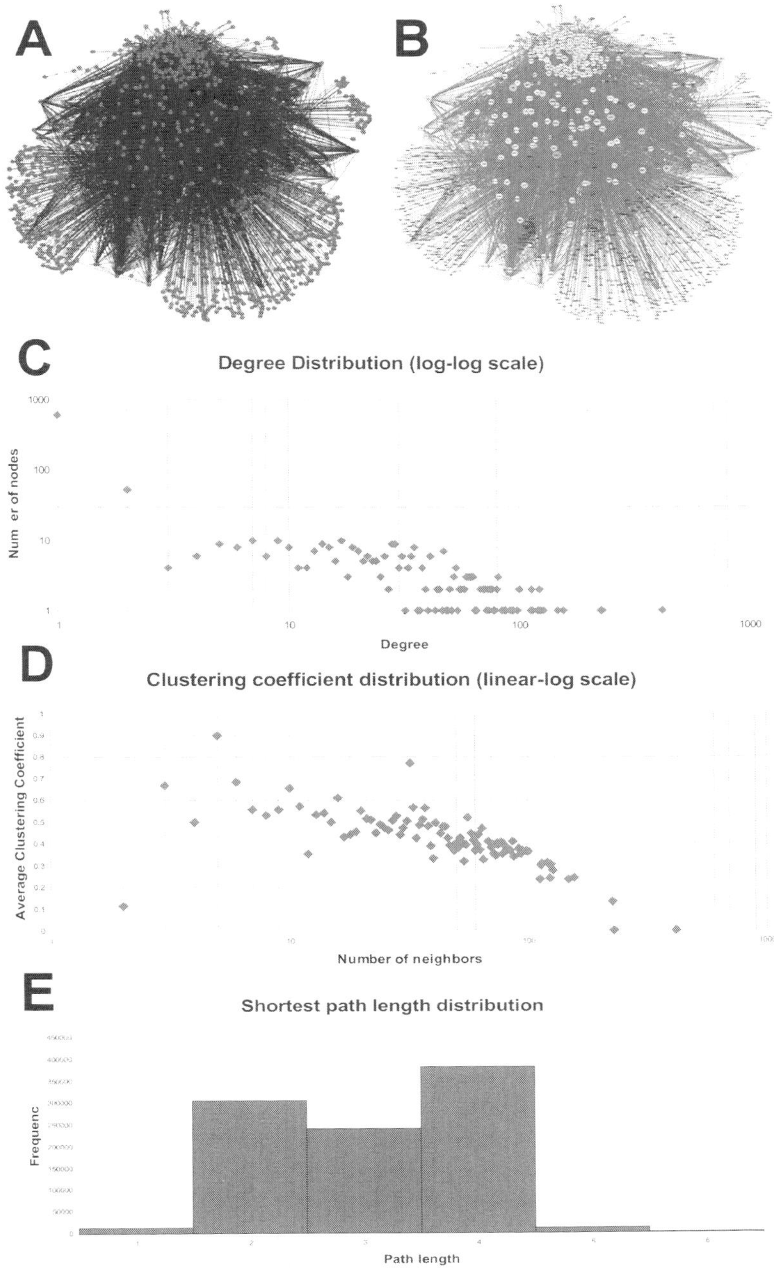

Figure 4: Gene regulatory network associated with proliferation in primary breast cancer. Gene Regulatory Network (GRN) inferred from differential gene

expression profiling in 1191 whole genome expression experiments for biopsy samples from breast cancer patients/controls (Baca-López et al. 2012). Panel A depicts the associated GRN. The relative importance of highly connected genes is not evident. Panel B depicts the same network, nodes are size-coded and color-coded according with their connectivity degree (big red nodes correspond with highly connected genes, whereas small green nodes are lowly connected genes). Some genes apparently stand-out as *relevant*, however the intricate network structure does not permit to tell indirect connections from direct ones, also a number of medium-level connected nodes add complexity to the analysis. In panel C we can see the connectivity degree distribution: no definite trend is evident -e.g. a power law, a stretched exponential, etc.- but the distribution remains somehow homogeneous for the range between a few connections and more than a hundred connections. In panel D we can see how the average clustering coefficient almost follow a power-law ($R^2=0.85$ for the power-law fit) indicating that there may be a hierarchy related with how nodes associate. This could be an indication that a number of not-so-strong interactions are present. Panel E presents the short path length distribution that in this case is a bimodal (with maxima at distances of 2 and 4 links) with a short tail.

If we now consider the MI-inference of a GRN network with the same gene expression profiling data as the one in Figure 4, but constrained by the absolute (i.e. $DPI_{tol}=0$) application of the Data Processing Inequality (as in Figure 1), the network depicted in Figure 5 is generated. Panels A-E are the same as in Figure 4. Panel A highlights the relative importance genes as master regulators. This is even more evident in panel B where the role of *MEF2C* (big red node) and *MNDA* (big dark-orange node) (known master regulators (Baca-López et al. 2012) becomes evident, with important but bounded inter-relationships with other genes. Panel C displays the degree distribution, a monotonic-decaying power-law like function highlighting the importance of a few highly connected nodes, whereas the vast majority of genes present low connectivity. Panel D shows the average clustering coefficient distribution that has obviously dropped to zero since all indirect interactions were pruned-out of the network. Interestingly enough, the Shortest-Path length distribution grossly remains a bimodal (with the same maxima at distances 2 and 4) that however exhibits a long tail, a clear indicator of diminished network navigability that results from eliminating shortcuts given by indirect interactions.

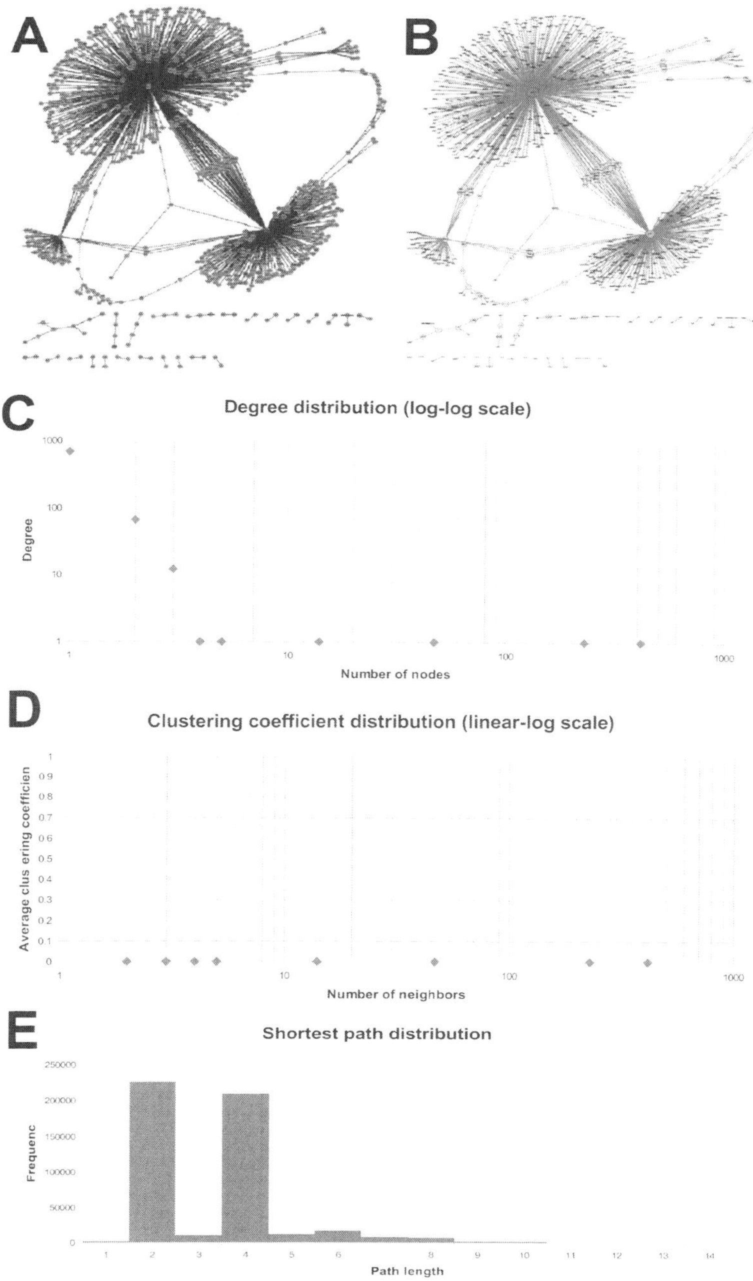

Figure 5: Gene regulatory network associated with proliferation in primary breast cancer after DPI pruning of indirect interactions. Panel A depicts the

Breast Cancer associated GRN (Baca-López et al. 2012) (same as in Figure 4)after eliminating all indirect interactions by means of the application of the Data Processing Inequality (DPI). The relative importance of highly connected genes is now more evident. The network is basically founded on the role of two major hubs (that are also connected by means of intermediate nodes) and a couple of medium-high connected nodes. Panel B depicts the same network, nodes are size-coded and color-coded according with their connectivity degree. Two major hubs appear corresponding to *MEF2C* (red node) and *MNDA* (dark orange node). These two genes have been recognized as *transcriptional master regulators* in breast cancer (Baca-López et al. 2012). In panel C we can see the connectivity degree distribution that now resembles a power law distribution (too few nodes, however to have a reliable statistic for the fit) indicating a relatively high importance of few nodes and a low importance for most nodes. In panel D we can see how the average clustering coefficient has drop to zero. This is a clear effect of pruning for indirect interactions, since the relative importance of a node is now given in terms of its direct connections and not because of neighbor-by-neighbor influence. Panel E presents the short path length distribution that again is a bimodal (with maxima at the same distances of 2 and 4 links) that however shows a somehow larger tail than in Figure 4. This may be due to the fact that navigability in the network was easier in the presence of indirect interactions.

Network Assessment in Social Networks

Scientific Collaboration Networks

In order to test the value of IT methods for community inference in the context of social networks, we describe a Scientific Collaboration Network (SCN from now on) based on the coauthorship history of the researchers of the National Institute of Genomic Medicine of Mexico (referred as INMEGEN). INMEGEN is one of the National Institutes of Health in Mexico and it was created in 2005, being so the second youngest institute of all 13.

The study of SCN dates back to the Erdös Number Project (Cardillo et al. 2006) and due to the accessability of data through the Internet, it has become lately a common place for those interested in social networks (Newman 2004). In this context, a very intuitive way of understanding scientific collaboration is by means of coauthorship, that is, when two scientists have worked together in one or more publications. In order to build the network, we used data of the publications reported by INMEGEN (as retrieved from *PubMed*) from 2005 to the beginning of 2012. The network includes collaborations among researchers of INMEGEN with themselves and with scientists from other institutions, as well as those collaborations between the latter as long as they have also coauthored publications with researchers from INMEGEN within the same network. The edges of the network are weighted according with their corresponding co-authorship probability (the value of every link depends on the strength of collaboration between two scientists -a MI-like function inferred from the number of collaborations with a maximum value of 11 and a minimum of 1-). This network is called *Network 1* (Figure 6). From *Network 1* we obtained a subnetwork (*Network 2* in Figure 7). *Network 2* is also a collaboration network of the researchers of INMEGEN but in which collaborators external to INMEGEN that otherwise are not connected with each other have been left out. Finally, by applying IT methods for network assessment to *Network 2*, and eliminating the weakest link between the nodes of a triangle, we obtained one last network, (*Network 3* shown in Figure 8). From *Network 3* we were able to identify four research groups and we built the collaboration network for each one of them. Groups were defined by the fact that all their members have coauthored at least two papers.

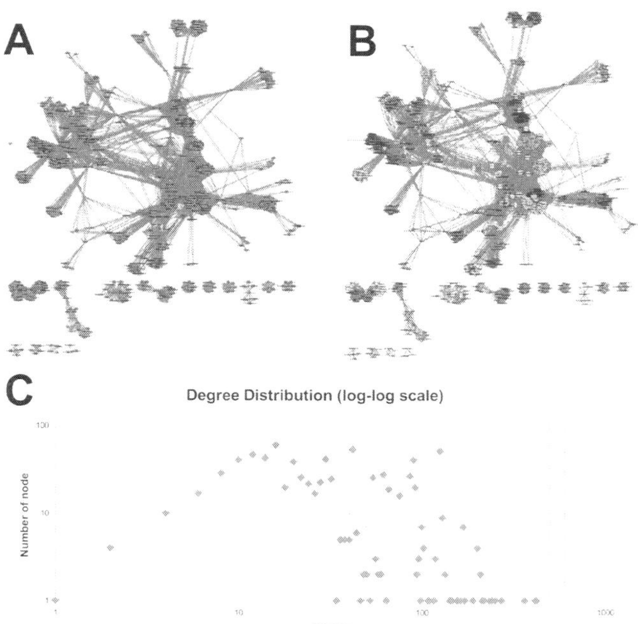

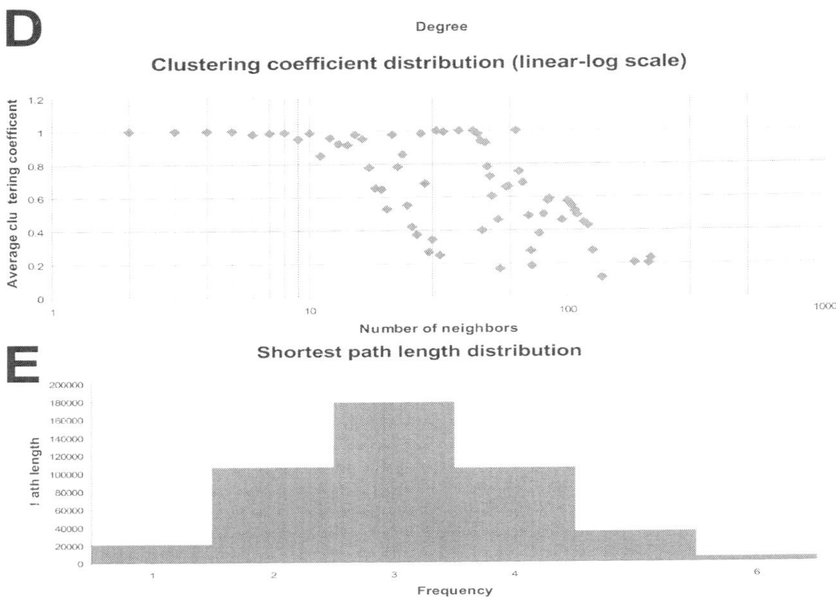

Figure 6: Scientific collaboration network of researchers at the National Institute of Genomic Medicine (INMEGEN) and their extended partners - network 1. Panel A presents a SCN that includes collaboration links among researchers of INMEGEN with each other and also with scientists from other institutions, also collaborations between the latter as long as they have also coauthored publications with researchers from INMEGEN within the same network. Panel B depicts the same network, nodes are size-coded and color-coded according with their connectivity degree. The presence of a couple of well-connected individuals (bigger red nodes) as well as a number of medium-high connected ones (orange mid-sized nodes) points out to the existence of some kind of hierarchic structure. In panel C we can see the connectivity degree distribution that shows a somehow anomalous behavior in the very low degree region, and then displays a typical power-law behavior. This anomaly (a very low number of barely connected nodes) may be due to incidental collaboration. In panel D we can see the average clustering coefficient that also presents a *left-hand* tail, most likely also due to incidental collaboration and after this a power-law like behavior. Panel E presents the short path length distribution, which is a quasi-symmetric unimodal with an average length of three steps.

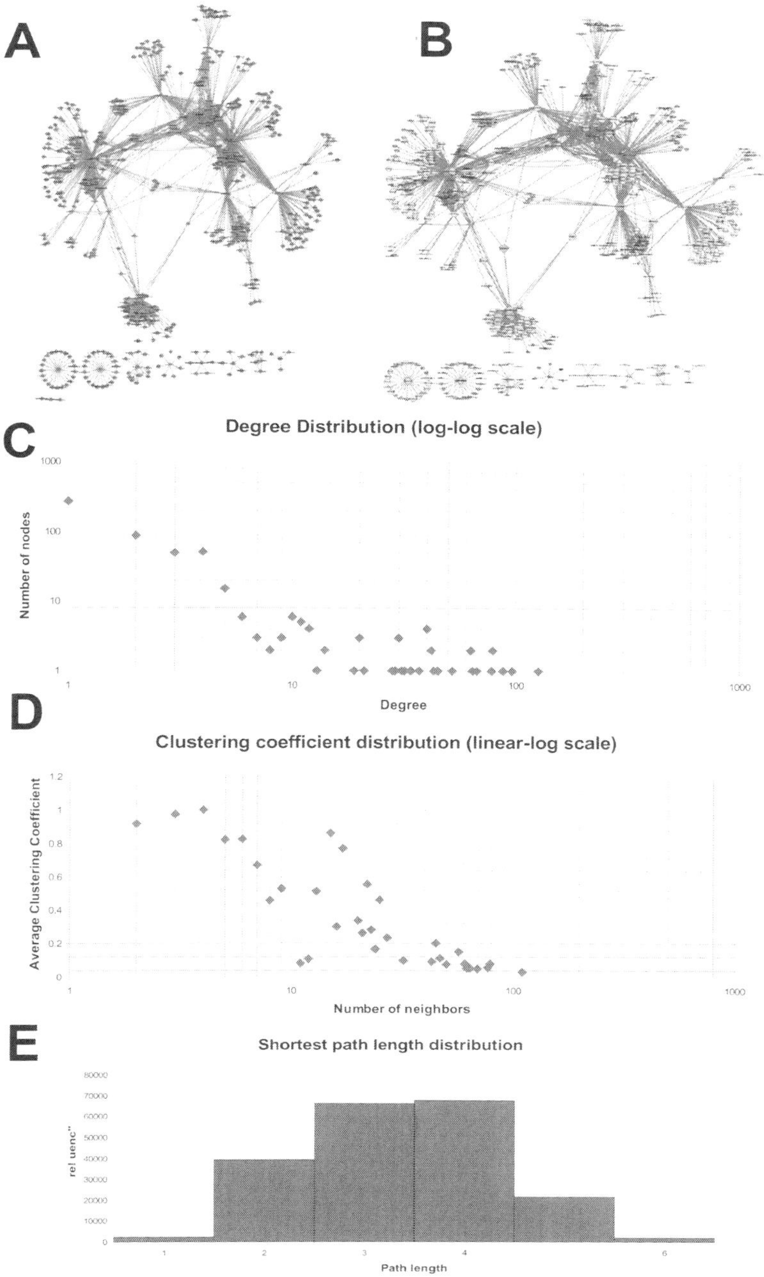

Figure 7: Scientific collaboration network of researchers at the National Institute of Genomic Medicine (INMEGEN) and their extended partners - network

1. Panel A presents a SCN that includes collaboration of INMEGEN researchers but in which collaborators external to INMEGEN -otherwise not connected with each other- have been left out. We can see that this network presents a topology resembling that of Network 1 but decimated in the number of links. Panel B depicts the same network, nodes are size-coded and color-coded according with their connectivity degree. The presence of a couple of well-connected individuals (bigger red nodes) as well as a number of medium-high connected ones (orange mid-sized nodes) points out to the existence of some kind of hierarchic structure. In this Network, the presence of localized hubs (that we may later identify as group leaders) is more evident. In panel C we can see the connectivity degree distribution that shows a power-law behavior with no further appearance of the incidental collaboration anomaly. In panel D we can see the average clustering coefficient displaying again a power-law like behavior. Panel E presents the short path length distribution, which is also a quasi-symmetric unimodal but with a less-defined expected value (between three and four steps) for the separation length.

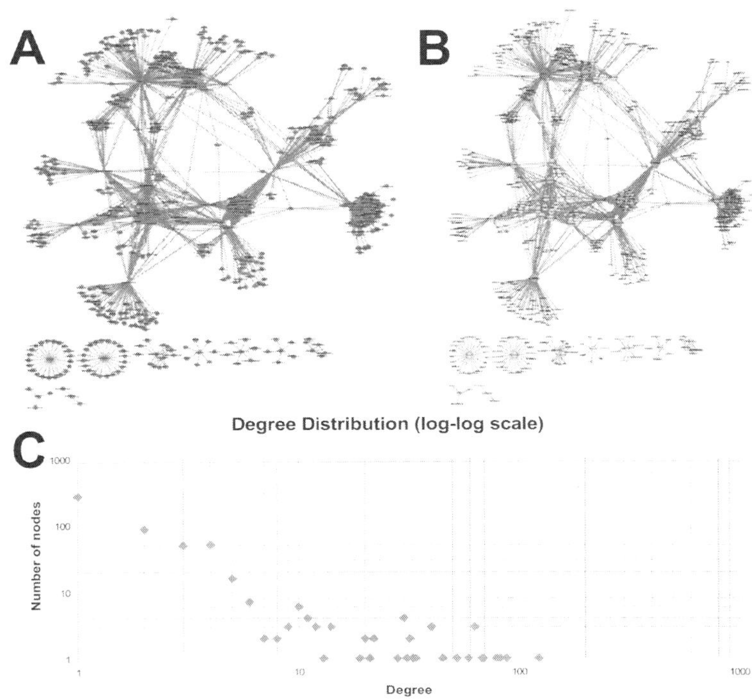

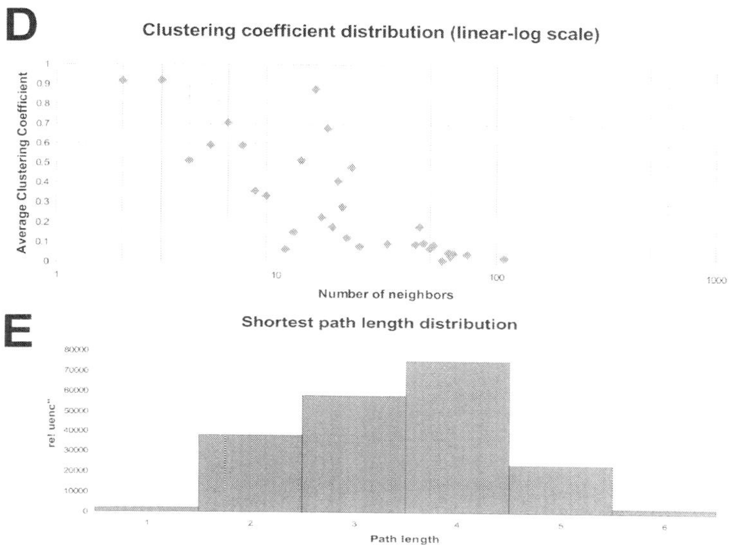

Figure 8: Scientific collaboration network of researchers at INMEGEN after DPI pruning of indirect interactions - network 3. Panel A presents a SCN similar to Network 2 but indirect interactions have been pruned-out, (same as in Figure 7) after eliminating all indirect interactions by means of the application of the DPI. Network 3 is similar to Network 2, i.e. the effect of eliminating *indirect* interactions has a small impact. In many triadic interactions (triangles in the network) two-out of three interactions presented the same value so that no edge was eliminated. Panel B depicts the same network, nodes are size and color-coded according to their connectivity degree. The presence of a couple of well-connected individuals (big red nodes) as well as a number of medium-high connected ones (orange mid-sized nodes) points out to some kind of hierarchic structure. Research groups are easily identified, yet a certain degree of collaboration between most large groups is found. In panel C the connectivity degree distribution shows a power-law like behavior but with a smaller high-degree tail than in Network 2. In panel D the average clustering coefficient displays again a power-law like behavior. The reason for finite (non-zero) clustering coefficients is that a number of triangles remained in the network for two of its edges presented equal strength of interaction. Panel E presents the short path length distribution, which is unimodal but with a sharp expected separation length of 4. It is noticeable that eventhough DPI-pruning did not eliminated all the triadic connections in the network, the effect of network as-

sessment to knock-out clear indirect interactions affected network navigability (average path length went from a clear value of three in Network 1, to between three and four in Network 2 and then to a clear value of three in Network 3).

We paid special attention to general values such as number of nodes *N*, clustering coefficient *C*, centrality degree, network centralization, and characteristic pathways $\langle l \rangle$. *Network 1* has $N=847$, a very high clustering coefficient $\langle C \rangle =0.925$ (see Figure 6C) and a network centralization of 0.221. The characteristic path length $\langle l \rangle$ is 3.09 (Figure 6E), and the average number of neighbors is 24.8. *Network 2* has $N=535$, a $\langle C \rangle =0.411$ (Figure 7C), and a network centralization of 0.199. The characteristic path length $\langle l \rangle$ is 3.36 (Figure 7E), and the average number of neighbors is 4.3. *Network 3*displayed the following values, $N=535$, $\langle C \rangle =0.341$ (Figure 8C), and a network centralization of 0.197. The characteristic path length $\langle l \rangle$ is 3.43 (Figure 8E), and the average number of neighbors is 4.1.

Due to the shortness of the characteristic path length and its clustering coefficient, the structure of *Network 1* is close to that of a *small-world* network, which is not at all strange to collaboration networks and to SCN, as has been noted before (Yousefi-Nooraie et al. 2008). When compared to *Network 2*, its clustering coefficient decreases abruptly compared to its value in *Network 1*, and the characteristic path length increased slightly. Given the difference between the clustering coefficients of these two networks, it lead us to suggest that external collaboration plays a great deal in defining the small-world structure of INMEGEN's collaboration network, and most of all, it gives it a closure that makes it easily navigable. By closure we mean that INMEGEN, being such a young institution, it may still depend, to a degree, on external collaborators. Such dependency might be fueled by the fact that INMEGEN doesn't provide clinical services which means that it doesn't have the possibility to systematically recruit research subjects from where to get biological samples, depending so, on other institutions for this purpose. This idea still remains to be tested and is part of our future work.

The differences between *Network 2* and *Network 3*, are minimal. *Network 3* was inferred by IT Methods for network assessment, even though, its clustering coefficient remained almost unaffected -decreased from $C=0.411$ in *Network 2*, to $C=0.341$ in *Network 3*- and the characteristic path length was also basically the same. This is an important result because it means that most of the clustering was due to external collaboration as it was noted already, and for *Networks 2* and *3*, triangle formation depends on having at least two nodes from INMEGEN.

IT Methods applied to *Network 2*, eliminated what was left from the 'unbalanced' triangles and the weakest link between two nodes in a triangle was deleted, generating *Network 3*. This method made possible to identify more clearly the existence of four communities that we could recognize as INMEGEN's main research groups, with a leader (laboratory or researcher leader) easily detectable. It is important to mention that there were other, rather small groups, that have not been articulated to the main network and with a small weight value in their collaborations links. These are hypothesized to be groups in process of consolidation.

In order to analyze the structure of the four groups, we created a subnetwork for each one, based on researchers that have collaborated in at least two publications. The subnetworks made also possible to recognize different collaborative strategies among them. Group 1 and 2, with an $N=79$ and $N=139$ respectively, were very similar, both displayed a high value for network centralization, over 0.750 and low clustering coefficient, having Group 2 the highest $\langle C \rangle = 0.356$. For these groups, there was a leader with a very high k, and one or two very close collaborators. The difference between first and second order collaborators was large, with the highest weight (corresponding approximately to $k=9$) between the leader and its first order collaborator and a lesser weight (implying $k=2$) with the lowest, with an average connectivity difference of 7. The structure of these two networks could be interpreted as having a tendency towards a star-shaped structure.

In group 3, with N=106, network centralization was not as high as in groups 1 and 2, although it remained important (Network centralization= 0.681). Noteworthy was the clustering coefficient over $\langle C \rangle = 0.500$. According to these numbers, group 3 behaves more like a group and the weight of collaborations is similar among its members.

Group 4 had the highest network centralization 0.938 as well as the highest clustering coefficient $\langle C \rangle = 0.746$. The network of group 4 depends on two members of INMEGEN on which all collaborations are fixed. The situation for group 4 is that from all, this is the extreme case of group behavior since there is a minimal connectivity degree difference of 1 between first and second order collaborators, that is, for most nodes, the weight of collaboration among each other is the same (weight 2) and only few have a weight of 1, and none of weight 3 or higher. It is important to mention that this group is different from the other three because is the newest and the smallest ($N=64$).

We created these networks -i.e. *Networks 1, 2, and 3*- in such a way because we wanted to have a complete image of the collaborative network of INMEGEN, and to assess INMEGEN's internal community structure. Overall, there are some emerging groups and senior researchers are still quite scarce. This may be the consequence of genomics being a new field (as compared, for instance to other biomedical disciplines), and INMEGEN is still in its infancy. If this is true, we will be able to see it in our future work when we compared INMEGEN's SCN with the networks of other National Institutes of Health, some of them with more than 50 years of history. We would also like to model the collaboration strategies followed by the different groups using agent based modeling.

CONCLUSIONS

When reconstructing the basic structure of a network, to being able to assess direct and indirect interactions among nodes can be very useful and informative. One important aspect when analyzing complex networks is to being able to distinguish and assess direct from indirect interactions (even more, with regards to local interaction levels that may shape the large scale structure and the functional features of such networks). In this paper we have shown how an application of simple theorems of information theory (Hernández-Lemus and Rangel-Escareño 2011) makes possible for researchers to grasp the nature of the links in a clear-cut way. We discuss this in the context of both biological and social networks. However, we believe that this same general method arguments may apply to any network inferred by means of mutual information measures, and to some extent to other networks inferred by other quantitative interaction measures. What is more, such a general method may serve to the general purpose of unveiling similarities and differences between networks that are as different as it is a transcriptional network from a collaborative one. It is important to recall that DPI-pruned networks may be considered along with non-DPI pruned (and even with relatives degrees of pruning as given by different values of DPI_{tol}) in order to assess for structural features of the network. This is specially relevant if one is to consider the *de-Novo* functional discovery of the role of specific individuals or the reassurance of already envisioned hypothesis along the same lines.

The material presented here is intended to show some useful features of the application of IT methods for network inference and assessment. The biological and social networks chosen here were of different nature: different in size and, more important, in structure. This may be seen by the fact that DPI-pruning of the biological network (that was much larger than the social one) resulted in a tree-like structure (with no triangles and hence with a null clustering coefficient), whereas in the social network, DPI-pruning uncover different structural properties due to the presence of undecidable cases where three edges were equally weighted. However, in both cases one important outcome of direct-vs-indirect interaction analysis resulted, i.e. the presence (or better the *visibility* of the presence) of *key players*: in one case transcription factor genes acting as master regulators and in the other researchers identified as *group leaders*. Both studies revealed some surprising elements -like the presence of novel master regulators, or some *in-consolidation* researchers that are really emerging as *emerging group leaders*- given the structure of the direct interaction networks.

The study of the relationship between topological structure and functional organization in complex networks is, of course, still in a very early stage of development. The problems and challenges that arise include not only the determination of direct and indirect interactions and the role that some privileged nodes may play in network structure and navigability. A thorough study may also considered the issues of community structure and local connectedness. In the case of networks probabilistically inferred from experimental data one may also take into account the role that inference errors, noise and asymptotics may play. Further on, since complex networks are often dynamically adaptive systems, driven both by their inner structure and their environmental constraints, the role that dynamic evolution, fluctuations and adaptability may play will most surely also be fundamental to understand their behavior. We can only envisage what lies in the future: too many challenges should still be tackled before reaching a complete understanding of the behavior complex networks. However, we believe that information theoretical concepts and tools may play a fundamental role when facing such complexities.

ACKNOWLEDGEMENTS

The authors gratefully acknowledge support by grants: 179431/2012 (CONACYT) and PIUTE10-92 (ICyT-DF) [Contract 281-2010], as well as federal funding from the National Institute of Genomic Medicine (México).

AUTHORS' CONTRIBUTIONS

EHL conceived the study, EHL and JMSG performed simulations, analyses and calculations. EHL and JMSG wrote the paper. All authors read and approved the final manuscript.

REFERENCES

1. Andrecut M, Kauffman SA: A simple method for reverse engineering causal networks. *J Phys Math Gen* 2006, 39: L647-L655. 10.1088/0305-4470/39/46/L01
2. Kauffman SA, Andrecut, M: Mean-field model of genetic regulatory networks. *New J Phys* 2006, 8:148. 10.1088/1367-2630/8/8/148
3. Assenov Y, Ramírez F, Schelhorn S, Lengauer T, Albrecht M: Computing topological parameters of biological networks. *Bioinformatics* 2008, 24: 282–284. 10.1093/bioinformatics/btm554
4. Barabási AL: The network takeover. *Nat Phys* 2012, 8: 14–16.
5. Baca-López K, Hidalgo-Miranda A, Mayorga M, Gutiérrez-Nájera N, Hernández-Lemus E: The role of master regulators in the metabolic/transcriptional coupling in breast carcinomas. *PLoS ONE*2012,7(8):e42678. 10.1371/journal.pone.0042678
6. Bader GD, Hogue CW: An automated method for finding molecular complexes in large protein interaction networks. *BMC Bioinformatics* 2003,4(1):2. 10.1186/1471-2105-4-2
7. Baker NE: Master regulatory genes; telling them what to do. *Bioessays* 2001,23(9):763–766. 10.1002/bies.1110
8. Baldazzi V, Ropers D, Markowicz Y, Kahn D, Geiselmann J, de Jong H: The carbon assimilation network in Escherichia coli is densely connected and largely sign-determined by directions of metabolic fluxes. *PLoS Comput Biol* 2010,6(6):e1000812. 10.1371/journal.pcbi.1000812
9. Bansal M, Belcastro V, Ambesi-Impiombato A, di Bernardo D: How to infer gene networks from expression profiles. *Mol Syst Biol* 2007, 3: 78.
10. Beltrán E, Valiente-Banuet A, Verdú M: Trait divergence and indirect interactions allow facilitation of congeneric species. *Ann Bot* 2012, 110: 1369–1376. 10.1093/aob/mcs089
11. Bickel PJ, Doksum KA: *Mathematical Statistics: Basic, Ideas and Selected Topics Vol. 1*. New Jersey: Pearson-Prentice Hall; 2007.
12. Brennecke J, Aravin AA, Stark A, Dus M, Kellis M, Sachidanandam R, Hannon GJ: Discrete small RNA-generating loci as master regulators of transposon activity in Drosophila. *Cell*2007,128(6):1089–1103. 10.1016/j.cell.2007.01.043

13. Cardillo A, Scellato S, Latora V: A topological analysis of scientific coauthorship networks. *Physica A* 2006,372(2):333–339. 10.1016/j.physa.2006.08.059
14. Chua HN, Ning K, Sung W-K, Leong HW, Wong L: Using indirect protein-protein interactions for protein complex prediction. *J Bioinform Comput Biol* 2008,6(3):435–466. 10.1142/S0219720008003497
15. Callaway RM, Howard TG: Competitive networks, indirect interactions, and allelopathy: a microbial viewpoint on plant communities. *Prog Bot* 2007, 68: 317–335. 10.1007/978-3-540-36832-8_14
16. Costello JC, Dalkilic MM, Beason SM, Gehlhausen JR, Patwardhan R, Middha S, Eads BD, Andrews JR: Gene networks in Drosophila melanogaster: integrating experimental data to predict gene function. *Genome Biol* 2009,10(9):R97. 10.1186/gb-2009-10-9-r97
17. Cover TM, Thomas JA: *Elements of, Information Theory*. New York: John Wiley & Sons; 1991.MATH
18. Crowley-Riddey L: An information-theoretic approach to finding community structure in networks. *Ph.D. Dissertation* 2009. Trinity College Dublin, School of Mathematics Trinity College Dublin, School of Mathematics
19. de Jong H: Modelling and simulation of genetic regulatory systems: a literature review. *J Comp Biol* 2002,9(1):67–103. 10.1089/10665270252833208
20. Dong W: Mutual Information: inferring tie strength and proximity in bipartite social network data with non-metric associations. 2011. M.Sc. Dissertation, University of Illinois at Urbana-Champaign, USA
21. Fowler JH, Christakis NA: The spread of obesity in a large social network over 32 Years. *N Engl J Med* 2007,357(4):370–379. 10.1056/NEJMsa066082
22. Christakis NA, Fowler, JH: Cooperative behavior cascades in human social networks. *Proc Natl Acad Sci USA* 2010,107(12):5334–5338. 10.1073/pnas.0913149107
23. Fleuret F: Fast binary feature selection with conditional mutual information. *J Mach Learn Res*2004, 5: 1531–1555.
24. Harrison MM, Li X-Y, Kaplan T, Botchan MR, Eisen MB: Zelda binding in the early drosophila melanogaster embryo marks regions subsequently activated at the maternal-to-zygotic transition. *PLoS Genet* 2011,7(10):e1002266. 10.1371/journal.pgen.1002266

25. Hernández-Lemus E, Rangel-Escareño C: The role of information theory in gene regulatory network inference. In *Information Theory: New Research*. Edited by: Deloumeaux P, Gorzalka JD. ematics Research, Developments Series, Nova Publishing; 2011:109–144.
26. Hernández-Lemus E, Velázquez-Fernández D, Estrada-Gil JK, Silva-Zolezzi I, Herrera-Hernández MF, Jiménez-Sánchez G: Information theoretical methods to deconvolute genetic regulatory networks applied to thyroid neoplasms. *Physica A* 2009, 388: 5057–5069. 10.1016/j.physa.2009.08.013
27. INFOTHEO: A collection of information theoretical tools based on several entropy estimatorsPublished 2012–12–10, Last accesed 2013–03–17 [http://cran.r-project.org/web/packages/infotheo/index.html] [] Published 2012-12-10, Last accesed 2013-03-17
28. Lefebvre C, Lim WK, Basso K, Dalla-Favera R, Califano A: A context-specific network of protein-DNA and protein-protein interactions reveals new regulatory motifs in human B cells. *Lect Notes Bioinform* 2007, 4532: 42–56.
29. Lefebvre C, Rajbhandari P, Alvarez MJ, Bandaru P, Lim WK, Sato M, Wang K, Sumazin P, Kustagi M, Bisikirska BC, Basso K, Beltrao P, Krogan N, Gautier J, Dalla-Favera R, Califano A: A human B-cell interactome identifies MYB and FOXM1 as master regulators of proliferation in germinal centers. *Mol Syst Biol* 2010, 6: 377. 10.1038/msb.2010.31
30. Liang KC, Wang X: Gene regulatory network reconstruction using conditional mutual information. *EURASIP J Bioinform Syst Biol* 2008, 253894. 10.1155/2008/253894
31. Madni AM, Andrecut M: Design and implementation of a gene network reverse engineering method based on mutual information. *J Integr Des Process Sci* 2007,11(3):55–68.
32. Mani KM, Lefebvre C, Wang K, Lim WK, Basso K, Dalla-Favera R, Califano A: A systems biology approach to prediction of oncogenes and molecular perturbation targets in B-cell lymphomas. *Mol Syst Biol* 2008, 4: 169.
33. Margolin AA, Nemenman I, Basso K, Wiggins C, Stolovitzky G, Dalla Favera R, Califano A: ARACNe: An Algorithm for the reconstruction of gene regulatory networks in a mammalian cellular context. *BMC Bioinformatics* 2006,7(Suppl I):S7.

34. Mislove AE: Online social networks: measurement, analysis, and applications to distributed information systems. *Ph.D. Dissertation* 2009. Rice University, Department of Computer Science Rice University, Department of Computer Science

35. Mislove AE, Viswanath B, Gummadi KP, Druschel P: You are who you know: Inferring user profiles in online social networks. *Proceedings of the 3rd ACM International Conference on Web Search and Data Mining WSDM'10* 2010, 251–260.

36. Nawrath J, Romano MC, Thiel M, Kiss IZ, Wickramasinghe M, Timmer J, Kurths J, Schelter B: Distinguishing direct from indirect interactions in oscillatory networks with multiple time scales. *Phys Rev Lett* 2010, 104: 038701.

37. Newman MEJ: The structure and function of complex networks. *SIAM Rev* 2003, 45: 167–256. 10.1137/S003614450342480

38. Newman, MEJ: Coauthorship networks and patterns of scientific collaboration. *Proc Nat Acad Sci*2004,101(Supl1):5200–5205.

39. Peng H, Long F, Ding C: Feature selection based on mutual information: criteria for max-dependency, max-relevance and min-redundancy. *IEEE Trans Pattern Anal Mach Intell*2005,27(8):1226–1238.

40. Sehgal MSB, Gondal I, Dooley L, Coppel R, Mok GK: Transcriptional gene regulatory network reconstruction through cross platform gene network fusion. *In Pattern Recognition in Bioinformatics, Lecture Notes in Computer Science* 2007, 4774: 274–285. 10.1007/978-3-540-75286-8_27

41. Shalizi CR, Thomas AC: Homophily and contagion are generically confounded in observational social network studies. *Sociol Meth Res* 2011,40(2):211–239. 10.1177/0049124111404820

42. Shannon CE, Weaver W: *The ematical, Theory of Communication.* Urbana: The University of, Illinois Press; 1949.

43. Subramanian A, Tamayo P, Mootha VK, Mukherjee S, Ebert BL, Gillette MA, Paulovich A, Pomeroy SL, Golub TR, Lander ES, Mesirov JP: Gene set enrichment analysis: a knowledge-based approachfor interpreting genome-wide expression profiles. *Proc Natl Acad Sci USA* 1554, 102: 5–15550.

44. Tsatskis I: Systemic losses in banking networks: indirect interaction of nodes via asset prices.*SSRN 2062174* 2012. [http://papers.ssrn.com/sol3/papers.cfm?abstract_id=2062174]

45. Tresch A, Beissbarth T, Saltmann H, Kuner R, Poustka A, Buness A: Discrimination of direct and indirect interactions in a network of regulatory effects. *J Comput Biol: J Comput Mol Cell Biol*2007,14(9):1217–1228. 10.1089/cmb.2007.0085

46. van Someren EP, Wessels LFA, Backer E, Reinders MTJ: Genetic network modelling.*Pharmacogenomics* 2002,3(4):507–525. 10.1517/14622416.3.4.507

47. Wells BS: Drosophila p53: A master regulator of DNA and tissue damage repair. 2009. Ph.D. Dissertation, Columbia University, Genetics, Department

48. Wu X, Liu Q, Jiang R: Align human interactome with phenome to identify causative genes and networks underlying disease families. *Bioinformatics* 2009, 25: 98–104. 10.1093/bioinformatics/btn593

49. Yan KK, Maslov S, Mazo I, Yuryev A: Prediction and verification of indirect interactions in densely interconnected regulatory networks. *arXiv preprint arXiv:0710.0892* 2007. [http://arxiv.org/abs/0710.0892]

50. Yousefi-Nooraie R, Akbari-Kamrani M, Hanneman RA, Etemadi A: Association between co-authorship network and scientific productivity and impact indicators in academic medical research centers: a case study in Iran. *Health Res Policy Syst* 2008, 6: 9. 10.1186/1478-4505-6-9

51. Yu H, Xia Y, Trifonov V, Gerstein M: Design principles of molecular networks revealed by global comparisons and composite motifs. *Genome Biol* 2006, 7: R55. 10.1186/gb-2006-7-7-r55

52. Zhang X, Zhao XM, He K, Lu L, Cao Y, Liu J, Hao JK, Liu ZP, Chen L: Inferring gene regulatory networks from gene expression data by path consistency algorithm based on conditional mutual information. *Bioinformatics* 2012,28(1):98–104. 10.1093/bioinformatics/btr626

53. Zhao K, Karsai M, Bianconi G: Entropy of dynamical social networks. *PLoS ONE* 2011,6(12):e28116. 10.1371/journal.pone.0028116

CHAPTER 13

REVEALING NETWORKS FROM DYNAMICS: AN INTRODUCTION

Marc Timme[1,2,3] and Jose Casadiego[1]

[1] Network Dynamics, Max Planck Institute for Dynamics and Self-Organization (MPIDS), D-37077 Gottingen, Germany

[2] Bernstein Center for Computational Neuroscience (BCCN) Gottingen, ¨ D-37077 Gottingen, Germany

[3] Faculty of Physics, Institute for Nonlinear Dynamics, University of Gottingen, ¨ D-37077 Gottingen, Germany

ABSTRACT

What can we learn from the collective dynamics of a complex network about its interaction topology? Taking the perspective from nonlinear dynamics, we briefly review recent progress on how to infer structural connectivity (direct interactions) from accessing the dynamics of the units. Potential applications range from interaction networks in physics, to chemical and metabolic reactions, protein and gene regulatory networks as well as neural

CITATION (APA): Timme, M., & Casadiego, J. (2014). Revealing networks from dynamics: an introduction. Journal of Physics A: Mathematical and Theoretical, 47(34), 343001. (36 pages)

circuits in biology and electric power grids or wireless sensor networks in engineering. Moreover, we briefly mention some standard ways of inferring effective or functional connectivity.

WHERE ARE YOU LINKED?

Relating Connection Topology of a Network to its Dynamics

Networks are everywhere. And most of them are dynamic. From networks of biochemical reactions that regulate the metabolism in the cells of our bodies to the neuronal circuits in our brains, from social ties forming networks of our friendships and collaborators to the power grids and internet that provide huge amounts of electric energy and information every second. All of these systems form networks of units that interact to yield complex, collective forms of functions—and all are crucial to our everyday life.

The interaction topology of complex networks strongly impact their collective dynamics and thus the function of entire systems. For many network dynamical systems, for instance in physics and biology, the dynamics of individual units becomes more and more accessible whereas their intricate web of interactions remains uncertain or even often largely unknown. As an example, many constituents of protein and gene interaction networks are well characterized but how they interact and which pathways are relevant for suitable functioning is not well understood [1, 2]. In neuroscience, the number of units from which one can simultaneously measure neuronal activity is increasing rapidly from a few units to hundreds of them [3]. Still, identifying the synaptic connections of a neuronal circuit by anatomical methods is mostly restricted to individual synapses and computer-aided reconstruction based on optical methods for more than two cells becomes available only since very recently [4–6]. In social networks, even in simple settings such as basic games, pairwise interactions are roughly understood, but often both the (temporally varying) interaction network and its collective consequences remain a riddle. Thus, reconstructing the structure of interaction networks from (only) the collection of local dynamical data constitutes a current open challenge, with applications across the natural and social sciences as well as engineering.

Network dynamics: forward versus inverse problem. Yet, the vast majority of research on network dynamics has focused on the ‹forward direction› of modeling and asked what types of collective dynamics emerge from a network of given topology. Researchers from the natural sciences and engineering systematically address the reverse questions—how to control a network or, more generally, how to design networks for a desired dynamics and thus function and how to infer topology from dynamics—now at a rapidly increasing pace: in particular in engineering, the design of systems for a specific function always was core [7] and with the systems becoming more complex, considering recurrently interacting networks becomes indispensable. Conversely, complex systems in physics and biology require a view on networked systems to understand how complex emergent phenomena contribute to (possibly optimal) system's dynamics or function [8–12]. Finally, also how one could perhaps redesign collective dynamics of networks, e.g., gene and protein networks [13, 14] poses challenges of frontier research.

The inverse problem of how to infer interaction topology from network dynamics constitutes the main topic of this review.

Aims and Options of Network Inference

What do we aim to infer? Before addressing any inference problem, we have to clarify what we actually want to find out about the networked system and which level of detail we are interested in, see table 1. For instance, we may want to know *effective* or *functional connectivity* not caring about individual interactions *per se* but only about statistical dependences which the entire set of interactions yield between pairs of units through the collective network dynamics. Inter-unit correlations and various information-theoretic measures have been devised to solve such problems. These often neglect the temporal dynamics as they use temporal averages or statistical distributions of observables. Moreover, effective connectivity may depend on the current collective state and function of a given system and thus the same physical network may display different effective topologies for different functions or in different states. We may alternatively want to know *structural connectivity*, i.e. which unit directly interacts with which other units (and how)?

Table 1: Levels of interest. Which properties of the connections are we interested in?

	Distinctions
Property of connection	...and examples thereof

Structural versus effective	*Direct interaction or statistical dependence?*
	EEG versus connectomics for neural 'connectivity'
Pure existence	*Presence or absence of links?*
	does one gene directly influence another?
Sign	*Positive, negative, or mixed-sign interaction?*
	phase-advancing or phase retarding (for coupled oscillators);
	activating or inhibiting (for gene and protein interactions);
Directedness	*Directed or undirected (bi-directed) interactions?*
	chemical versus electrical neuronal synapses
Type of interaction	*Continuous time or discrete time; linear versus nonlinear?*
	chemical versus electrical synapses
	diffusive versus nonlinear coupling
Time scales	*Instantaneous, temporally localized or extended?*
	slot communication (in mobile phones) versus genetic interactions
Spatial scales	*Local, global, non-local, bounded?*
	message broadcasting, from wireless networks to cell tissue

In this review, we focus on these direct physical interactions and address various levels of detail. We also briefly present basic methods to infer different types of effective connectivity. By construction, such a review cannot be complete, also because the field is currently developing at a breathtaking pace. We therefore select specific reconstruction methods from those that are commonly used, appear promising for the future of the field, or have been recently developed and form the basis of current research.

What can we learn about the connections of a network from accessing the units' dynamics? Mathematically, inferring the connectivity constitutes a high-dimensional inverse problem and various methods have been devised to address this question. Every inference method starts from different levels of pre-knowledge about the system and has its own aims what to reconstruct, cf table 1. We may be interested in whether the interactions are directed or undirected, in whether or not a link exists, in the sign of the interaction dynamics, the type of links (e.g., electric versus chemical synapses, diffusive versus nonlinear coupling), the strengths and the temporal and spatial scales of interactions etc.

Structural connectivity may be very different from effective or functional connectivity (figure 1). On the one hand, high correlations may exist between two units that are not directly connected but only influenced by each other via a third unit they both directly interact with. In general, such indirect interactions may be induced not only by one third node, but equally by the entire collective dynamics of a network. On the other hand, even a strong direct interaction between two units does not necessarily mean that their dynamics is highly correlated; correlations could be submerged, e.g., by external noise or recurrent inputs the two units receive, e.g., from two distinct other parts of the network, or even from outside of it.

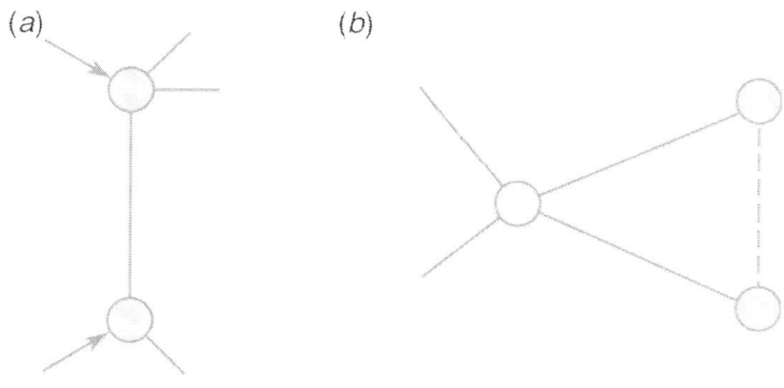

Figure 1: Diverse dynamic impact of structural links on effective connectivity. (*a*) Structural links may not be detectable by certain correlation measures due to strong independent driving signals (e.g. noise). For example, strong inputs along two links (from within or outside the network, marked by arrows) may decorrelate the dynamics of the two nodes (gray disks) although they directly interact. (*b*) Common input may create effective link without the structural link present. For example, common input from a third node (open circle) may create effective connectivity (dashed line) between two nodes (gray disks) that are not

directly connected. In both examples, (*a*) and (*b*), it may depend on the entire collective state of the network (and the external inputs it may receive) whether or not an effective connection is detected.

In general, effective and structural connectivity are related in a highly non-trivial way. In fact, a number of counter-intuitive phenomena have been observed in various systems. For instance, recent work on coupled oscillator networks highlight that under certain conditions noise may aid to reconstruct structural connectivity from correlation-based effective connectivity [15]; and one may detect small-world features in the functional connectivity even if it is derived from randomly connected dynamical systems without any specific small-world features [16].

Where do we start from? It is important to clarify which knowledge about the networked system we presuppose. Is the collective dynamics known to be simple such as converging to a fixed point of concentrations in biochemical networks or to a limit cycle in coupled oscillator networks? Or do we expect more complex, irregular, and perhaps unpredictable, chaotic or random types of dynamics? Can we change the dynamics of units by interfering with the system or can we just observe? Does the research question require a global analysis in state space or do we focus on a specific dynamical state where local analysis may suffice? We should answer these questions, among others, before using or developing any inference method—to achieve reconstruction at the level we need with best quality and minimal efforts, both experimentally and computationally.

Here we review recently developed approaches to inferring structural connectivity of a network from accessing its collective dynamics. The presented approaches assume various levels of pre-knowledge about the system and may or may not require the observer to interfere with the system. The article is structured as follows. We first clarify in section 2 what we mean by a network dynamical system, taking the view of continuous-time dynamics described by coupled ordinary differential equations. In sections 3–5, we explain three principally different classes of methods to obtain information about the structure of the network topology from dynamical quantities.

Section 3 illustrates basic theoretical approaches based on measuring and evaluating the response of a network to external perturbations or driving. As the response depends on both the external driving signal and the interaction topology of the network, sufficiently many driving-response experiments yield information about the entire network topology. A second class of methods sets up a model copy of the original system (section 4) and adapts the coupling matrix of the model so as to synchronize its dynamics with the original. If synchronization is achieved, the topology obtained for the model is taken as a proxy for the original. Finally, a set of direct methods (section 5) rely on measuring time series (or features thereof), evaluating temporal derivatives, and exploiting smoothness assumptions to find solutions to an optimization problem given by the restrictions by data.

We briefly comment on technical issues (section 6) and mention basic core approaches for identifying effective connectivity (section 7). These approaches rely on simple linear correlation, maximum entropy principles and related statistical inference methods. Finally, we provide an outlook (section 8) where we highlight current challenges, point out aspects sometimes overlooked and show potential research paths towards uncovering more of the topology of the networks that surround us.

This review on purpose is short but self-contained. It is intended for researchers mainly in the physical and biological sciences and engineering and assumes basic knowledge of dynamical systems and probability theory. Let us start with the details.

NONLINEAR DYNAMICS OF NETWORKS

Systems of Ordinary Differential Equations Describing Networks

Throughout the main part of this review, we consider networks of units assumed to be described by systems of ordinary differential equations. Discrete time maps coupled to a network are not discussed but typically approaches similar to those presented here are viable in slightly modified forms. We discuss specific issues for systems of pulse-coupled units, such as spiking neurons, in section 5. These are formally hybrid systems, i.e. mixtures of continuous-time and discrete time systems.

Assuming that interactions occur between pairs of coupled units, a generic network dynamical systems is given by

$$\frac{\mathrm{d}}{\mathrm{d}t}\boldsymbol{x}_i = \boldsymbol{f}_i(\boldsymbol{x}_i) + \sum_{j=1}^{N} J_{ij}\boldsymbol{g}_{ij}(\boldsymbol{x}_i, \boldsymbol{x}_j) + \boldsymbol{I}_i(t) + \boldsymbol{\xi}_i(t) \tag{1}$$

where $i, j \in \{1, 2, ...N\}$, $\boldsymbol{x}_i(t) = [x_i^{(1)}(t), x_i^{(2)}(t), \ldots, x_i^{(D)}(t)]^{\mathsf{T}} \in \mathbb{R}^D$

describes the state of the ith unit at time $t \in \mathbb{R}$, and the functions $f_i : \mathbb{R}^D \to \mathbb{R}^D$ and $g_{ij} : \mathbb{R}^D \times \mathbb{R}^D \to \mathbb{R}^D$ mediate intrinsic and interaction dynamics of the D-dimensional units, respectively. The term $Ii(t)$ represents a vector of external driving signals (possibly random) and $\xi i(t)$ symbolically represents external noise. Finally, the J_{ij} define the interaction topology, in the simplest setting in terms of the adjacency matrix A such that $J_{ij} = A_{ij} = 1$ if there is a direct physical interaction from unit j to i and $J_{ij} = A_{ij} = 0$ otherwise. In general, units› interaction may be higher order, e.g. requiring terms like $hijk(xi, xj, xk)$ added to the right-hand side of (1). For instance in gene and protein interaction networks, a protein (the so-called transcription factor, say unit k) is directly influencing the rate of transcription of a gene (say, unit j) to a DNA segment (say unit i). We do not treat such terms here explicitly. Their relevance for network dynamical systems is discussed in [17].

For some methods to infer effective connectivity, the functional form of (1) does not directly enter the inference argument, other methods can be extended to include higher order terms explicitly. We comment on higher order terms where appropriate (cf also section 5).

Rescaling, Simplifications, and Common Interactions

Some a priori technical issues. Considering (1) as our basic level of description, in case the dimension D_i of the local dynamical system i depends on unit i, we would just consider the maximal occurring dimension $D = \max_{i \in \{1,\dots,N\}} D_i$ and for each unit ignore the $D - D_i$ dummy variables. This is done purely for notational simplification. Note further that in general, the quantity J_{ij} is a $D \times D$ matrix of coupling strengths $J_{ij}^{dd'}$, but that for many paradigmatic model systems, only one of these elements is non-zero, i.e. $J_{ij}^{dd'} = J_{ij}^{dd'} \delta_{d,d_1} \delta_{d',d_2}$ such that $J_{ij} = J_{ij}^{d_1 d_2}$ is sufficient to describe the influence of unit j on i.

Given a dynamical system of the form (1), the right-hand side is determined up to $N \times D$ additive constants C_i^d and one overall multiplicative constant C_0. Shifting the state variables x_i^d to co-moving frames and rescaling time enables us to set $C_i^d = 0$ for all i and all d and $C_0 = 1$ without loss of generality.

For simplicity of presentation, we furthermore describe the methods below as if they were for networks of coupled one-dimensional units only. Often, different dimensions may be treated independently during reconstruction.

We thus take the coupled equations

$$\frac{\mathrm{d}}{\mathrm{d}t}x_i = f_i(x_i) + \sum_{j=1}^{N} J_{ij}g_{ij}(x_i, x_j) + I_i(t) + \xi_i(t) \tag{2}$$

as our basic network characterization we start from, where now the variables x_i and x_j and the functions f_i and g_{ij} are treated as real scalars.

Common Interaction Functions. Only non-trivial coupling terms that are not identically zero,

$$\frac{\partial g_{ij}}{\partial x_j} \neq 0 \tag{3}$$

for the relevant domain of arguments actually contribute to interactions, so we assume all coupling functions g_{ij} for which $J_{ij} \neq 0$ to be non-trivial in this sense. A broad range of systems exhibits diffusive coupling such that

$$g_{ij}(x_i, x_j) \propto (x_j - x_i) \tag{4}$$

which is a special case of coupling functions

$$g_{ij}(x_i, x_j) = \tilde{g}_{ij}(x_j - x_i) \tag{5}$$

that depend on state differences (e.g. phase differences for coupled oscillators) only. The simplest non-trivial form of interaction is linear coupling,

$$g_{ij}(x_i, x_j) = A_{ij}x_j, \tag{6}$$

and does not depend on the dynamical variable of the unit it influences.

We have now set the stage to dive into specific inference approaches.

DRIVING-RESPONSE EXPERIMENTS

One idea of inferring network topology is to measure the collective response of a network dynamical system to driving by external signals. For instance, if a system exhibits a stable collective state (e.g. fixed point or periodic orbit, cf figure 2), it will relax back to that state after a transient input (pulse), if the latter is not too strong (such as to not kick the system out of the basin of attraction of the stable state) [18–22]. Here, the input signal (driving) effectively changes the initial condition of the system, leaving the system features (given by the local and interaction functions and their parameters) the same. Similarly, a stable state of the system will generically move in state space (and keep qualitatively the same stability properties) in response to sufficiently weak, temporally constant external perturbations. These kinds of perturbations effectively creates a non-identical but similar systems with different parameters determined by the driving signal.

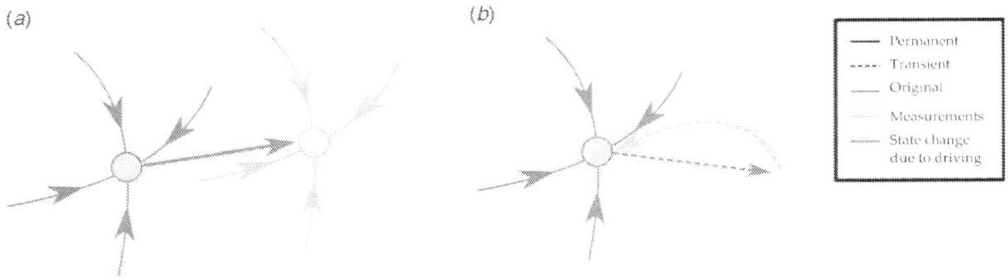

Figure 2: Stable state approaches to network inference. (*a*) Constant external driving signal (parameter change) moves the stable fixed point (gray) in state space to another position (blue). The difference vector v (red) depends on both the driving signal and the network topology such that several measurements of v under different driving conditions yields information about the topology. (*b*) Same as (*a*) but for transient perturbative signal: After the driving signal is switched off, the dynamics relaxes back (dashed blue trajectory) towards the original fixed point (gray). Now, the relaxation trajectory contains information about the topology.

Both the relaxation dynamics and the shift in state space in general depend not only on the external signal (which unit is perturbed, how and how strongly, i.e. known quantities), but also on the (unknown) interaction topology of the network. Each collective response of the system to an external perturbation yields a restriction on the network topology such that sufficiently many driving-response experiments may reveal the entire topology. In this section, we present the main ideas underlying several related driving-response approaches.

Restrictions from Local Fixed Point Analysis

Recent efforts for developing methods to identify a network's topology have emerged from the need to understand biological, in particular gene regulatory networks [1, 23–25]. Such networks consist of genes and proteins that interact with each other within a cell [26, 27]. These interactions in particular control (indirectly) the rates at which genes are transcribed into mRNA and the regulatory features emerge via the interactions and generally do not follow from the single-gene level [28]. Gene regulatory networks and other reaction networks, e.g., in chemistry or population dynamics, are often modeled by nonlinear differential equations

$$\dot{z}_i = f_i(z_1, z_2, \ldots, z_N; \boldsymbol{p}, \tilde{\boldsymbol{I}}), \tag{7}$$

describing the rate of change of the expected numbers (or concentrations) $z_i(t)$ of entities (e.g. genes, atoms and molecules or individual organisms) at time t in terms of their dependence on the number of other entities [27]4. Here $\boldsymbol{p}$ is a set of parameters and $\tilde{\boldsymbol{I}}$ represents a set of external perturbations directed to the entities. The z_i are typically positive real numbers but mathematically there is no restriction for them to also be negative. Under the assumption that such a system is close to a steady state $z^* = (z_1^*, z_2^*, \ldots, z_N^*)$, a stable fixed point where $f_i(z_1, z_2, \ldots, z_N; \boldsymbol{p}, \tilde{\boldsymbol{I}}) = 0$ for all i, the dynamics for perturbations $x_i(t) = z_i(t) - z_i^*$ from steady state concentrations of such nonlinear models may be approximated to first order in the x_i by

$$\dot{x}_i = \sum_{j=1}^{N} J_{ij} x_j + I_i(t), \tag{8}$$

where the local Jacobian $J_{ij} = (\partial f_i / \partial z_j)(\boldsymbol{z}^*)$ is the effective interaction matrix given the steady state and $I_i(t)$ is assumed to be an external perturbation linearly coupling to deviations of variables x_i. We remark that when deriving (8) from (7) we implicitly use the relation

$$I_i(t) = \sum_j \tilde{I}_j(t) \frac{\partial f_i}{\partial \tilde{I}_j}\bigg|_{z=z^*, \tilde{I}=0} + \mathcal{O}(\tilde{I}_i \tilde{I}_j) \tag{9}$$

ignoring the second and higher order terms.

Driving the System Constantly to Move a Stable State (Changing Parameters)

The method presented now effectively moves the parameters and thus in particular the fixed points of a system. Originally intended for gene and protein interaction networks, Gardner and coworkers [1, 29] have explicated that reconstruction is possible, at least for small networks. We here discuss two approaches of using the general form (8) to reconstruct the interaction network, i.e. the J_{ij}. The first is based on driving the system (8) by sufficiently small, temporally constant driving forces $I_i(t) = Ii_{,} m$ to new fixed points $x^*_{i,m} \neq 0$ close to the original one $x^*_i = 0$. As the original fixed point was structurally stable, the new fixed point will generically exist and have qualitatively the same stability properties if the *Ii m* are small enough (figure 2(*a*)). This is because solutions of (7), in particular fixed point solutions, typically vary continuously with changing parameters (here: changing *Ii m* from zero) and there is no bifurcation close to the parameters yielding a generic stable fixed point y^*_i.

The observed values assumed *at* each new steady state together with the (known) driving signals provide information about the interaction topology J_{ij}. Performing several perturbation experiments $m \in \{1, ..., M\}$ yield $N \times M$ equations

$$\sum_{j=1}^{N} J_{ij} x^*_{j,m} = -I_{i,m}, \tag{10}$$

one for each experiment m and for each unit i.

After an arbitrary number M of experiments, equation (10) in matrix form becomes

$$JX = Y, \tag{11}$$

where $J \in \mathbb{R}^{N \times N}$ represents the connectivity among the units, $X \in \mathbb{R}^{N \times M}$ the steady state values with $X_{i,m} = x^*_{i,m}$, and $Y \in \mathbb{R}^{N \times M}$ the perturbations *Yi m* = −*Ii m* that we assume to be known. This matrix equation restricts the connectivity J given the measured data X and the input perturbations I. The matrix equations constraining the full network topology J can be split into N equations

$$\boldsymbol{J}_i X = \boldsymbol{Y}_i, \tag{12}$$

one for each input connectivity $\boldsymbol{J}_i := (J_{i,1}, \ldots, J_{i,N}) \in \mathbb{R}^{1 \times N}$ of a unit i. Thus, the same set of data X restrict all the sets of units providing interactions to $i \in \{1, ..., N\}$ but the data $\boldsymbol{Y}_i = (Y_{i,1}, \ldots, Y_{i,M})^{\mathsf{T}}$ are unit dependent. This reduction to N individual equations also admits to split the computational effort for solving them. The problem becomes trivially parallelizable because for different i, these restrictions (12) are independent in the sense that reconstruction of the input coupling strengths to each unit i can be performed without taking care of input coupling strengths of other units $k \neq i$.

Observe Relaxation to Stable State after Transient Driving (Changing Initial Conditions)

A second approach assumes that the quantities y_i (and thus the x_i) are perturbed such that at time t_0 we have $x_i(t_0) = x_i^{(0)}$ and the transient dynamics $y_i(t)$ of relaxation back to the original fixed point y_i^* (and thus x_i^*) are observed at a sequence of times $t_m > t_0$, $m \in \{1, \ldots, M\}$. This yields the same type of equation (11), but now with the $Y_{i,m} = -\hat{\dot{x}}_{i,m}$ being estimates of the derivatives $\dot{x}_i(t_m)$. We remark that these derivatives may be estimated in various ways, each of them requiring a resolution of the measured data on sufficiently small time scales, cf section 5.1. In this second approach, the different times the transient dynamics is measured replaces the different driving experiments in the first approach. Finally, both approaches can of course be combined, several experiments evaluated at several time points, again yielding the same form of restrictions (11).

Solving the Restricting Equations

How can we finally obtain the coupling elements J_{ij} and thus the interaction network? In principle, solving the matrix equation (11) yields the interaction matrix J as a function of the known data X and Y. One may naively assume that it is directly solvable once the number of experiments equals the number of units in the network, $M = N$. However, this problem can be numerically ill-conditioned [30] for large N, such that the result is not reliable. In addition, as also stressed in [1], the results may be sensitive to noise in the measured data.

A way to overcome this problem is by performing (many) more experiments than nodes available, $M \gg N$, thus *over-determining* (11). In general, due to noise and measurement inaccuracies, this yields the system (12) to be inconsistent such that there is no vector ***Ji*** that satisfies all constraints. It will still be possible to find a robust approximation $\hat{\boldsymbol{J}}_i$ that minimizes the error between the predicted dynamics $\boldsymbol{J}_i X$ and the actual dynamics $\boldsymbol{Y}_i$ for a given node. Specifically, this error function may be modeled as

$$E_i(\hat{\boldsymbol{J}}_i) = d(\boldsymbol{Y}_i, \hat{\boldsymbol{J}}_i X), \tag{13}$$

where the distance measure $d(\boldsymbol{v}, \boldsymbol{w}) = \|\boldsymbol{v} - \boldsymbol{w}\|_p^p$ between two vectors $\boldsymbol{v}, \boldsymbol{w} \in \mathbb{R}^M$ is commonly defined in terms of the pth power

$$d(\boldsymbol{v}, \boldsymbol{0}) = \|\boldsymbol{v}\|_p^p = \left(\sum_{m=1}^{M} v_m^p\right) \tag{14}$$

of an L_p-norm with $p \geq 1$, due to its convexity properties. This guarantees that any local minimum of E_i is also a global minimum [31]. Particularly, the L_2-minimization criterion,

$$E_i(\hat{\boldsymbol{J}}_i) = \|\boldsymbol{Y}_i - \hat{\boldsymbol{J}}_i X\|_2^2, \tag{15}$$

is of great importance because it has an analytical solution for its extremum. Equating to zero the derivatives of the error function with respect to the matrix elements, $\frac{\partial}{\partial J_{ik}} E_i(\hat{\boldsymbol{J}}_i) \stackrel{!}{=} 0$, yields an analytical solution to L_2 error-minimization given by

$$\hat{\boldsymbol{J}}_i = \boldsymbol{Y}_i X^{\mathsf{T}} (X X^{\mathsf{T}})^{-1}. \tag{16}$$

Evaluating such equations for all $i \in \{1, ..., N\}$ yields the complete reconstructed network $\hat{J}$. This mathematical form of minimum L_2-norm solution is implemented in many mathematical packages (e.g. as the mrdivide function in Matlab [32] or the LeastSquares function in Mathematica [33]). We explicate that the obtained off-diagonal terms $\hat{J}_{ij}$ serve as the best estimate (in the procedural sense using the L_2-minimization above) for the coupling constants J_{ij} ; at the same time, the diagonal elements J_{ii} are not relevant for the network topology because the influence of these terms on the dynamics of unit i is physically indistinguishable from an intrinsic drive to i included in the local dynamics specified by $f(x_i)$ in (2), cf equation (3).

Experimentally, it is in principle possible to over-determine a system of equations by performing repeated measurements on the network until the condition $M \gg N$ is achieved. Nevertheless, it may often seem unsuitable for large networks due to the large number of experiments that would be required.

So, if the size of the network is an issue or the number of available measurements insufficient to over-determine the system, we have $M < N$. Assuming that the network is sparse (i.e. that each unit is connected with a small number of others and thus many connection strengths are $J_{ij} = 0$), may still yield the collection of all network links. This implies that several J_{ij} are effectively set to zero, therefore decreasing the amount of unknown coefficients to be solved for. It leaves us with the problem of finding which links are actually present and which are not. We present two related options to do so.

Sparse Solution with a Bounded Connectivity per Unit

If an upper bound for the number of links $K_i < M$ is known, we may assume only $M < N$ experiments are available and we have a rough idea of how many nodes (at most) are connected to a particular node. In particular, assume that the number of incoming connections for node i is given by at most $K_i < M$ [1]. This assumption shifts the system (12) from having more unknowns than constraints, $M < N$, to have more constraints than unknowns, $M > K_i$, therefore, implicitly over-determining the system. It means that out of the N nodes present in the network only K_i of them are chosen to be part of the system of equations for node i. Such assumption may be done when there is some *a priori*information about the network's connectivity and dynamics.

Specifically, the system of equations (12) may in principle be rewritten as

$$\boldsymbol{B}_i Z_i = \boldsymbol{Y}_i, \tag{17}$$

where $\boldsymbol{B}_i \in \mathbb{R}^{1 \times K_i}$ is the reduced connectivity vector for node i that contains the coupling strengths for the selected nodes and $\boldsymbol{Z}_i \in \mathbb{R}^{K_i \times M}$ is a matrix that contains the states of such nodes. If we knew which K_i of the $N - 1$ possible connections actually contributed, we could use equation (16) to solve (17) using L_2-minimization yielding

$$\hat{\boldsymbol{B}}_i = \boldsymbol{Y}_i Z_i^T (Z_i Z_i^T)^{-1}, \tag{18}$$

where $\hat{\boldsymbol{B}}_i$ is the best approximation to $\boldsymbol{B}i$.

Yet, it so far remains unclear which of the $\binom{N}{K_i} = \frac{N!}{K_i!(N-K_i)!}$ possible combinations of incoming connections is best suited for reproducing the dynamics of i. In principle, $\hat{\boldsymbol{B}}_i$ may be calculated for each combination of K_i genes, and the combination that yields the smallest value of the L_2 norm $\|\hat{\boldsymbol{B}}_i\|_2$ in (18) may be chosen as the best estimate for J_i. The efficiency of such a procedure relies in number K_i of interactions per node. Hence, choosing a proper K_i aiming to recover the largest number of real interactions with the smallest number of false positives is a key factor to achieve a successful topological reconstruction.

Maximizing the Sparseness of the Connectivity Matrix

If $M < N$ and the K_i are unknown, cannot be estimated or there are too many of them (making the combinatorial search practically impossible) maximizing the number of zero entries in $\boldsymbol{J}$, (i.e. minimizing the K_i and thus maximizing sparseness) may be a way to solve (12). This approach is particularly useful if the only *a priori* knowledge about the network›s connectivity is some sparsity.

For general matrix equations

$$Ay = \boldsymbol{b}, \tag{19}$$

where $A \in \mathbb{R}^{m \times n}, \boldsymbol{y} \in \mathbb{R}^{n \times 1}$ and $\boldsymbol{b} \in \mathbb{R}^{m \times 1}$, singular value decomposition (SVD) of A according to

$$A = U \Sigma V^{\mathsf{T}}, \tag{20}$$

yields an analytic solution

$$\boldsymbol{y} = V \tilde{\Sigma} U^{\mathsf{T}} \boldsymbol{b} + V \boldsymbol{c}, \tag{21}$$

where $\tilde{\Sigma} = \Sigma^{\mathsf{T}}(\Sigma\Sigma^{\mathsf{T}})^{-1}$that parametrizes the space of all solutions through the vector $c \in \mathbb{R}^{n \times 1}$with $c_i = 0$ for $i \in \{1, \ldots, r\}$ and $r = Rank(\boldsymbol{A})$.

In our reconstruction problem, we are seeking to maximize the number of zero entries in $\boldsymbol{J}$ based on solving the restricting equations (12) for $\boldsymbol{J}i$ as we solved (19) for $\boldsymbol{y}$. Consider the transpose

$$X^{\mathsf{T}} \boldsymbol{J}_i^{\mathsf{T}} = Y_i^{\mathsf{T}} \tag{22}$$

of (12). The analogous SVD-based solution then reads

$$\boldsymbol{J}_i^{\mathsf{T}} = V \tilde{\Sigma} U^{\mathsf{T}} \boldsymbol{Y}_i^{\mathsf{T}} + V \boldsymbol{c}, \tag{23}$$

where $U \in \mathbb{R}^{M \times M}, V \in \mathbb{R}^{N \times N}, \tilde{\Sigma} \in \mathbb{R}^{N \times M}$ and $\boldsymbol{c} \in \mathbb{R}^{N \times 1}$ is a vector of remaining coefficients parametrizing the solution space. Thus, the set of all possible solutions for $\boldsymbol{J}i$ is given by (23). The goal now is to pick the sparsest solution from this set. Therefore, equation (23) may be posed as the overdetermined ($M > N - r$) problem

$$V \tilde{\Sigma} U^T \boldsymbol{Y}_i^T = -V \boldsymbol{c}. \tag{24}$$

Minimizing the L_1 error

$$E_i(\boldsymbol{c}) = \| V \tilde{\Sigma} U^T \boldsymbol{Y}_i^T + V \boldsymbol{c} \|_1, \tag{25}$$

yields a sparse solution [31]. However, unlike the L_2 minimization, L_1 minimization has no analytical solution, so choosing an appropriate iterative algorithm to solve it is essential. The Barrodale Roberts algorithm [34] provides a particularly fast solver that has been vastly used in the field of network reconstruction [10, 12, 29, 35].

Remarks. *The core equations (50) also provide the option to reconstruct network connectivity via maximizing sparseness of the network and there is a particular relation to what is known as compressive sensing cf, e.g. [36].*

In general, linearization of dynamical equations, e.g. linearizing in state variables close to fixed points, often well approximates nonlinear dynamics. This seems to hold for gene regulatory networks [29] as well as in models of Drosophila segmentation networks [37] and may thus be of general use across systems. For gene and protein interaction networks, often single genes are selected for perturbations in an experiment, with the danger of providing non-generic restrictions in (12). Finally, for some systems increasing the number of experiments may reduce the resulting computational costs such that this trade-in may be considered.

Driving the System's State to a Fixed Point

One may also infer network structure by externally driving the system to a fixed point and shifting component values of the fixed point for individual units [38, 39]. As before, the differences between pairs of steady-state responses are analyzed. Let us describe the network as

$$\dot{x}_i = f_i(x_i) + \sum_{j=1}^{N} A_{ij} g_{ij}(x_i, x_j) + I_i, \tag{26}$$

where the $A_{ij} \in \{0, 1\}$ are the entries of the adjacency matrix specifying only if an interaction from j to i is present ($A_{ij} = 1$) or not ($A_{ij} = 0$), the $g_{ij}(x_i, x_j)$ are the coupling functions from $j \in \{1, 2, ..., N\}$ to i, and I_i is the driving signal applied to unit i. It was demonstrated by Yu and Parlitz [38] that under driving signals

$$I_i = -(x_i - \hat{x}_i)\theta, \tag{27}$$

with sufficiently large gain factor $\theta \in \mathbb{R}$ and Lipschitz continuous f_i and g_{ij}, the network may be driven to a globally stable fixed point $\boldsymbol{x}^* := (x_1^*, x_2^*, \ldots, x_N^*)^\mathsf{T} \in \mathbb{R}^N$ that is arbitrarily close to a predetermined point $\hat{\boldsymbol{x}} := (\hat{x}_1, \hat{x}_2, \ldots, \hat{x}_N)^{\mathsf{T}} \in \mathbb{R}^{\mathbb{N}}$, independent of the initial conditions [38] . At such fixed point we have

$$(x_i^* - \hat{x}_i)\theta = f_i\left(x_i^*\right) + \sum_{j=1}^{N} A_{ij} g_{ij}\left(x_i^*, x_j^*\right) \tag{28}$$

for all i. To understand how the network responds to changes in $\hat{\boldsymbol{x}}$, let us define

$$\Delta_i := f_i(\hat{x}_i) + \sum_{j=1}^{N} J_{ij} g_{ij}(\hat{x}_i, \hat{x}_j) - \left[f\left(x_i^*\right) + \sum_{j=1}^{N} J_{ij} g_{ij}\left(x_i^*, x_j^*\right) \right]. \tag{29}$$

hence, equation (28) may be rewritten in terms of Δi as

$$\left(x_i^* - \hat{x}_i\right)\theta = f_i(\hat{x}_i) + \sum_{j=1}^{N} A_{ij} g_{ij}(\hat{x}_i, \hat{x}_j) - \Delta_i. \tag{30}$$

The main idea at this point is to check whether unit k couples to unit i by evaluating how x_i^* reacts to the shifting of x_k^* through $\hat{x}_k$. Therefore, let us set

$$\hat{x}_j = \begin{cases} \hat{x}_k & \text{if } j = k \\ 0 & \text{if } j \neq k, \end{cases} \tag{31}$$

and evaluate (30) at this point, yielding

$$x_i^*\theta = A_{ik} g_{ik}\left(0, \hat{x}_k\right) - \Delta_{ik} + f_i(0) + \sum_{j \neq k}^{N} A_{ij} g_{ij}\left(0, 0\right). \tag{32}$$

Shifting the same component twice to $\hat{x}_k^{(1)}$ and $\hat{x}_k^{(2)}$, resp., fixes a reference frame and thereby yields equations characterizing the difference between responses of a given unit i to a shifted fixed point component for unit k. It results in

$$\left[x_{i,2}^* - x_{i,1}^*\right]\theta = A_{ik}[g_{ik}(0, \hat{x}_{k,2}) - g_{ik}(0, \hat{x}_{k,1})] + [\Delta_{ik,1} - \Delta_{ik,2}], \tag{33}$$

a condition that may be rewritten as

$$S_{ik}\theta = A_{ik}\eta_{ik} + \lambda_{ik}. \tag{34}$$

We remark that the differences $[\Delta_{ik,1} - \Delta_{ik,2}]$ in (33) are not known but the general form (34) may be used to reveal whether they are zero, $\lambda_{ik} = 0$, or not: given that we are dealing with entries of the adjacency matrix, we may infer two possible outcomes from (34), whether system k is coupled to i or not. Specifically,

$$S_{ik}\theta = \begin{cases} \eta_{ik} + \lambda_{ik} & \text{if unit } k \text{ coupled to } i \\ \lambda_{ik} & \text{if not.} \end{cases} \tag{35}$$

It was also demonstrated by Yu and Parlitz [38] that Δi decreases with θ if f_i and g_{ij} are Lipschitz continuous. This permits to discriminate whether there is a coupling between a pair $k \rightarrow i$. Especially, when θ is sufficiently large, the $|S_{ik}\theta|$ values may be classified into sets $\mathbb{I}_0$ and $\mathbb{I}_1$, non-coupled and coupled sets, respectively. To construct such sets, Yu and Parlitz [38] propose to:

- For fixed k, organize the $|S_{ik}\theta|$ values in an ascending order, i.e., the values should be arranged into a new series z where $zk_j < zk_{j+1} < ...$ that defines the indexing j of the zk_j.
- Establish the critical values of each set. In this case, the critical values j_c and jc_{+1} define the end and the beginning of $\mathbb{I}_0$ and $\mathbb{I}_1$, respectively. Yu and Parlitz suggest to find j_c by requiring the distance between any element from $\mathbb{I}_1$ with respect to zi_1 to be larger than twice the size of $\mathbb{I}_0$, $z_{k,j} - z_{k,1} \geqslant 2(z_{k,j_c} - z_{k,1})$ for all $j > j_c$.

Finally, by performing this process on every unit, the topology of the network may be reconstructed.

Remarks. *The approach relies on the feasibility of (i) perturbing the systems in a specific manner, and (ii) measuring the steady states, suggesting that it is model independent to a large extent in that it does not in principle require knowledge of local dynamics or coupling functions (yet it requires these to be Lipschitz continuous). These features may make the approach of interest under certain conditions where only little pre-knowledge about the system is available. Yet the approach requires substantial control over the system, in particular, the option of externally driving every unit (independently) constitutes a major requirement. The study [38] does not state how indirect actions are treated, for instance, how is an indirect effect from unit k via k' onto i distinguished from direct interactions from k to i? Possibly, driving k may indirectly affect i only weakly and this potentially second order contribution could be treated in a perturbative way.*

Distributed Perturbations to Collective Periodic Dynamics

The approaches presented above (sections 3.1–3.3) required the existence of fixed points either in the original system or in the presence of sufficiently strong external driving. Yet, more complex dynamics prevails in a large range of biological, physical or artificial systems. The second most simple invariant dynamics are periodic orbits and often arise as limit cycles of coupled oscillatory units, thus asking for a generalization beyond simple fixed point approaches. Even more complex dynamics, e.g. collective chaos, is treated by a direct approach below in section 5.1.

Is it possible to infer network topology from driving-response experiments also for oscillator networks? Below we positively answer this question, at the same time showing that distributed driving signals not precisely targeted to one or a few units are at least equally appropriate to infer network topology. Several theoretical model studies of coupled oscillators [40–45] have shown that the response of single units in a network to constant or periodic driving signals as well as the transient dynamics of synchronization depend on the network topology. Some recent works [43, 44] helped us to understand specific quantitative influence of structural features on the response and how the network response provides some information about the structure (and the driving signal). For instance, the magnitude of responses seem to decay exponentially with distance from the driving node [43], and the coarse-scale connectivity among connected components may qualitatively determine to which degree network dynamics is coordinated globally [44]. Further developing such insights, a follow-up work [10] presents a method of reconstructing network topology from systematic measurements of network responses to temporally constant, distributed driving signals in coupled phase-oscillator networks.

The basic idea is that any network displaying a stable invariant dynamics, not just fixed points, yield a specific response to a given perturbation as a consequence of the network's topology and the perturbation itself [40, 42, 44], cf figure 3. If the perturbations are small, the invariant set is typically *qualitatively* unchanged and only slightly moved in state space. Keeping track of which driving signals resulted in which responses, we can collect evidence about the interactions among units in a network. Sufficiently many repetitions of appropriate driving-response experiments then yield the network›s topology.

Figure 3: Topology revealed by driving a stable periodic orbit. As for fixed point approaches discussed above (section 3.1 and figure 2), the difference vector v (red) depends on both the driving signal and the network topology such that several measurements of v under different driving conditions yields information about the topology.

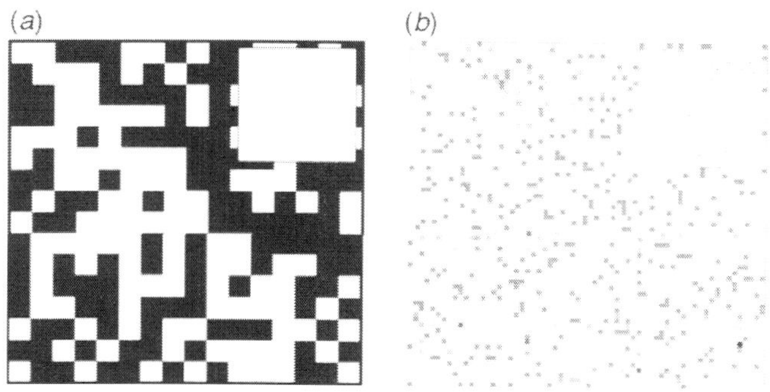

Figure 4: Revealing network topologies from response dynamics for directed networks of phase-locked Kuramoto oscillators. Variables evolve according to $\dot{x}_i = \omega_i + k^{-1}\sum_{j=1}^{N} J_{ij}\sin(x_j - x_i)$, with random frequencies $\omega i \in [0.1, 1]$ and $k = 8$ directed interactions of strengths $J_{ij} = k^{-1}$, randomly selected for each unit. Panels show reconstructed coupling matrices J for (a) $N = 16$, and $M = 32$, and (b) $N = 64$, and $M = 32$. The matrices are gray-coded from white ($\hat{J}_{ij} = 0$) to black $(\hat{J}_{ij} = \max_{i'j'}\{\hat{J}_{i'j'}\})$. Insets: element-wise absolute difference $|J_{ij}^{\text{derived}} - J_{ij}^{\text{original}}|$, plotted on the same scale.

Weakly coupled limit cycle oscillators are well-characterized by ignoring (in the long time limit) amplitude responses to coupling and by modeling them as phase-oscillators with coupling via their phase differences only. A method to infer network topology for coupled phase-oscillators with arbitrary stable, phase-locked dynamics has been presented in [10]. One key observation is that the phase differences (yet not the phases themselves) in such systems converge with time and that comparing differences of phase differences among different driving conditions yield restrictions to network topology. The network dynamics is given by

$$\dot{\phi}_i = \omega_i + \sum_{j=1}^{N} J_{ij} g_{ij}(\phi_j - \phi_i) + I_{i,m}, \tag{36}$$

where $\phi_{i(t)}$ and ωi are the phase and natural frequency of oscillator i, respectively, J_{ij} the connection strength from oscillator j to i and $I_{i,m}$ is a temporally constant driving signal applied to i during the experiment m. We assume that in the absence of driving, $I_{i,m} \equiv 0$, the network is in a phase-locked state where $\dot{\phi}_j - \dot{\phi}_i = 0$ for all i, j. We remark that one, several or all units may be perturbed during each given experiment, such that driving can be arbitrarily distributed and effectively changes the frequencies of the driven oscillators. As for the approaches relying on fixed points (section 3.1), the existence of a stable periodic orbit (and thus in particular a phase-locked state) implies that sufficiently small constant perturbations yield an (only slightly moved and slightly different) stable periodic orbit.

If for a given driving condition m, the dynamics becomes phase-locked, the phase differences

$$\Delta_{ij,m}(t) = \phi_{j,m}(t) - \phi_{i,m}(t) \tag{37}$$

become constant in time, $\Delta_{ij,m}(t) \rightarrow \Delta^*_{ij,m} := \lim_{t\rightarrow\infty} (\phi_{j,m}(t) - \phi_{i,m}(t))$ because all oscillators move at the same collective frequency

$$\Omega_m = \omega_i + \sum_{j=1}^{N} J_{ij} g_{ij}(\phi_{j,m} - \phi_{i,m}) + I_{i,m}. \tag{38}$$

Hence, if the network is perturbed by a sufficiently small driving signal, the original phase-locked state (for $I_{i,m} \equiv 0$) is slightly moved such that $|\Delta^*_{ij,m} - \Delta^*_{ij,0}| \ll 1$ and there is a small difference between the perturbed and non-perturbed collective frequencies Ωm and Ω_0. Defining the effective frequency difference $D_{i,m} := \Omega m - \Omega_0 - I_{i,m}$ of oscillator i, and approximating the arbitrarily nonlinear coupling functions g_{ij} by a first order Taylor expansion around $\Delta^*_{ij,0}$ we obtain

$$D_{i,m} = \sum_{j=1}^{N} \hat{J}_{ij}\theta_{j,m} \tag{39}$$

where $\theta j\ m := \phi_{j\ m} - \phi_{j,0}$ is the phase shift and $\hat{J}$ is the Laplacian matrix of the network given by

$$\hat{J}_{ij} = \begin{cases} -J_{ij}g'_{ij}\left(\Delta^*_{ij,0}\right) & \text{for } i \neq j \\ \sum_{k,k\neq i} J_{ij}g'_{ik}\left(\Delta^*_{ik,0}\right) & \text{for } i = j. \end{cases} \tag{40}$$

Now, identifying the matrices $X_{i,m} = D_{i,m}$ and $Y_{i,m} = \theta_{i,m}$ we have reduced the problem of identifying network topology using distributed perturbations in systems of limit cycle oscillators to solving the same linear algebraic equation (11).

As remarked in previous sections, several experiments are necessary in order to perform the reconstruction of $\hat{J}$. Therefore, from repeated measurements for different conditions it is possible to rewrite equation (39) in the form (11) in terms of $Y = D \in \mathbb{R}^{N\times M}$ and $X = \Theta \in \mathbb{R}^{N\times M}$ representing the differences between collective dynamics and phase shifts for each of the N systems during the M experiments.

Now, for sufficiently many experiments, i.e. $M \geq N$, the reconstruction may be accomplished in principle but may be ill-conditioned numerically. In addition, for large N the many required experiments may not be practical. if this condition is not fulfilled, methods like the *setting a maximum connectivity per system* or *maximizing the sparseness of the connectivity matrix* (section 3.2.2), may be applied in order to make the problem of retrieving $\hat{J}$ an overdetermined problem.

As shown in [10], we may compare how accurate our prediction is by defining $J_{\max} := \max_{i'j'}\left\{\left|J^{\text{derived}}_{i'j'}\right|, \left|J^{\text{original}}_{i'j'}\right|\right\}$, and using a relative difference defined as

$$\Delta J_{ij} := \frac{1}{2J_{\max}}\left|J^{\text{derived}}_{ij} - J^{\text{original}}_{ij}\right|, \tag{41}$$

where $\Delta J_{ij} \in [0, 1]\ \forall\, i, j$. In addition, the quality of reconstruction Q_α may be posed as the fraction

$$Q_\alpha := \frac{1}{N^2}\sum_{i,j} H((1-\alpha) - \Delta J_{ij}) \in [0, 1] \tag{42}$$

of connection strengths which are assumed to be correct. Here $\alpha \leq 1$ is a constant employed to set the required accuracy for predictions and H is the Heaviside function, $H(x) = 1$ for $x \geq 0$ and $H(x) = 0$ for $x < 0$. For instance, $\alpha = 0.95$ means that the derived matrix has a normalized relative error (41) of at most 5%. Moreover, we may estimate the minimum number of experiments

$$M_{q,\alpha} := \min \{M | Q_\alpha (M) \geqslant q\} \tag{43}$$

required for a reconstruction with a quality level q and with a prediction accuracy α. Figure 5illustrates these measures for random networks of phase-locked Kuramoto oscillators for several random topologies and parameters.

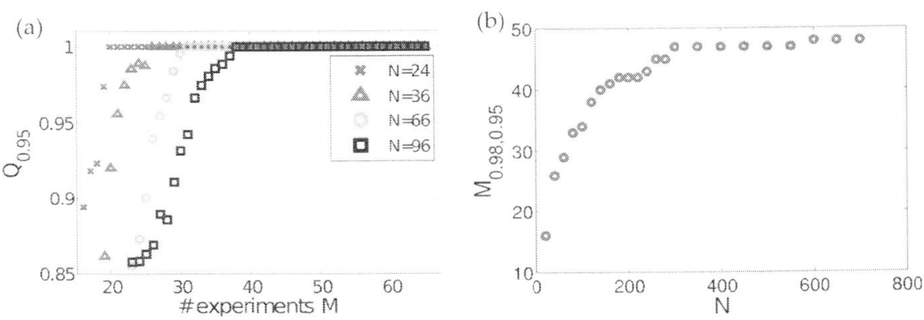

Figure 5: Quality of reconstruction and number of required experiments for reconstructing directed networks. Phase-locked Kuramoto oscillators with dynamics defined as in figure 4with $k_i = 10$ random incoming connections per node. (*a*) Quality of reconstruction at $\alpha = 0.95$ for $N = 24$ (×), $N = 36$ (⍰), $N = 66$ (◯) and $N = 96$ (□). (*b*) Minimum number of experiments required for a reconstruction of quality level $q = 0.98$ and accuracy $\alpha = 0.95$.

The driving response method in principle may be applied to a broad variety of problems involving stable dynamics. A model analogous to equation (39) could be inferred as long as the systems may be linearized around a stable state, allowing to retrieve the topology from the network responses as above. Yet, there may be practical problems. For instance, even for perturbations induced by constant driving signals, the invariant solution resulting from perturbations to more complex periodic orbits or other stable invariant sets may be describable only by time dependent quantities (and not, e.g., temporally constant phase differences), limiting the approach suggested above to specific classes of systems.

Features and Restrictions

One common advantage of the approaches presented above is that their required computational effort scales well (weaker than linearly) with system size N such that at least moderately large systems appear accessible (cf figure 5). At the same time, the approaches are relatively simple to realize because they do not require knowledge in higher mathematics or computational approaches beyond a basic standard.

A possible route of generalization is to combine some of the above approaches. For instance, one may first drive a system to a stable fixed point as in section 3.1 and then apply small perturbations around that new point as in section 3.3.

Yet, all these approaches require the researchers to be able to access (measure and drive) the dynamics of all units in the system. Moreover, the local dynamics as well as the (approximate) form of interactions typically need to be at least partially known. The collective dynamics suitable for the driving-response approaches described above also need to be simple, in fact to exhibit a stable fixed point or periodic orbit or to admit the system to be driven to such as state. Finally, the presented inference of the existence of physical interactions and their functional form [46] seems well understood for networks of phase-oscillators, where perturbations in oscillation amplitude decays on faster time scale than the relaxation of phases. It thus remains an open problem how to use a driving-response approach to properly infer structural network connectivity of coupled oscillators in systems, where the amplitude degrees of freedom play a role or are even dominant. More generally, systems exhibiting more complex dynamics, such asynchronous chaotic activity, bifurcations, multistability or other prevalent features of high-dimensional, nonlinear systems, currently still prevent network reconstruction by the methods presented above.

These requirements severely restrict the range of applicability in praxis to simple, well-accessible systems only. In particular for biological systems such as neural circuits or gene interaction networks, dynamics are typically more complex, systems are large and it is still hard to implement controlled large-scale driving experiments on the single-unit level. Direct methods (section 5) that do not rely on driving the system seem to offer viable directions towards reconstructing networks with such more complex dynamics as well. A currently open question of research constitutes how to exploit recorded time series from only a fraction of the units.

COPY-SYNCHRONIZATION: ADAPTING A MODEL COPY

Another way of reconstructing network topology of a given network is by adapting the topology of a second, model system, a *network copy,* such that it synchronizes with the original system. The idea is to update the model topology continuously until the copy system exhibits a dynamics identical to the original system; the rationale is that the final topology of the copy is likely to be the original topology [47].

Specifically, consider an (original) system of the form

$$\dot{x}_i = f_i(x_i) + \sum_{j=1}^{N} J_{ij} g_j(x_j), \tag{44}$$

where f_i and g_{ij} are known and assumed to be Lipschitz continuous. This original system can in principle have arbitrary dynamics. Now, let us propose a model copy

$$\dot{y}_i = f_i(y_i) + \sum_{j=1}^{N} K_{ij} g_j(y_j) + I_i, \tag{45}$$

where y_i represents the state of the copy system, $I_i(t)$ are control feedback signals and $K_{ij}(t)$ is current coupling strength in the test system. In order to synchronize the copy to the original system, both the feedback signals $I_i(t)$ and the coupling strengths $K_{ij}(t)$ evolve in time and depend on the states of the actual and the copy system. Defining the synchronization error

$$e_i = y_i - x_i \tag{46}$$

we adapt the coupling strengths in the model copy according to

$$\dot{K}_{ij} = -\gamma_{ij} g_j(y_j) e_i \tag{47}$$

and

$$I_i = -\alpha e_i \tag{48}$$

where $\alpha > 0$ and $\gamma ij > 0$. We remark that here the local dynamics $f_i(.)$ as well as the coupling function $g_{ij}(.)$ need to be known in order to set up the test copy. It was proven in [47] that under feedback signals (48) with sufficiently large α, the synchronization error e_i decreases in time, $\dot{e}_i \leqslant 0$ for all i such that the two systems converge to a synchronized state. The rationale is that after synchronization, the copy topology equals that of the original network, $K_{ij} \approx J_{ij}$, cf figure 6. With minor modifications on the control signals I_i this method admits to reconstruct networks and sub-networks in the presence of disturbances and modeling errors as well [47].

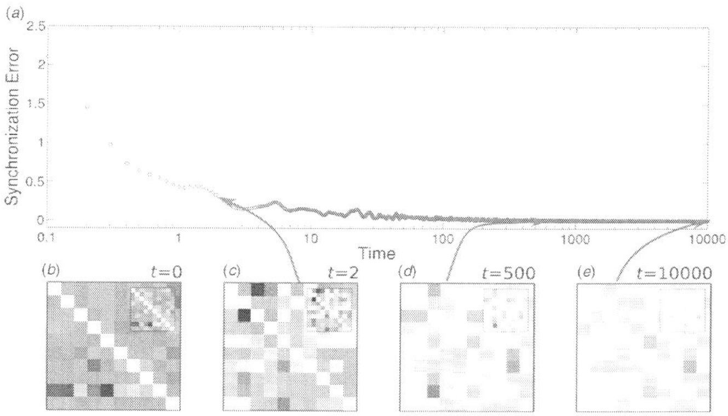

Figure 6: Revealing network topology via copy-synchronization. Dynamics of reconstructed network for directionally coupled Kuramoto oscillators $\dot{x}_i = \omega_i + k^{-1}\sum_{j=1}^{N} J_{ij}\sin(x_j - x_i)$ with random frequencies $\omega i \in [-1, 1]$, random coupling strengths $J_{ij} \in [0.5, 1]$, $N = 10$ and $k = 3$ random incoming connections per node. (*a*) Dynamics of synchronization error between the original and the copy network measured as $e(t) = \sqrt{N^{-1}\sum_i e_i^2(t)}$. (*b*), (*c*), (*d*) and (*e*) show the inferred topologies at $t = 0$, $t = 2$, $t = 500$ and $t = 10^4$. The matrices are gray-coded from white ($J_{ij} = 0$) to black ($J_{ii} = 1$). The insets depict the element-wise absolute differences $|J_{ij}^{\text{derived}} - J_{ij}^{\text{original}}|$, plotted on the same scale.

We remark that to the best of our knowledge, there is no guarantee that the resulting topology of the copy system actually reflects the one of the original network. In particular, symmetries might lead to disambiguities.

Further, the method based on copy-synchronization is model dependent such that knowing the intrinsic and coupling functions of units is vital, as in several parts of section (3). Its applicability has been explicated for sample networks of up to $N = 16$ nodes, but it remains unclear how to handle large networks as of the order of N^2 links K_{ij} need to be co-evolved in time and a bound of convergence times is currently missing. At the same time, the copy approach does not require perturbations to the original systems, so experimental access to it need not include driving access to its units. Interestingly, Yu *et al* [47] highlight that the method may be useful to track changes in a network's connectivity in real time.

DIRECT APPROACHES

In the previous two sections, we have introduced methods to infer network connectivity by either interfering with the system (driving-response approaches, section 3) or by setting up and synchronizing a second, model system (section 4). Both classes of methods work if certain requirements are met (in particular, the option to actively drive the system or the option to synchronize the copy with the original system, respectively). It remains a challenge to infer network topology from dynamics without such requirements.

Reconstruction by Purely Observing Dynamics[5]

Methods based on copy-synchronization (section 4) assume that the local dynamics f_i as well as the interaction functions g_{ij} in (1) are known and the g_{ij} do not depend on the state of unit i, $g_{ij}(x_j)$. Inferring network connectivity, i.e. the J_{ij}, then relies on the construction of a second, model network, with dynamics governed by equation (1) and network parameters J'_{ij} that are tuned to that of the real network by an error minimization procedure. As noted recently [12], one may solve the same reconstruction problem with significantly reduced efforts and reduced requirements by evaluating the states and their time derivatives directly from the time series recorded from the original system. In particular, such a simple direct method [12] removes the need to set up and synchronize a second, copied, system.

The idea is as follows. If the local dynamics and the coupling functions are known, their evaluations at different times are also known from recorded time series and the only remaining unknown parameters in equation (1) are the coupling strengths, which are to be determined. Specifically recording a time series $x_i(t_m)$ at sufficiently closely spaced times $t_m \in \mathbb{R}$ and estimating the temporal derivatives6 of it yields the dynamics of the ith unit given by

$$\dot{x}_i(t_m) = f_i(x_i(t_m)) + \sum_{j=1}^{N} J_{ij} g_{ij}(x_i(t_m), x_j(t_m)). \tag{49}$$

If there are M such times, $m \in \{1, ..., M\}$, we have M equations of the form

$$\dot{x}_{i,m} = f_{i,m} + \sum_{j=1}^{N} J_{ij} g_{ij,m} \qquad (50)$$

with abbreviations $\dot{x}_{i,m} := \dot{x}_i(t_m)$, $f_{i,m} := f_i(x_i(t_m))$ and . Repeated evaluations of the equations of motion (49) of the system at different times t_m thus comprise a simple and implicit restriction on the network topology J_{ij} as follows: writing $Y_{i,m} = \dot{x}_{i,m} - f_{i,m}$ and the matrix $X_i = (g_{ij,m})_{j,m} \in \mathbb{R}^{N \times M}$, these equations constitute the matrix equation

$$\boldsymbol{Y}_i = \boldsymbol{J}_i X_i \qquad (51)$$

where $\boldsymbol{Y}_i \in \mathbb{R}^{1 \times M}$ and $\boldsymbol{J}_i \in \mathbb{R}^{1 \times N}$ is the ith row of the interaction matrix J, comprising the vector $(J_{ij})_{j \in \{1, ..., N\}}$ of all input coupling strengths to unit i.

The main restricting equations (51) again have the same form as the generic restrictions (11) and thus may be solved analogously. In [12], Euclidean L_2-norm minimization was used to infer the topology. Numerical tests show that reconstruction works well for transient as well as attractor dynamics, for simple as well as complex, chaotic units and collective states, and even in the presence of noise that substantially alters the dynamics and thus the recorded time series. Figure 7 illustrates successful reconstruction in the presence of various levels of noise.

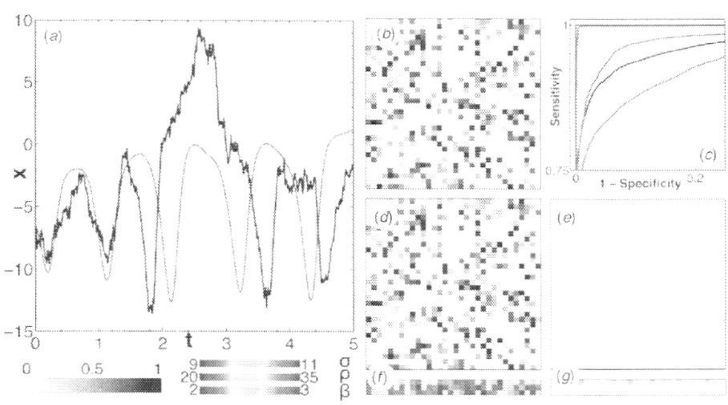

Figure 7: Reconstructing a random network complex chaotic dynamics. Example of N=32 Lorenz oscillators with dynamics given by $\dot{x}_i = \sigma_i(y_i - x_i) + \sum_{j=1}^{N} J_{ij}(x_j - x_i) + \xi_i^{(x)}$, $\dot{y}_i = x_i(\rho_i - z_i) - y_i + \xi_i^{(y)}$, $\dot{z}_i = x_i y_i - \beta_i z_i + \xi_i^{(z)}$, with unknown parameters $\forall i, j$ $\{J_{ij}, \sigma_i, \rho_i, \beta_i\}$ in the presence of external Gaussian white noise ξ_i of amplitude λ. (*a*) Dynamics of a unit in the absence (blue) and presence (black)

of noise with $\lambda = 5$. Original (*b*) and reconstructed (*c*) connectivity, inferred parameters (*f*), and their errors compared to the original topology and parameters (*e*) and (*g*). (*c*) Receiver operating characteristics (ROC) of reconstruction under noise-free (blue) and noisy measurements (black and red) for $\lambda \in \{0.1, 1, 10\}$. Adapted from [12].

It is clear that certain types of dynamics do not allow topology inference. For instance, if all local dynamics are identical $f_i \equiv f$, all coupling functions are identical, $g_{ij} \equiv g$, and collectively the units are identically synchronous, i.e. all in the same states at the same times, there is no information about the connection topology one can possibly obtain from (only) recording time series, because for *all*strongly connected topologies7, the collective dynamics would be identical.

Further theoretical considerations show that following the same lines of analysis as above, all parameters occurring linearly in the local dynamics of coupling functions, can also be reconstructed by the same error minimization based on equation (11). For instance for the (fictional) coupling function $g_{ij}(x, y) = a\sin(xy) - \exp(b + y - x)$, the parameters a and b need not be known but can be inferred as well (up to one constant prefactor for each pair (i, j) of nodes), because both a and $\exp(b)$ occur linearly in the equations of motions (49). In many physical systems, parameters actually do occur linearly. These include, for instance, the dynamics of coupled classical electric LCR circuits and the strengths of diffusive (4) or direct linear coupling (6). Moreover, widely used model systems for networks of neurons, such as leaky integrate-and-fire (LIF) and quadratic integrate-and-fire neurons [50] or networks of coupled chaotic systems such as Rössler or Lorenz systems [51, 52] exhibit all or at least most parameters occurring linearly or affinely. For all such systems, local dynamics f_i and coupling functions g_{ij} may not or not completely be required to be known in advance.

Such a direct inference method [12] for network topology from complex collective dynamics thus serves as a simple starting strategy for connectivity reconstruction in a broad range of networked systems.

Pulse-coupling: Reconstruction from Spike Patterns

Networks of spiking neurons or other pulse-coupled units are formally hybrid systems [53], because the continuous-time flow is interrupted at certain time points, where events (e.g. spike sending or reception) modify the dynamics [54, 55] via discrete time maps.

To start addressing the reconstruction problem for such hybrid systems, researchers have considered one of the simplest model networks of pulse-coupled units based on LIF models. Such network models are most broadly used in mathematical analysis and computational modeling. Each unit in such a network has one state variable, its membrane potential, roughly emulates leaky capacitive properties observed for membranes of biological neurons and is complemented with a threshold where a pulse (action potential or spike [56]) is artificially created and the membrane potential is reset. Although these models lack certain biological details, such as a natural spike generating mechanisms, they are simple enough to be studied analytically and they have been useful for furthering our understanding how the information is processed among neurons [57].

Explicitly, the membrane potential $V_i(t)$ of a LIF unit i changes as in an RC circuit (resistor and a capacitor connected in parallel) according to

$$\dot{V_i} = \gamma_i(R_i I_i - V_i) + S_i(t), \qquad (52)$$

where R_i, γi and I_i are the membrane resistance, inverse membrane time constant and the external current applied to neuron i, respectively. Once the potential of a unit j crosses a threshold, $V_j(t) \geq VTj$, the potential is reset to $V_j(t^+) = VR\, j$ and the unit emits a pulse that it transmitted to the neuron›s post-synaptic neighbors [58]. The time of this pulse sending event is remembered as the unit's mth spike time $tj\ m := t$. The collection of such pulses then define the actions onto post-synaptic units $i$such that the quantity

$$S_i(t) = \sum_{j=1}^{N} \sum_{k \in \mathbb{Z}} J_{ij}\delta(t - t_{j,m} - \tau_{ij}), \qquad (53)$$

in (52) represents the pulses that unit i receives from the rest of the network. Here, J_{ij} and τij are the synaptic coupling strength and the synaptic transmission delay from unit j and i, respectively. Furthermore, δ(.) is the Dirac delta distribution modeling a potential response kernel that is much faster than all other time scales involved.

The main question we address now is whether and how the network connectivity, as specified by the matrix of coupling strength J_{ij} can be reconstructed if the pattern of spikes times $(t_{j,m})_{j\in\{1,\dots,N\},m\in\mathbb{Z}}$is given? We remark that we do not assume access to the natural state variables, the membrane potentials $V_i(t)$, which may not be observable, but only to the spike times that are generated by the dynamics of these potentials. This difference constitutes the main novelty of the approaches presented in this subsection compared to those for continuous time state variables presented before.

First observe that (52) has an explicit solution [35] given by

$$V_i(t) = R_i I_i (1 - e^{-\gamma_i(t-t_0)}) + V_i(t_0) e^{-\gamma_i(t-t_0)} + \sum_{j=1}^{N} \sum_{t_0 < t_{j,m} + \tau_{ij} \leqslant t} J_{ij} e^{-\gamma_i(t-t_{j,m}-\tau_{ij})}. \tag{54}$$

if the time interval $[t_0, t)$ lies in between two subsequent spikes of neuron i. The first sum is taken over all neurons while the second is taken over those spikes generated by the other neurons $j \neq i$ that affect i in the time interval $[t_0, t)$.

Based on this solution, we now present two distinct approaches to infer network connectivity, one direct and exact inference method assuming oscillatory units and one coarse approximate method based on stochastic optimization.

Direct Topology Inference Assuming Oscillatory Units

Van Bussel *et al* [35] assumes that all the parameters of (54), besides the synaptic couplings J_{ij}, are known. The rationale is that delays and membrane time constants as well as a unit's equilibrium potential $R_i I_i$ may be estimated beforehand. How could these coupling strengths and thus the structural network topology be inferred? If the currents I_i are sufficiently large such that $R_i I_i > VT_{.}i$the units are oscillatory such that they create spike even without recurrent inputs from the remaining network. After crossing the threshold $VT_{.}i$, unit i sends a spike to its post-synaptic neighbors and is reset to a resting value $VR_{.}i$, so that the unit's state at the boundaries of each inter-spike interval $[ti_{.}k_{-1}, ti_{.}k)$ is determined and one may take $t_0 = ti_{.}k_{-1}$ and $t = ti_{.}k$. However, these threshold crossings may be induced in two manners at $t = ti_{.}k_{-1}$, (*a*) by incoming excitatory spikes from other neurons such that $V_i(t^-) < VT_{.}i$ but $V_i(t^-) + \sum_{\{j:\exists m: t_{j,m}+\tau_{ij}=t\}} J_{ij} > V_{T,i}$ or (*b*) by the intrinsic oscillation of the unit such that $V_i(t^-) = VT_{.}i$ without simultaneously incoming spike(s). In both cases, the membrane potential is at reset value immediately after sending a spike. So identifying $t_0 = ti_{.}k_{-1}$ in (54), we have

$$V_i(t_0) := V_i(t_{i,k-1}) = V_{R,i}. \tag{55}$$

If the next spike is oscillation-induced (*b*), the membrane potential approaches its threshold value $VT_{.}i$ continuously such that in addition

$$V_i(t^-) := \quad V_i(t_{i,k}^-) = V_{T,i} \tag{56}$$

where we now identified $t = ti\,k$ in (54). Thus, each oscillation-induced spike at some $t = ti\,k$ implies a linear restriction of the form (54) for the coupling strengths J_{ij} by equating $[t_0, t) = [ti\,k_{-1}, ti\,k)$.

For M different inter-spike intervals obeying (55) and (56), this provides a linear system of equations restricting the coupling matrix. However, as van Bussel *et al* [35] remark, consecutive intervals may display similar patterns, such that it is often advisable to select $M > N$ inter-spike intervals sufficiently separated in time, to minimize correlations between intervals and thus numerical inaccuracies. For each i, defining the subset of inter-spike intervals

$$D_{i,m} := (t_{i,l_m-1}, t_{i,l_m}) \tag{57}$$

and

$$T_{i,m} = t_{i,l_m} - t_{i,l_m-1} = |D_{i,m}| \tag{58}$$

equation (54) may be rewritten as

$$X_i \boldsymbol{J}_i^T = \boldsymbol{Y}_i, \tag{59}$$

where $\boldsymbol{J}_i := (J_{ij})_{j \in \{1,\dots,N\}}$ and $X_i \in \mathbb{R}^{M \times N}$ and $\boldsymbol{Y}_i \in \mathbb{R}^{M \times 1}$ are given by

$$(x_i)_{mj} = \sum_{t_{j,k}+\tau_{ij} \in D_{i,m}} J_{ij}\, \mathrm{e}^{-\gamma_i (t_{i,l_m} - t_{j,k} - \tau_{ij})} \tag{60}$$

and

$$Y_{i,m} = V_{T,i} - R_i I_{i,m}(1 - \mathrm{e}^{-\gamma_i T_{i,m}}) - V_{R,i}\, \mathrm{e}^{-\gamma_i T_{i,m}}. \tag{61}$$

Repeating the same process for all units i yields the topology of the whole network. As shown in section 3.2, the overdetermined problem, $M > N$, and the undetermined, $M \ll N$, may be solved minimizing the L_2-norm or maximizing the sparseness of the network, respectively. A major limitation is that the presented approach requires a large amount of prior knowledge about the system, such as the time delays between units and their time constants, among others. Given this knowledge, the approach is capable of inferring large networks of neurons as the computation time, as well as the number M of required inter-spike intervals to be evaluated scales linearly with system size [35, 59].

Stochastic Optimization of All Systems Parameters

In a complementary study, Makarov *et al* [60] showed how stochastic optimization of all system parameters

$$p_i = \left(\gamma_i, R_i, I_i, \tau_{ij}, J_{ij}\right)_{i,j\in\{1,\dots,N\}} \tag{62}$$

given the spike trains for all neurons in the network may yield a network topology roughly consistent with the actual one.

The idea is to evaluate the predicted inter-spike interval durations $\bar{T}_{i,m}(p)$ from a test set of parameters and to optimize those parameters to reproduce a given (measured) spike train as closely as possible. Thus, minimizing the difference between the predicted and measured spike trains is vital for this method. The authors in particular applied their idea also to recordings of biological neurons.

As the equation (54) yields transcendental relations for the $\bar{T}_{i,m}(p)$, their estimates need to be determined numerically. Thus, finding

$$p_i^* = \arg\min_{p_i} E_i(p_i) = \arg\min_{p_i} \sum_{m=1}^{M} (T_{i,m} - \bar{T}_{i,m}(p_i))^2 \tag{63}$$

yields the best (in Euclidean distance norm) solution for the set of parameters.

Briefly, to find the minimum p_i^* in (63) one must explore the parameters space. This means that the solution is found through iteratively choosing random values for ***pi*** and comparing the value of (63) for consecutive iterations. By relating the changes in (63) with the changes in ***pi*** one may establish directions in which the minimum may be found by gradient descent. However, it is also advisable to change directions in the parameters space randomly. Mainly, because there may better solutions for p_i^* in regions of the parameters space where they are not expected to exist [61]. Makarov *et al* [60] thus applied stochastic optimization for searching the minimum.

As a strong requirement, this method needs the number of recorded spikes to be $M \gg 2N$; as noted in [60], robust regression models are more suitable to handle this kind of problems and special care with the inter-spike intervals must be taken as, e.g., spike bursts may lead to false intervals.

Features and Limitations

The stochastic optimization approach provides a generic approach in finding best fitting parameters and thus here, potentially a well fitting network; yet, it is computationally demanding and simultaneously requires many recordings compared to the size of the network. In contrast, the direct approach assumes a large amount of pre-knowledge, in particular regarding the unit's parameters. In addition, by requiring to pre-process the data sets (i.e. choosing appropriate time intervals), the minimization problem is no longer required to deal with transcendental relations. This leads to a considerable increase on the computational performance and it is the key factor that makes the method suitable for large networks. We remark that still, both methods are currently not suitable in our opinion to reliably infer network topology from real recordings of spike data, because of several reasons: For instance, other network-external sources of spike generation, stochastic fluctuations due to intrinsic noise and the highly specific conditions required for reconstruction still limit the methods applicability. Finally, both approaches assume linearity of interactions, yet it is known that interactions can be nonlinear, e.g. due to dendritic spikes in response to sufficiently strong, simultaneous inputs to single dendritic branches of a neuron [62–64].

TECHNICAL ISSUES

Let us briefly comment on four technical issues related to the structural inference approaches discussed above. We mention how some of them may be directly transferred to settings, where discrete time dynamics describes the system (section 6.1), discuss issues when measuring data from real, e.g. biological networks (section 6.2), remark on L_1 versus L_2 minimization (section 6.4) and finally discuss what happens for hyper-networks, where more than two-point interactions influence the collective state (section 6.5).

Discrete Time Maps

If the dynamical system is described as a network of coupled discrete time maps instead of by coupled differential equations (2), approaches similar to the ones presented above are often viable in slightly modified form. For instance, the core equation becomes (for one-dimensional local units)

$$x_i(t+1) = f_i(x_i(t)) + \sum_{j=1}^{N} J_{ij} g_{ij}(x_i(t), x_j(t)) + I_i(t) + \xi_i(t) \tag{64}$$

where now the time $t \in \mathbb{Z}$ is an integer. Here, the direct method of subsection 5.1 is immediately applicable, even with the additional advantage that no temporal derivatives need to be estimated such that an observed time series $(x_i(t))i_{\in\{1,\ldots,}N_{\},}t_{\in\{1,\ldots,}M_{\}}$ enters the inference equations without approximations. Of course, that time series may still contain substantial measurement errors that propagate into any solution of the inference problem.

Sampling Dynamics of Real Networks

When sampling real life phenomena, measurement constraints may arise due to the nature of the given phenomena, making the study of such a major challenge. For instance, in *EEG analysis*, a group of sensors is employed to measure the brain activity. These recordings help in finding functional connections among regions in the brain. Basically, each sensor measures the activity of a population of neurons below its active area for a certain amount of time. Then, by analyzing the correlations among the time series of each sensor, the interaction network is constructed. Nevertheless, as remarked by Bialonski in [65], special considerations regarding the *spatial* and *temporal sampling*must be accounted for. A simple constraint is that sensors placed too close to each other, may record overlapping activity from neighboring populations, thus increasing the difficulty to discern between *direct* and *indirect* interactions. Moreover, if the size and frequency of the sampling fails to capture the intrinsic time scales, false functional connections in network may be constructed. Clearly this example is particularly relevant for inferring effective connectivities, as discussed in section 7. As an example regarding the time domain, gene and protein interaction networks often still have a very limited number of sampling points available [66, 67] such that a network inference problem may become drastically under-determined. Additional information such as the known existence of certain interactions in such a network (but not the existence of others) sometimes is available but how precisely to use this is currently unknown and requires further method development. So briefly, when studying real networks, special technical considerations should be accounted for to actually use what it is measured (the data) for what one wants to know about the system, cf section 1.2.

L_1 Norm Minimization as a Linear Program

Given the optimization problem

$$\min_{\boldsymbol{y}} \|A\boldsymbol{y} - \boldsymbol{b}\|_1 , \tag{65}$$

where $A \in \mathbb{R}^{M\times N}$, $\boldsymbol{y} \in \mathbb{R}^N$ and $\boldsymbol{b} \in \mathbb{R}^M$, (65) may be posed as [31]

$$\min_{s} \mathbf{1}^{\top} s \quad \text{s.t.} \begin{array}{l} Ay - b \preceq s \\ Ay - b \succeq -s \end{array}, \tag{66}$$

where $\mathbf{1} \in \mathbb{R}^{M}$ is a vector of ones and $\preceq$ and $\succeq$ denote entry-wise comparison and s is an auxiliary variable. To solve (66), one has to solve the linear program

$$\min_{x} \tilde{c}^{\top} x \quad \text{s.t.} \quad \tilde{A}x = \tilde{b}, \tag{67}$$

where

$$x = \begin{bmatrix} y \\ s \end{bmatrix}, \quad \tilde{A} = \begin{bmatrix} A & -I \\ -A & -I \end{bmatrix}, \quad \tilde{b} = \begin{bmatrix} b \\ -b \end{bmatrix}, \quad \tilde{c} = \begin{bmatrix} \mathbf{0} \\ \mathbf{1} \end{bmatrix} \tag{68}$$

and $I \in \mathbb{R}^{M \times M}$ and $\mathbf{0} \in \mathbb{R}^{N}$ are the identity matrix and a vector of zeroes, respectively. The advantage of posing problem (65) as (67) is that the latter can be easily solved in a standard way by implementing any solver for linear programs (e.g. the linprog function in MATLAB [32]).

L_2 versus L_1 Norm Minimization

The need to choose a minimization scheme may turn the reconstruction problem into a great challenge, because different schemes may yield different solutions, thus forcing us to discern which scheme is best suited for our purposes in specific reconstruction problems. As illustrated in [1, 10, 29, 35, 59, 60, 68] these differences between minimizers may be exploited in certain situations. For instance, the L_2 minimizer finds the closest solution in the L_2-norm and moreover has an analytic solution. Yet, given the nature of the minimizers (check [68]), an L_1 minimizer is suited for finding particular sparse solutions and it is more robust to outliers than L_2, so it might be seen as more useful for applications. However, minimizing an L_1 norm is computationally more costly compared to the $L_2$and it may have more than one solution [31]. We note that $\|p\|_2 \leq \|p\|_1$ for any vector $p \in \mathbb{R}^{N}$.

Hypernetworks: Beyond Two-point Interactions

We have explicitly excluded networks with more than two-unit interactions from our general mathematical description (2) or dynamical networks with temporally changing connections. Those may occur, for instance in networks of computers, or other communication networks of engineering, where inputs from several units that give input to the same other unit are nonlinearly processed (for instance, multiplied) to change the latter units' state. Similarly, non-additive dendritic interactions in neurons [20, 63], where two simultaneously received synaptic inputs in close spatial proximity initiate so-called dendritic spikes and thereby a nonlinearity [62], naturally imply three-point interactions.

We remark that direct three-unit and higher order interactions (i.e. three-term and higher order products such as $x_i^d x_j^{d'} x_k^{d''}$ where i, j, k are mutually different) are omitted in (1) because they refer to dynamical systems on hypernetworks, thus going beyond the scope of this review. Such third order and higher order terms are not covered by (1), firstly, because the notation stays much simpler without them, but secondly and more importantly, because there seem to be few major theoretical results on reconstructing such systems, if any, that may be or become of practical use. Yet, in social, communication and information networks, such questions may soon become of interest as 'big data' are pouring in.

A brief introduction to the dynamics of complex hypernetworks is given in [17].

EFFECTIVE CONNECTIVITY: CORRELATION-BASED METHODS

Linear Correlation and Covariance

One common way of quantifying effective connectivity is to measure scalar time series $\boldsymbol{x}(t) = (x_1(t), ..., x_N(t))$, $t \in \{t_1, ...t_M\}$ from the N units of a system (spike rates of neurons, expression levels of genes, ...), compute their (temporal) averages $\mu i = \langle x_i(t)\rangle\ t$ and variances $\sigma_i^2 = \langle (x_i(t) - \mu_i)^2 \rangle_t$ and from those compute the covariance matrix

$$\mathrm{Cov}_{ij} := \frac{\langle (x_i(t) - \mu_i)(x_j(t) - \mu_j)\rangle_t}{\sigma_i \sigma_j}. \tag{69}$$

Here the averages are time averages

$$\langle y_i(t)\rangle_t := M^{-1} \sum_{m \in \{1,...,M\}} y_i(t_m). \tag{70}$$

If multiple measurements are available, averages may be taken over ensembles of experiments rather than (or in addition to) temporal averaging.

The obtained covariance matrix is then often thresholded, either just choosing a heuristic value or according to some measure of statistical significance (against an appropriate null hypothesis) and the resulting non-zero values interpreted either as 'strength' of effective connectivity (weighted connectivity matrix) or just as existence (adjacency matrix) of a functional link between units. Sometimes, the value of a threshold is systematically varied and changes in resulting connectivity evaluated. Correlation, covariance and other linear algebra measures are commonly used, often complemented by nonlinear operations such as thresholding, to analyze, for instance functional magnetic resonance imaging (fMRI) or other spatio-temporal data [69]. We note that correlation measure may be (mathematically) seen in a number of different ways [70].

Entropy Maximization

Entropy measures the uncertainty associated with a given probability distribution and constitutes a key quantity in *information theory* [71]. Given the probabilities of a set of events, the entropy measures how uncertain, on average, the occurrence of an event is; or in other words, how much information, on average, one obtains by measuring the occurrence of that event knowing the probability distribution of events. Reversely, given a collection of (observed) data points, we can choose probabilities to maximize the entropy. Such a distribution, known as a *maximum entropy probability distribution,* would be the least biased distribution possible and any other would require further assumptions on the nature of the problem [72]. For a network dynamical system, i.e. systems of coupled dynamical units, we can ask what the effective interactions are such that the probability distribution that best describes the data (averages and correlations) has maximum entropy.

Specifically, the principle of maximum information from Bayesian statistics postulates the 'most likely' probability $\rho(\boldsymbol{x})$ of measuring $\boldsymbol{x}$ given the same type of time series data $\boldsymbol{x}(t) = (x_1(t), \ldots, x_N(t))$ is the one maximizing the information obtained from measuring one more state $\boldsymbol{y}$. More precisely, the goal is to find the probability $\rho(\boldsymbol{x})$ that maximizes Shannon entropy

$$S = -\sum_{\boldsymbol{x}} \rho(\boldsymbol{x}) \ln \rho(\boldsymbol{x}) \tag{71}$$

under the constraints that the first and second moments are consistent with those estimated from the data,

$$\langle x_i \rangle_\rho := \sum_{x} x_i \rho(\boldsymbol{x}) \stackrel{!}{=} \langle x_i(t) \rangle_t \tag{72}$$

and

$$\langle x_i x_j \rangle_\rho := \sum_{x} x_i x_j \rho(\boldsymbol{x}) \stackrel{!}{=} \langle x_i(t) x_j(t) \rangle_t . \tag{73}$$

By restricting ourselves to these conditions (and no others), we here focus on pairwise interactions and neglect three-point and higher order coupling that arise in hypernetworks (cf section 6.5 and [17]). This yields the probability distribution of the form

$$\rho(\boldsymbol{x}) = Z^{-1} \exp\left(\sum_{i=1}^{N} h_i x_i + \frac{1}{2} \sum_{i=1}^{N} \sum_{j=1}^{N} \hat{J}_{ij} x_i x_j \right)$$

where $Z = \int_{\mathbb{R}^N} \exp\left(\sum_{i=1}^{N} h_i x_i + \frac{1}{2} \sum_{i=1}^{N} \sum_{j=1}^{N} \hat{J}_{ij} x_i x_j \right) d^N x$ is a normalization constant and h_i and $\hat{J}_{ij}$ are parameters to be chosen such that (C.3) and (C.4) hold. The quantities $\hat{J}_{ij}$ are interpreted as effective couplings between units i and j.

If the matrix of covariances between the N time series is

$$C_{ij} = \langle x_i(t) x_j(t) \rangle_t - \langle x_i(t) \rangle_t \langle x_j(t) \rangle_t . \tag{74}$$

the effective coupling matrix is given by its inverse

$$\hat{J} = C^{-1} . \tag{75}$$

This means that the best available probability distribution (i.e. that yielding the maximum information) is given by second order effective coupling strengths determined by (but by no means identical to) the linear correlation matrix.

This concept is applied to a range of different systems, in particular in biology, including gene networks [73], protein networks [74] and neural circuits [75]. We comment on an approach originally suggested by Bialek and coworkers [75, 76] for revealing in how far two-point interactions characterize the coupled dynamics of neural circuits in retina. In fact, a systematic study of neural activity in the retina of larval tiger salamander and guinea pig revealed that 90% of the multi-information (which measures all correlative dependences in a system [76]) is covered by pairwise correlations only [75]. The authors took this finding as a sign that pairwise interactions well characterize the full network dynamics and that higher order interactions may be neglected. Specifically, they temporally discretized neural responses into '1' or '0' states depending on whether a neuron was or was not active during each considered time interval of generically 20 ms. Thus, the state of the entire (observed) network at each time interval is given by an N-dimensional word composed of the binary components of the N neurons. As sometimes done for effective connectivity, researchers often tend to go beyond what Bialek and coworkers concluded and interpret the effective coupling strengths (75) as actual physical interactions of experimentally observed units. However, it is typically not clear how correlative dependences yield information about direct physical interactions and such that such interpretations are not justified without substantial further knowledge about the system.

Further in Finding Effective Connectivity

We would like to emphasize that there is a multitude of additional approaches for finding effective or functional connectivities. For instance, a range of methods are based on information-theoretic measures such as mutual information [77], transfer entropy [78] and Granger causality [79, 80] and extensions thereof. In addition, there are various methods relying on Bayesian statistics or explicit or implicit modeling or extended regression such as used in generalized linear models [81]. In particular in biological sciences, such statistical methods have been used early on even at times not many or not very reliable data were available, see for instance [82] for an early study regarding effective connectivity in neural circuits.

There are a number of open challenges regarding both precision of such methods and the interpretation of the respective results. For instance, fMRI experiments of brain areas may rely on differences in blood oxygen level (so-called BOLD signals) as observables but often aims to relate actual dependences in neural spiking activity. The reasoning here is that the cell metabolism and thus the blood oxygen consumption in a local region (typically one cubic millimeter) of the brain is larger the more action potentials per time are generated by the (10^4–10^6) neurons in that region so that such approaches are not undisputed [83]. Moreover, the terms effective connectivity, functional connectivity and structural or anatomical connectivity are sometimes not well defined, used in different meanings across studies. There are even overlapping definitions of non-structural forms of connectivity, e.g. for functional connectivity, effective connectivity etc. Here we did not delve into historic waters and used the term 'effective connectivity' for all forms of connectivity that is not structural. Sometimes, effective connectivity is even taken as an indication for structural connectivity of physical interactions. For instance, distinguishing direct interactions from indirect influences may be an important issues (cf figure 1) that is not yet fully clarified [84].

Finally, we mention that for neural circuits [85] have devised a statistical method to find the couplings J_{ij} that optimize the likelihood that a class of integrate and fire models generates the spike trains observed in experiments. This statistical method assumes the same model class as the approach for inferring structural connectivity presented above (section 5.2) and its goal indeed is finding the (most likely) structural connectivity.

As a conclusive warning, we remark that in particular different methods exit for obtaining effective connectivity given the *same* data set; the results each provide information about specific features of the system: exactly those defined by the method. Some might almost coincide with others, some might be congruent with and some may well be contradicting a given structural connectivity (cf [86]).

Conclusion and open questions

This review focuses on how structural connectivity of networks may be inferred from dynamical features of the networks' nodes. It is on purpose on an introductory level and (given that the area of network reconstruction is rapidly growing simultaneously in different fields, from gene and neural networks to engineering systems) by construction only highlights some selected approaches, most of which based on a perspective of considering the collective nonlinear network dynamics. We provide basic approaches about finding effective or functional connectivities of a network from time series as a brief complement and to get a taste for essential differences in perspective. One main distinction between approaches for identifying qualitatively different types of connectivity is that structural inference, aiming to reconstruct real physical interactions, take into account the time evolution of a system. In contrast, finding effective connectivity is often based on a stationarity assumption and uses distributions of states, neglecting all or parts of their temporal evolution. Relating observable, possibly effective and physical connectivity, is at the heart of current interdisciplinary research [15, 87]. As also mentioned in the Introduction and briefly discussed in section 7, functional versus structural inference if often not well distinguished and, in particular in early publications, the qualitative difference in approach was sometimes not even mentioned.

The methods and approaches presented in this review aim to tell whether or not and how strongly units in a network directly interact with each other. This is in distinction to the entire field of system's identification [7] where the aim is to identify the rules underlying a dynamical system from time series and predict its (future) dynamics based on this identification. Systems identification for multi-dimensional systems, and thus in particular for large networks, seems intractably hard because even if the 'real' system is almost (but not entirely) perfectly reconstructed, predicting their future dynamics can be impossible due to chaos (sensitive dependence on initial conditions), exponentially many different collective states and uncontrolled external influences, with all these factors becoming typical for multi-dimensional complex systems. At the same time, as in part illustrated in this review figure 5, learning the existence of strengths of interactions only (and not the precise form of dynamics) may well be successful for much larger systems. We thus speculate that novel methods complementing those of systems identification may yield further insights into the interaction networks of various complex systems.

A number of key issues are not discussed in this review but still are sometimes pressing for progress research, in particular with respect to applications to real world settings. We list a few.

- How much information can we actually access (cf [88])? Can we observe all the units of a network? Can we observe all dynamical variables (dimensions) of each unit? In which sense may it make sense to seek connectivity of nodes that are not observed?
- What do we know a priori about the system? Are the functional forms of network interactions known? What can we say about stability or instability of the dynamics? Perhaps the dynamics even exhibits a complex mixture of stable and unstable dynamics [9, 89, 90].
- What are reasonable (or possible) number of experiments or measurements that can be done and what is a clever trade-off against the required computational efforts that may depend on the quality and quantity of those measurements [23].
- Which of the structural features are actually the most relevant and which are even possible. In biochemical and gene regulatory networks, electric circuits and power grids, for instance, many interaction links (or their absence) may have been identified by other methods not related to network collective dynamics. Under

these circumstances, can we improve existing approaches to take such information into account?

- How can we address genuinely stochastic dynamics, e.g. in excitable systems, possibly even induced by small number fluctuations—a pathway of though that links to non-equilibrium physics, cf [91]?
- What if network connectivities change, for instance by synaptic plasticity in neural circuits [92–94] or other forms of adaptation, e.g. in social networks [95].
- ...think of your own, there are many more questions with great opportunity for scientific progress in a range of fields ...

Answers to all these questions will specifically restrict or enhance the space of optional networks consistent with recorded data. From an abstract point of view, these modify the form and dimensionality of the set of all possible networks and points to a direction that still addresses network dynamics as an inverse problem but goes beyond network reconstruction. Given all restrictions, perhaps we can design or control a network to robustly exhibit a specific functionality. Network design and control form two additional branches in the theory of network dynamical systems, with currently highly active research, from biological sciences to engineering [11, 96–99]. Such approaches might soon be extremely influential and thought-provoking, when, e.g., the neural and biochemical networks in our bodies as well as our infrastructure networks surrounding us can not only be reconstructed, but even controlled and specifically engineered. We recommend a view point by Freeman Dyson for a very practical taste [100].

ACKNOWLEDGMENTS

We thank Srinivas Gorur Shandilya for help during project initiation and Raoul-Martin Memmesheimer, Mor Nitzan, Dirk Witthaut, David Gross, Matthias Wendland, Sarah Hallerberg and Hinrich Arnoldt for valuable discussions and constructive comments on parts of the manuscript. Partially supported by the International Max Planck Research School (IMPRS) Physics of Biological and Complex Systems (JC) the Ministry for Education and Science (BMBF), Germany, through the Bernstein Center for Computational Neuroscience, grant no. 01GQ1005B (MT) as well as by a grant by the Max Planck Society to MT.

REFERENCES

1. Gardner T, di Bernardo D, Lorenz D and Collins J 2003 Science 301 *102–5*
2. Palsson B 2006 Systems Biology: Properties of Reconstructed Networks (Cambridge: Cambridge University Press)
3. Brown E, Kass R and Mitra P 2004 Nature Neurosci. 7 *456–61*
4. Briggman K and Denk W 2006 Curr. Opin. Neurobiol. 16 *562–70*
5. Helmstaedter M, Brigmann K and Denk W 2011 Nature Neurosci. 14 *1081–8*
6. Sporns O, Tononi G and Koetter R 2005 PLoS Comput. Biol. 4 *e42*
7. Maier M and Rechtin E 2000 The Art of Systems Architecting 3rd edn (Boca Raton, FL: CRC Press)
8. Prinz A, Bucher D and Marder E 2004 Nature Neurosci. 7 *1345–52*
9. Rabinovich M, Huerta R and Laurent G 2008 Science 321 *48–50*
10. Timme M 2007 Phys. Rev. Lett. 98 *224101*
11. Memmesheimer R M and Timme M 2006 Physica *D* 224 *182–201*
12. Shandilya S and Timme M 2011 New J. Phys. 13 *13004*
13. Francois P and Hakim V 2004 Proc. Natl Acad. Sci. USA 101 *580–5*
14. Slusarczyk A, Lin A and Weiss R 2012 Nature Rev. Genet. 13 *406–20*
15. Ren J, Wang W, Li B and Lai Y 2010 Phys. Rev. Lett. 104 *058701*
16. Hlinka J, Hartman D and Palus M 2012 Chaos 22 *033107*
17. Timme *M et al 2014 in preparation*
18. Timme M, Wolf F and Geisel T 2002 Phys. Rev. Lett. 89 *258701*
19. Wiley D, Strogatz S and Girvan M 2006 Chaos 16 *015103*
20. Memmesheimer R 2010 Proc. Natl Acad. Sci. USA 107 *11092–7*
21. Jahnke S, Memmesheimer R and Timme M 2008 Phys. Rev. Lett. 100 *048102*
22. Menck P, Heitzig J, Marwan N and Kurths J 2013 Nature Phys. 9 *89–92*
23. Tegner J, Yeung M, Hasty J and Collins J 2003 Proc. Natl Acad. Sci. USA 100 *5944–9*
24. di Bernardo D, Thompson M, Gardner T, Chobot S, Eastwood E, Wojtovich A, Elliott S, Schaus S and Collins J 2005 Nature Biotechnol. 23 *377–83*

25. Faith J, Hayete B, Thaden J, Mogno I, Wierzbowski J, Cottarel G, Kasif S, Collins J and Gardner T 2007 PLoS Biol. 5 *54–66*
26. de Jong H 2002 J. Comput. Biol. 9 *67–103*
27. Lee W-P and Tzou W-S 2009 Briefings Bioinform. 10 *408–23*
28. Levine M and Davidson E 2005 Proc. Natl Acad. Sci. USA 102 *4936–42*
29. Yeung M, Tegner J and Collins J 2002 Proc. Natl Acad. Sci. USA 99 *6163–8*
30. Stoer J and Bulirsch R 1993 Introduction to Numerical Analysis (New York: Springer)
31. Boyd S and Vandenberghe L 2009 Convex Optimization (Cambridge: Cambridge University Press)
32. MATLAB7 2004 (Natick, MA: The Mathworks Inc.)
33. Mathematica80 2010 (Champaign, IL: Wolfram Research Inc.)
34. Barrodale I and Roberts F 1974 Commun. ACM 17 *319–20*
35. van Bussel F, Kriener B and Timme M 2011 Front. Comput. Neurosci. 5 *3*
36. Wang W-X, Lai Y-C, Grebogi C and Ye J 2011 Phys. Rev. X 1 *021021*
37. von Dassow G, Meir E, Munro E and Odell G 2000 Nature 406 *188–92*
38. Yu D and Parlitz U 2010 Phys. Rev. E 82 *026108*
39. Yu D 2010 Automatica 46 *2035–40*
40. Kori H and Mikhailov A 2004 Phys. Rev. Lett. 93 *254101*
41. Kawamura Y, Nakao H, Arai K, Kori H and Kuramoto Y 2008 Phys. Rev. Lett. 101 *024101*
42. Zanette D 2004 Europhys. Lett. 68 *356–62*
43. Radicchi F and Meyer-Ortmanns H 2006 Phys. Rev. E 73 *036218*
44. Timme M 2006 Europhys. Lett. 76 *367–73*
45. Arenas A, Diaz-Guilera A and Perez-Vicente C 2006 Phys. Rev. Lett. 96 *114102*
46. Levnajic Z and Pikovsky A 2011 Phys. Rev. Lett. 107 *034101*
47. Yu D, Righero M and Kocarev L 2006 Phys. Rev. Lett. 97 *188701*
48. Timme M and Wolf F 2008 Nonlinearity 21 *1579–99*
49. Diestel R 2010 Graph Theory (Heidelberg: Springer)

50. Izhikevich E 2007 Dynamical Systems in Neuroscience: The Geometry of Excitability and Bursting(Cambridge, MA: MIT Press)
51. Rössler O 1976 Phys. Lett. A 57 *397*
52. Lorenz E 1963 J. Atmos. Sci. 20 *130*
53. Ashwin P and Timme M 2005 Nonlinearity 18 *2035–60*
54. Cessac B and Vieville T 2008 Front. Comput. Neurosci. 2 *1–20*
55. Kielblock H, Kirst C and Timme M 2011 Chaos 21 *025113*
56. Rieke F, Warland D, de Ruyter van Steveninck R and Bialek W 1997 Spikes: Exploring the Neural Code (Cambridge, MA: MIT Press)
57. Burkitt A 2006 Biol. Cybern. 95 *1–19*
58. Dayan P and Abbott L 2001 Theoretical Neuroscience (Cambridge, MA: MIT Press)
59. van Bussel F 2010 Topological optimization in network dynamical systems PhD Thesis Georg-August-Universität Göttingen
60. Makarov V, Panetsos F and de Feo O 2005 J. Neurosci. Meth. 144 *265–79*
61. Gentle J, Haerdle W and Y Mori E 2004 Handbook of Computational Statistics (Berlin: Springer)
62. Ariav G, Polsky A and Schiller J 2003 J. Neurosci. 23 *7750–8*
63. Memmesheimer R and Timme M 2012 PLoS Comput. Biol. 8 *e1002384*
64. Jahnke S, Timme M and Memmesheimer R 2012 Phys. Rev. X 2 *041016*
65. Bialonski S 2012 Inferring complex networks from time series of dynamical systems: pitfalls, misinterpretations, and possible solutions PhD Thesis Rheinische Friedrich-Wilhelms-Universität Bonn
66. Bansal M, Belcastro V, Ambesi-Impiombato A and di Bernardo D 2007 Mol. Syst. Biol. 3 *78*
67. Hecker M, Lambeck S, Toepfer S, van Someren E and Guthke R 2009 Biosystems 96 *86–103*
68. Napoletani D and Sauer T 2008 Phys. Rev. E 77 *026103*
69. Erdhardt E, Rachakonda S, Bendrick E, Allen E, Adali T and Calhoun V 2011 Hum. Brain Mapp. 322075–95
70. Rodgers J and Nicewander W 1988 Am. Stat. 42 *59–66*
71. Cover T and Thomas J 2006 Elements of Information Theory (New York: Wiley)

72. Jaynes E 1957 Phys. Rev. 106 *620–30*
73. Braustein A, Pagnani A, Weigt M and Zecchina R 2008 J. Stat. Mech. P12001
74. Tkacik G, Schneidman E, Berry M and Bialek W 2006 arXiv:q-bio. NC/0611072
75. Schneidman E, Berry M, Segev R and Bialek W 2006 Nature 440 *1007–12*
76. Schneidman E, Still S, Perry M and Bialek W 2003 Phys. Rev. Lett. 91 *238701*
77. MacKay D 2003 Information Theory, Inference and Learning Algorithms (Cambridge: Cambridge University Press)
78. Schreiber T 2000 Phys. Rev. Lett. 85 *461–4*
79. Granger C 1969 Econometrica 37 *424–38*
80. Ladroue C, Guo S, Kendrick K and Feng J 2009 PLoS One 4 *e6899*
81. Pillow J, Shlens J, Paninski L, Sher A and Litke A 2008 Nature 454 *995–9*
82. Aertsen A M, Gerstein G L, Habib M K and Palm G 1989 J. Neurophysiol. 61 *900–17*
83. Horwitz B 2003 Neuroimage 19 *466–70*
84. Nawrath J, Romano M, Thiel M, Kiss I, Wickramsinghe M, Timmer J, Kurths J and Schelter B 2010 Phys. Rev. Lett. 104 *38701*
85. Cocco S, Leibler S and Monasson R 2009 Proc. Natl Acad. Sci. USA 106 *14058–62*
86. Lünsmann *B et al 2014 in preparation*
87. Barzel B and Barabási A 2013 Nature Biotechnol. 31 *720–5*
88. Su R, Wang W and Lai Y 2012 Phys. Rev. E 85 *065201*
89. Ashwin P and Timme M 2005 Nature 436 *36–37*
90. Schittler Neves F and Timme M 2012 Phys. Rev. Lett. 109 *178701*
91. Metzner P, Dittmer E, Jahnke T and Schütte C 2007 J. Comput. Phys. 227 *353–75*
92. Morrison A, Aertsen A and Diesmann M 2007 Neural Comput. 19 *1437–67*
93. Tetzlaff C, Kolodziejski C, Timme M, Tsodyks M and Wörgötter F 2013 PLoS Comput. Biol. 9e1003307

94. Bassett D, Wymbs N, Porter M, Mucha P, Carlson J and Grafton S 2011 Proc. Natl Acad. Sci. USA 1087641
95. Skyrms B and Pemantle R 2009 A Dynamic Model of Social Network Formation Adaptive Networks ed T Gross and H Sayama (Dordrecht: Springer) pp 231–51
96. Memmesheimer R and Timme M 2006 Phys. Rev. Lett. 97 *188101*
97. Li H 2009 Reliability-based dynamic network design with stochastic networks PhD Thesis Delft University
98. Liu Y, Slotine J and Barabasi A 2009 Nature 473 *167–73*
99. Preciado V, Zavlanos M and Jadbabaie A 2012 arXiv:1112.0765
100. Dyson F 2005 The Darwinian Interlude (Cambridge, MA: MIT Press)
101. Yan X and Su X G 2009 Linear Regression Analysis: Theory and Computing (Singapore: World Scientific)
102. Lange K 2010 Numerical Analysis for Statisticians (New York: Springer)
103. Strang G 1993 Am. Math. Mon. 100 *848–55*

INDEX

D

E